Nanotechnology and Energy

Nanotechnology and Energy

edited by

Kaufui V. Wong

PAN STANFORD PUBLISHING

Published by

Pan Stanford Publishing Pte. Ltd.
Penthouse Level, Suntec Tower 3
8 Temasek Boulevard
Singapore 038988

Email: editorial@panstanford.com
Web: www.panstanford.com

British Library Cataloguing-in-Publication Data
A catalogue record for this book is available from the British Library.

Nanotechnology and Energy

ISBN 978-981-4774-24-6 (Hardcover)
ISBN 978-1-315-16357-4 (eBook)

Cover image: Close-up of a big boulder of natural turquoise mined and displayed in the Southwest USA

Printed in the USA

Contents

Preface

Nanotechnology (technology at the nanoscale dimensions, that is, <100 nm) is at present a very active and dynamic area of interdisciplinary research in physics, chemistry, biology, and engineering. Nanotechnology has the potential to renovate a wide selection of scientific and technological fields. Since we are in an energy–water–food nexus era, which is not going to let up on any of its associated problems, this book *Nanotechnology and Energy* is timely. That there are limited textbooks/reference books in this field and there are generations of practicing engineers worldwide who have not learned about this field makes the publication of this book imperative.

This book is about contribution of some of the fundamentals of fluid and thermal sciences to nanoscience and nanotechnology and their significant applications to date. It covers the materials and structure side of nanotechnology as a matter of completeness and in review. The monograph will be appropriate for use in an engineering or science course. It is designed as a textbook for a whole course at the senior undergraduate level and first-year graduate level and also for a general readership. It will not only be helpful for engineering students but also for those practicing engineers who did not hear about nanotechnology in their college days. The topic has a universal appeal because it is new compared to other sciences, with significant applications already proven and commercialized, and is therefore an essential read for all engineers.

I acknowledge with thanks the academic journals in which all chapters have been previously published as articles and the perceptiveness of Mr. Stanford Chong and Ms. Jenny Rompas of Pan Stanford Publishing who recognized the technological/scientific and public service value of my work.

Kaufui V. Wong

Summer 2017

Chapter 1

Nanoscience and Nanotechnology

Kaufui V. Wong
Department of Mechanical and Aerospace Engineering, University of Miami, Coral Gables, FL 33124, USA
kwong@miami.edu

Nanoscience, the conduct of physical systems when confined to nanoscale (<100 nm) dimensions together with the physical phenomena that take place at the nanoscale, is at this time one of the most active and quickly evolving areas of interdisciplinary research in physics, chemistry, biology, and engineering. Nanotechnology, the application of these properties and phenomena, has the possibility to transform a great variety of scientific and technological fields. This, and that a few technological fields have already seen the results of such transformation, seem to be the driving factors for nanoscience.

Physics, chemistry, and biology faculty research groups have participated in pivotal roles for developing and progressing the state of the art of nanofabrication and applying the ensuing tools and proficiencies in a great variety of nanoscience and nanotechnology research endeavors. These areas include the understanding, operating, and applying materials and happenings at the nanoscale. Because nanoscale science and technology are at the forefront of

Nanotechnology and Energy
Edited by Kaufui V. Wong

ISBN 978-981-4774-24-6 (Hardcover), 978-1-315-16357-4 (eBook)
www.panstanford.com

a lot of today's engineering and medical research, there is overlap between these doings and others mentioned in this chapter.

All the subsequent chapters of this book are published academic papers, which means that they were peer-reviewed and deemed state of the art or timely at the time of publication. The second chapter of this book is devoted to the discussion of how to introduce the topic of nanotechnology in the basic sciences and, hence, into the curricula of engineering students. Most engineering programs in the United States of America still do not have an undergraduate degree with a major in nanotechnology, as it was the case in 2007 when this second chapter of this book was published. However, it is possible that the progress in this direction could be sure, if slow.

The next two chapters are about the properties of nanofluids, which are macrofluids that contain nanoparticles. Nanofluids are prominent in nanotechnology applications in the energy area because they have proven themselves to be very useful in many engineering applications. The fifth chapter is about nanoscience and nanotechnology of solar cells. The sixth chapter is about the applications of nanofluids, both current and future. This published journal article of 2012 is the most highly cited publication of the author. In other words, it is the most popular among researchers in nanotechnology, and possibly one of the most helpful for researchers in nanofluids.

The seventh chapter is on the innovative use of ultrasound thermometry to study natural convection of opaque nanofluids. The eighth chapter is regarding the impact of nanotechnology on the automobile industry, an important sub-area of mechanical engineering.

The ninth chapter is a review of the use of carbon nanotubes in renewable energy applications and environmental protection and remediation. The tenth chapter is a 2016 journal article on nanotechnology and electric batteries.

The electronic properties of nanoscale structures and the advance of nanoscale gadgets founded on these properties constitute a major sub-area of nanoscience and nanotechnology. Projects include a range of research efforts on both Si- and carbon-based nanoelectronic systems that have the possibility of spreading and enhancing the capabilities of current electronic technologies. The use of nanotechnology in extending the life of batteries together with the use of innovative materials is another important sub-area. A whole chapter has been devoted to this subject in this book, as previously mentioned.

Nanomagnetics, the study and application of magnetic materials in the nanoscale, is another area of research interest. The relatively new discovery that the electron-spin-polarized current flowing to or from a thin-film ferromagnet can reversibly switch the magnetic orientation of another nearby nanomagnet by a "spin-transfer" process is introducing the possibility of an innovative method for ultra-high-density information storage. Another possibility is the development of new nanoscale components for high-frequency electronics. Research concerned with the injection and manipulation of electron spin in normal metal and semiconductor nanostructures could point to the advance of quantum computer elements as well as to a number of other "spintronics" projects. The manipulation and application of the optical properties of materials at the nanoscale is a facet of the labors by physics and engineering faculty groups to invent devices and systems for future photonics and optical communication uses.

The science and technology of nanofluids is critical for the energy sciences. Four chapters in this book are devoted to this important topic. All the work has been undertaken and published by Professor Vafai and his research group. Their work is republished here with Prof. Vafai's permission, and the permission of Elsevier, the publisher of the original journal articles. The first two chapters of the four look at the thermophysical properties of nanofluids and the performance of cylindrical heat pipes. The next two chapters examine the mixed convection of nanofluids, as well as the boiling phenomenon of nanofluids.

Nanoelectromechanical system (NEMS) research is about developing the ability to fabricate ultra-small electromechanical devices to the advancement of powerful and multipurpose nanoscale sensors and systems for an extensive range of electronic, optical, and biological applications. These research endeavors encompass work on single biomolecule detection with a nanofabricated "lab-on-a-chip" and the invention of very high-frequency NEMS mechanical oscillators for sensor and optical switching applications.

An important characteristic of the nanoscience and nanotechnology is the extensive range of research activities, the goals of which are to invent new tools and techniques for the characterization and study of materials at the nanoscale and for the manufacture and processing of nanoscale gadgets and systems. These empowering nanoscale sciences and technologies include, in addition to the development of innovative approaches for ultra-high-resolution nanolithography and materials processing, the development of prevailing, new scanned probe tools for the measurement of electronic and magnetic properties at the nanoscale. Nano-characterization research encompasses the advancement of the application of analytical scanning transmission electronic microscopy techniques for determining the electronic structure of interfaces of heterogeneous materials with resolution that goes down to nanometers and below.

It is not the purpose of this chapter to list the many and varied areas of research in nanoscience and nanotechnology, but only the significant ones that may have an impact on nanotechnology and energy, whether at present or in the future.

Chapter 2

Introduction of Nanotechnology in the Basic Energy Sciences

Kaufui V. Wong[a] and Pablo A. Garcia[b]
[a]*Department of Mechanical and Aerospace Engineering, University of Miami, Coral Gables, FL 33124, USA*
[b]*Department of Civil, Architectural and Environmental Engineering, University of Miami, Coral Gables, FL 33124, USA*
kwong@miami.edu

Nanotechnology is, according to some authoritative sources, the new wave of the immediate future in science and technology. Many universities in the world do not have this topic taught at the college level in engineering.

The objective of this chapter is to prepare teaching material in nanotechnology that could be used in a couple of lectures to be incorporated in a junior-level course of thermodynamics or other basic energy science course. The aim of preparing these modules is an effort to introduce our current undergraduate engineering

---

This chapter has been reprinted from the *Proceedings of IMECE2007,* 2007 ASME International Mechanical Engineering Congress and Exposition, Kau-Fui V. Wong and Pablo A. Garcia, Introduction of nanotechnology in the basic energy sciences, Copyright 2007, with permission from ASME.

Nanotechnology and Energy
Edited by Kaufui V. Wong

ISBN 978-981-4774-24-6 (Hardcover), 978-1-315-16357-4 (eBook)
www.panstanford.com

students to some important concepts in nanotechnology. The modality of modules is so that many other college programs in the world can adopt these modules and incorporate in their curriculum easily; the result is that the education of our future engineers will be more complete and our engineers more competitive in the world market.

The current work concentrates on the discussion and elaboration of the topics to be covered in nanotechnology in two one-and-a-half-hour lectures. Introductory items include the differences between nanoscience and macroscience as well as atomic-level science theory. It cannot just be a survey of where nanotechnology has affected our modern lives. It should contain enough basics so that future courses/topics can build on these preliminary foundation lectures.

The topics will tie in two existing lectures with nanotechnology applications. First, the principles of heat transfer relating to thermodynamics will be introduced. Second, an overview of nanotechnology will be presented. Third, students will be given an in-depth look at the various heat transfer processes. Finally, the nanotechnology applications relating to heat transfer shall be discussed, including a focus on nanofluids.

This chapter introduces some of the fundamental ideas regarding nanotechnology that relates to the energy sciences as it may be presented to a junior-level engineering class. This will include a discussion of some of the properties of nanoparticles, the synthesis of nanoparticles and the various materials used in fabrication, as well as the importance of nanofluids to complex thermal energy systems based on current and past research.

After laying out the lesson plan integration in the current course curricula, several sample problems are presented in the appendix that would lend the students a greater grasp of the new material.

An engineering upperclassman should understand the fundamentals of nanoscale engineering. Initial educational evaluation data based on a couple of lectures alone gave an average concept inventory, ([1] C. I., Foundation Coalition, 2007) score (geared for the two lectures) of 61.4% for a class of about 20 students (Spring, 2007) in thermodynamics; the corresponding improvement in the average C. I. score for about 10 students (Summer, 2007) was from 43.7% to about 60%.

2.1 Introduction

Nanotechnology represents the new wave of science and technology. Although it was first mentioned in the mid-1970s, it has recently been applied successfully and in many disciplines. In previous centuries, the development of machines progressed civilized nations into technical civilizations with capabilities extending beyond physical strength. Nanotechnology has the potential to move science into the scale of the building blocks of matter where properties can be altered. Nanotechnology also has incredible potential in terms of its energy applications in all branches of engineering, particularly in mechanical and electrical.

In order to be able to develop the field and integrate it successfully into engineering and energy sciences, it is important for it to be introduced at the educational level. The most beneficial scenario would include the introduction of nanotechnology lessons at the earliest practical level, starting in high school. The creative manipulation of matter at the macroscale and nanoscale is significantly different, and in order for a quicker development of the field, it is imperative to develop a spatial understanding of the nanoscale. The addition of a nanoscale perspective in the classroom with respect to all related areas will lead to a better understanding of fundamentals.

Currently, a growing number of educational institutions around the world, particularly in Korea, Australia, and Europe, offer baccalaureate degrees in nanoscience and nanotechnology. Although there has been an increase in students moving into the physical and engineering sciences due to the research by the National Nanotechnology Initiative, only one university in the United States, the University of Central Florida, offers a nanotechnology-related degree and a few others have begun to integrate nanotech courses into the engineering curricula. Although there continues to be steady engineering-related technological improvements, the lack of initiative to enhance and revise the current educational content being given to engineering students places this growth in a perilous position. Nanotechnology has the potential to increase the efficiency of many engineering processes and solve various scientific obstacles that have obstructed the development of engineering design capabilities for many years.

Since nanotechnology is a multidisciplinary field involving engineering, the importance of the engineer to the expansion of the field is undoubted. Applications of nanotechnology will involve nanoscale systems to which engineering methods of investigation and problem solving will be significant. One expected application of nanotechnology is to the computer and microchip industry, where millions of devices must function simultaneously and without fail. Engineers have a background in which they understand the principles of macroscopic phenomena, can define transport coefficients, and frequently optimize processes and products of all sorts. Added to this, nanotechnological progress would be defined by the ability to use it to manipulate matter under control, and integrate it into the manufacture of material structures, devices, and systems. Engineering methods would, thus, be required in order to develop tools and processes that could measure, calibrate, design, and manufacture nanostructures that would eventually be self-replicating [2].

The objective of this chapter is to prepare teaching materials relating to the field of nanotechnology that can be introduced in a couple of lectures in a junior-level thermodynamics course. Although this is hardly a complete education in nanotechnology, it represents a starting point from which, progressively, new courses may be introduced, and be made a requirement for attaining an undergraduate degree in all branches of engineering. The topics presented can be easily extended throughout educational programs around the country.

The topics introduced are formatted for two sections. The first section includes a discussion of the principles of heat transfer with regards to thermodynamics. This is a standard topic that is covered in undergraduate thermodynamics classes. However, due to its intrinsic relation to nanotechnology, it is important to introduce it prior to a discussion on nanotechnology. This will help the students to maintain the heat transfer processes, particularly convection and conduction, fresh in the their minds. It will also make it easier for the students to make mental connections with the study material. Once these have been presented, nanotechnology will be introduced. Nanotechnology is an interesting field of study due to the many physical and thermal properties between nanomaterials and micromaterials.

The remainder of the chapter focuses on the differences between nanoscience and microscience. There are quantum mechanical effects when dealing with systems on the nanoscale that are not present when decreasing from macro to micro, particularly the effect of shape on electrical and thermal properties. Additionally, certain changes in physical properties are unprecedented, such as the increased surface-area-to-volume ratio in nanomaterials due to the large proportion of constituent atoms residing at the grain boundary, which distinguishes it from the generally coarse grain of particles at or larger than the microscale [3]. Additionally, several properties have been found to "reverse" at the nanolevel. Most notably, several insulators such as silicon become conductors; gold, a stable material, becomes combustible; and inert materials such as platinum become catalysts. All aspects of nanotechnology should be covered while still following through with the course objectives that have been a part of thermodynamics instruction. Therefore, it is important to touch upon the different techniques for creating nanofluids, their advantages and disadvantages. Additionally, pivotal to the creation of nanofluids are also different materials that are used for production.

The second section consists mainly of a discussion of heat transfer applications. Mechanical and electrical engineers have to deal with heat transfer commonly in design. Many of the common applications of heat transfer in the workplace or the lab can be discussed, so the students can better understand the connection between education and implementation of principles and study. Engineering students benefit not only from understanding the scientific basis behind nanofluids, as well as the current technological achievements in the field, but also from understanding how such engineering innovations have an economic impact on the market. There is an increasing level of global competition in the engineering market, and within the field of heat transfer, there is consistent research into the development of heat transfer fluids with enhanced heat transfer properties.

Following this, the effects on nanotechnology developments in heat transfer applications will be discussed. Several case studies, such as those discussed here, can be presented to students in order to show them the wealth of research and relevant scientific quandaries have come about as a result of the growth of nanotechnology. This is an important aspect of this section, because it allows the students to see what questions remain, and what new questions have come about as more research is conducted.

There are many mechanical obstacles in which nanotechnology will serve a part in resolving. Of special concern to energy engineers is the decrease in efficiency when too much heat is rejected in a thermal process. This is the case with the internal combustion engine (ICE). Gasoline-fueled ICEs typically have an efficiency varying from 20% to 37%. Most of the energy loss is due to heat rejection, and economically speaking, an improvement in this figure would have a significant impact. Until now, attempting to improve the efficiency of the ICE has resulted in loss of performance. However, with nanotechnological applications, catalysts that are nanofabricated can help improve the efficiency of the ICE due to their increased resistance to carbon monoxide reactions.

In an increasingly globally connected world, engineering students will face a more competitive job market, and thus it is more important for their educational background to be on par or ahead of other students. With the addition of nanotechnology, on any scale, to an educational curriculum, engineering students will be better prepared and more competitive. Additionally, as the field expands, there will also be an increase in the number of nanotechnology workers that will be necessary to keep it growing, an estimated 2 million in the next 10–15 years [2].

2.2 Basic Energy Sciences

2.2.1 Macroscale versus Nanoscale

The concept of nanoscale is first introduced. In macroscale, micro-scale, and atomic-scale, the building blocks of matter are taught to be the same. In other words, a molecule of oxygen is identical to another molecule of oxygen. A lump of pure water in macroscale is also identical to another lump of pure water. In nanoscale, the building blocks are not all the same. They are a function of the size and shape. Five 5 nm football-shaped clay particles have different characteristics/properties from a 25 nm spherically shaped particle of the same clay material, for instance. This fundamental difference has to be driven home. In thermodynamics, the property tables for various fluids play an important part in engineering calculations. Corresponding property tables in nanoscale engineering are not relevant at this point of discussion.

2.2.2 Nanofluids and Heat Transfer

Thermal conductivity is a measure of how quickly heat flows through a substance. The defining equation, Fourier's law, in one dimension is

$$k = \frac{-(Q/A)\chi}{\Delta T} \tag{2.1}$$

where Q is the heat flow rate, A is the surface area, χ is the thickness, and ΔT is the change in temperature [4]. The units for conductivity are Btu/h-ft-°F or W/mK. The thermal conductivity of a heat transfer fluid is an important topic because of its effects on the technological development of heat transfer equipment such as electronics, HVAC&R (heating, ventilating, air conditioning, and refrigeration), and transportation. Therefore, it is an essential application for thermal management.

Experimental results have shown that the thermal conductivity of nanofluids is significantly higher than that of base fluids and the predicted values of existing models. Solids, used in creating nanoparticles, have a greater thermal conductivity than liquids. Carbon nanotubes, for example, have a thermal conductivity of 3000 W/mK, while water has a thermal conductivity of 0.613 W/mK. However, since nanoparticles are costly to manufacture, nanofluids typically have a low volume fraction.

Of the subfields of nanotechnology, an important one is heat transfer in nanofluids. The subject may be introduced along with the principles of heat transfer, specifically conduction and convection. The research subject of nanofluids is an important and promising one for various reasons. When nanoparticles of higher thermal conductivities are added to heating and cooling fluids, the thermal conductivities of the resulting nanofluids seem to be significantly increased. Nanofluids have a high thermal conductivity, stability, and prevent clogging in micro-channels [5]. As opposed to microparticles, which have many clogging problems, metallic nanoparticles are small enough as to behave like a molecule of the liquid and pass through micro-channels, making it useful for high-heat load applications. Nanoparticles stay suspended for greater amounts of time than their larger (milli- or microscale) counterparts, and below a certain size (a critical size), the particle suspension becomes indefinite. They also have better rheological properties [6]. The average size of the nanoparticles suspended in nanofluids is just under 50 nm. Research in nanofluids may lead to the design of more efficient cooling systems and, thus, smaller designs. In the laboratory, Ma et al. [7] combined nanofluids in an oscillating heat pipe (OHP) in order to develop a high-performance cooling device. Applications of such a use of nanofluids could be in the computer industry, where microchips can be cooled with an OHP in order to reduce the likelihood of overheating and burning out due to I^2R power losses.

In heat exchangers, where heat is transferred between two fluids at different temperatures that, depending on the type of heat exchanger, may or may not be in contact, there is also an opportunity for significant improvement with the use of nanofluids. In conventional heat exchangers, the only way to increase the rate of heat transfer is through mechanically increasing the pumping power of the fluids into the heat exchanger. However, by taking the fluid and suspending nanoparticles within it, the thermal conductivity will increase. Therefore, this will cause an increase in the rate of heat transfer without the need for increased pumping power, which leads to significant savings to the operator [3].

In order for heat transfer to be more energy efficient, it must be minimized in size with an increase in heat flux to compensate. Common heat transfer fluids such as water, mineral oil, and ethylene glycol have poor heat transfer properties when compared to solids, particularly metals. This represents the primary obstacle in the miniaturization of heat exchangers. Nanofluids seem to be the solution to this problem. As compared to microparticle suspensions, nanofluids cause significantly less erosion in pipelines and do not suffer from a noteworthy pressure drop caused by momentum transfer [8]. The heat transfer coefficient of a nanofluid is given by Eqs. (2.2) and (2.3).

$$Nu_{\mathrm{nf}} = \frac{h_{\mathrm{nf}} D}{k_{\mathrm{nf}}} \tag{2.2}$$

$$h_{\mathrm{nf}} = \frac{q}{T_{\mathrm{w}} - T_{\mathrm{f}}} \quad (2.3)$$

where Nu_{nf} is the Nusselt number of the nanofluid (dimensionless), h_{nf} is the heat transfer coefficient (W/m^{2}K), D is the inner diameter of the tube (m), k_{nf} is the thermal conductivity of the nanofluid, q is the heat flux (W/m^2), T_{w} is the temperature of the tube wall (K), and T_{f} is the temperature of the base fluid (K) [8]. Experimental results have shown that the flow velocity and volume fraction are also variables that affect the heat transfer coefficient [6]. As with other historical equations, the Dittus–Boelter equation, used in fluid mechanics to calculate the surface coefficient of heat transfer for turbulent fluid flow in round pipes, cannot be used to obtain the Nusselt number of the nanofluid. When pressure drop occurs, a nanofluid typically exhibits a friction factor roughly equal to that of water at the same flow velocity and is independent of volume fraction. This finding suggests that in applications, pumping power will not be sacrificed by the nanofluid and the single-phase friction factor can be used [8]. This is given by Eq. (2.4):

$$\lambda_{\mathrm{nf}} = \frac{\Delta P_{\mathrm{nf}} \cdot D \cdot 2g}{Lu_{\mathrm{m}}^2} \quad (2.4)$$

where λ_{nf} is the friction factor, ΔP_{nf} is the pressure drop of the nanofluid (Pa), D is the pipe diameter, g is the acceleration of gravity (~9.81 m/s^2), L is the length of the test tube (m), and u_{m}^2 is the mean velocity squared.

Heat transfer through convection can be increased through different methods. These are boundary conditions, varying the flow geometry, or enhancing the fluid thermal conductivity [9]. Suspensions prior to the creation of nanoparticles frequently create greater flow resistance and, in some cases, erosion. This is because the large size and density of the particles prevent them from settling into suspension. Since heat transfer occurs at the surface of the particle, surface area is an important variable in potential heat transfer applications. As discussed previously, the decrease in size from micro to nano brings an increase in surface area with respect to volume. Although the volume has decreased enough to prevent the particle from settling quickly, the increased surface area makes nanoparticles ideal for heat transfer applications.

Several models are used to predict the thermal (effective) conductivity of nanofluids. The Maxwell–Garnett model (MG) for loading of spherical particles is given by

$$k_{\text{eff}}/k_{\text{f}} = \frac{k_{\text{p}} + 2k_{\text{f}} - 2\phi(k_{\text{f}} - k_{\text{p}})}{k_{\text{p}} + 2k_{\text{f}} + 2\phi(k_{\text{f}} - k_{\text{p}})} \tag{2.5}$$

Here k_{eff} is the effective thermal conductivity of a liquid with particle suspension, k_{f} is the thermal conductivity of the fluid, k_{p} is the thermal conductivity of particles, and ϕ is the volume fraction of particles. The renovated Maxwell (R-MG) model is given by

$$k_{\text{eff}}/k_{\text{f}} = \left[\frac{k_{\text{pe}} + 2k_{\text{f}} - 2(k_{\text{pe}} - k_{\text{f}})(1+\gamma)^3\phi}{k_{\text{pe}} + 2k_{\text{f}} - 2(k_{\text{pe}} - k_{\text{f}})(1+\gamma)^3\phi}\right] \tag{2.6}$$

where h is the layer thickness, a is the particle radius, $\gamma = \frac{h}{a}$, and k_{pe} is the thermal conductivity of the particle. This model yields better predictions because it takes into account the increased volume fraction by considering the interfacial layer, where layering of liquid molecules at the solid-particle surface occurs. The interfacial layer of a single spherical particle is shown in Fig. 2.1. These models, however, have been unable to predict the thermal conductivity of a fluid at the nanoscale.

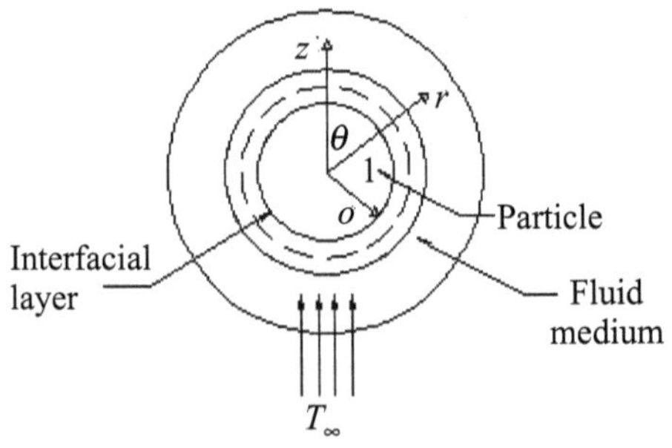

Figure 2.1 Graphical representation of a spherical nanoparticle and the interfacial layer that separates it from the fluid medium.

The increase in thermal conductivity is not due solely to the addition of nanoparticles to the base fluid, however, because conventional models have consistently underpredicted the experimental results. For low volume fractions, it is about one order greater than predicted by the classical models. Many models have been proposed to explain this phenomenon. The Hamilton and Crosser (HC) model for a suspension containing spherical particles is given by Eq. (2.7).

$$\frac{k_{\mathrm{eff}}}{k_{\mathrm{f}}}=\frac{\varepsilon_{\mathrm{p}}+2-2(1-\varepsilon_{\mathrm{p}})\phi}{\varepsilon_{\mathrm{p}}+2+(1-\varepsilon_{\mathrm{p}})\phi} \tag{2.7}$$

In this equation, $\varepsilon_{\mathrm{p}} = k_{\mathrm{p}}/k_{\mathrm{f}}$ is the ratio of the particle's thermal conductivity to the base fluid's, and $\Phi=\frac{4n\pi r_{\mathrm{p}}^{3}}{3}$ is the original volume fraction of nanoparticles in the nanofluid; n is the number of nanoparticles per volume of fluid, and r_{p} is the radius of the nanoparticle. By observing the equation, it is obvious that there is no term that accounts for the particle size [5]. Again, while this model predicts the thermal conductivity of macro- and microfluids relatively well, it has fared poorly when compared to experimental determinations of k_{eff} for nanofluids. One of the main reasons for the discrepancy between the classical models and the effect of nanoparticles on a fluid's thermal conductivity lies at the interfacial layer (or nanolayer for nanofluids). The layer acts as a thermal bridge between the base liquid and the nanoparticle, contributing to the increased thermal conductivity. Since liquid molecules adopt structured, solid-like layers at solid surfaces, their thermal conductivity lies between the smaller value for the bulk liquid and the larger value for the solid particle [10].

More recent models that have shown better results have included more parameters. These acknowledge the discontinuous temperature gradient at the interfacial boundaries where molecules exhibit different thermophysical properties and concentration. However, there is still a need for a model that quantifies the thickness of the nanolayer and its thermal properties.

Some models consider a nanoparticle and its attached nanolayer as a complex nanoparticle. As a result of the random thermal motion (Brownian motion) in the surrounding liquid molecules, the complex nanoparticles experience restless motion. Brownian motion is a continuous-time stochastic process that can be explained by a probability density. Therefore, this motion causes microconvection of the base fluid and enhances its overall heat transport. However, Brownian motion remains a controversial contributing factor among scientists because of the difficulty in quantifying Brownian effects on nanoparticle dispersion as well as the degree to which it affects the thermal conductivity of a nanofluid.

Ren et al. [5] described four heat transfer modes that account for the overall heat flux in the nanofluid, which is given by

$$\langle \vec{q}_{\mathrm{t}} \rangle = \langle \vec{q}_{\mathrm{f}} \rangle + \langle \vec{q}_{\mathrm{e,p}} \rangle + \langle \vec{q}_{\mathrm{e,l}} \rangle + \langle \vec{q}_{\mathrm{e,c}} \rangle \tag{2.8}$$

The subscripts t, f, p, l, and c represent the total heat flux, the heat flux through the base liquid, the heat flux through the nanoparticle, heat flux through the nanolayer, and the heat flux due to micro-convection, respectively. The subscript e denotes that the term is a nano-related term. The schematic for the overall heat flux is shown in Fig. 2.2. In order to analyze the thermophysical behavior of the interfacial nanolayer, it is necessary to analyze the factors that affect it. These are the suspended nanoparticle, the base fluid, and the relationship between the two.

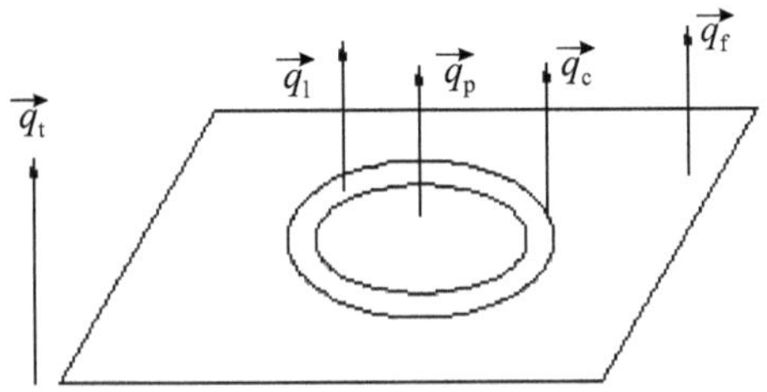

Figure 2.2 Schematic of the four heat transfer modes that make up the total heat flux in the nanofluid.

The most common method of measuring effective thermal conductivity of a nanofluid is the transient hot-wire method [9]. In this method, wires with a small diameter are immersed within a fluid and are used as resistance thermometers to measure electrical resistance power dissipation. The thermal conductivity is simply found from graphing the change in temperature over time and calculating the slope. This is a preferred method because it eliminates many of the effects of convection.

2.2.3 Heat Transfer Studies

Experimental data have yielded certain quantifiable relationships that may allow future models to accurately predict thermal conductivity and, thereby, understand how to control it. The motion of the particles is dependent on temperature, and accordingly, the thermal conductivity of the nanofluid also exhibits a strong relation to temperature. Additionally, as the thickness of the nanolayer increases, the effective thermal conductivity does as well. This is an important experimental discovery as it would suggest that modifying the nanolayer structure would significantly increase the effective thermal conductivity of the nanofluid. Experimental data have also supported the statement that as the size of the suspended particles increases, the nanolayer and thermal movement effects diminish.

Wen and Ding [11] conducted an experimental study that analyzed the convective heat transfer coefficient of λAl_2O_3 nanoparticles, which were suspended in deionized water in laminar flow through a copper tube. The experimenters concentrated on the entrance region of the particles. The experimental apparatus consisted of a straight copper tube, about 970 mm in length, 4.5 mm inner diameter, and 6.4 mm outer diameter for the test section. A silicon rubber flexible heater with a 300 W capacity kept a constant heat flux boundary condition, and a maximum flow rate was maintained by using a peristaltic pump with adjustable rotational speed. Various concentrations of nanoparticles were used and were stabilized by a dispersant, sodium dodecylbenzene sulfonate. A schematic of the experimental setup is shown in Fig. 2.3.

Experimental results found that the heat transfer coefficient varied with the Reynolds number and the volume fraction of particles inside the deionized water. The Reynolds number is a measure in fluid mechanics, which is the ratio of inertial forces to viscous forces within a fluid. Overall, the heat transfer coefficient of the fluid was significantly increased by the presence of nanoparticles, causing a decrease in the boundary layer and along the tube length.

Similarly, Yang et al. [12] studied the convective heat transfer coefficient of graphite nanoparticles dispersed in a liquid in laminar flow in a horizontal tube heat exchanger. Two different nanofluids were used, and the nanoparticles had disc-like aspect ratios in order to minimize the increase in fluid viscosity. Several parameters such as Reynolds number, volume flow rates, and fluid temperature were kept within specific ranges. As shown in Fig. 2.4, the study was intended to look at the effects of the aforementioned variables on the convective heat transfer coefficient.

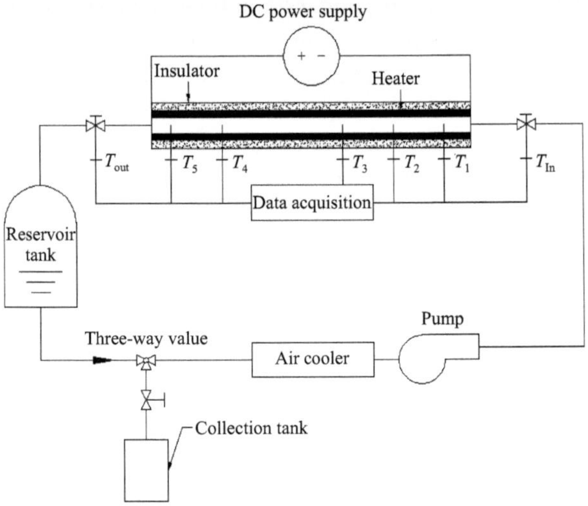

Figure 2.3 Schematic of the Wen and Ding experimental apparatus and setup [11].

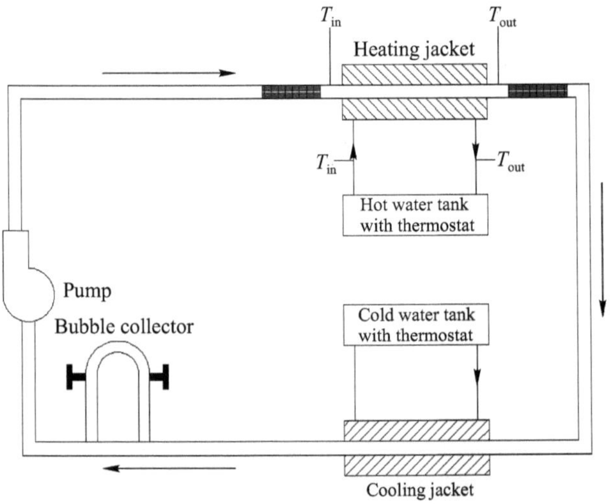

Figure 2.4 Schematic of the experimental apparatus by Yang et al. [12].

After mathematical analysis of the experimental data, similar results were obtained when compared to Wen and Ding's study. Increasing the Reynolds number and the particle volume fraction caused an increase in the heat transfer coefficient of the nanofluid.

2.2.4 Synthesis of Nanofluids

In addition to the principles and facts regarding nanofluids, it is important for engineering students to understand how to create a nanofluid in order to improve efficiency of the process. In order to create a nanofluid, several basic criteria must be met. The particle suspension must be even and stable, with a low amalgamation of particles. A fatty acid is commonly used as a dispersant to cover the nanoparticles and reduce the amalgamation. Additionally, there should be no chemical change to the fluid. Nanofluids can be produced in two ways: the single-step and the two-step method [9]. The single-step, direct evaporation approach is called the VEROS (vacuum evaporation onto a running oil substrate) technique and is shown in Fig. 2.5. The liquid is contained in a cylinder that is perpetually rotated to maintain a thin layer of fluid above a resistively heated crucible. The cooling system cools the liquid to prevent an increase in vapor pressure due to evaporation radiation. Nanophase powders in the vapor phase are condensed directly into the fluid. The single-step technique minimizes the agglomeration of nanoparticles but only works with low vapor pressure fluids. In this method, the nanoparticles are manufactured first and are then dispersed into the fluid with techniques such as ultrasound and high shear. The two-step method is used widely in the creation of nanofluids in view of the available commercial nanopowders manufactured by various companies. However, although this technique works well with oxide nanoparticles, it is not as effective with denser, metal nanoparticles [3].

Direct evaporation is a physical vapor deposition technique, meaning that a parent material in the solid phase, such as wires, sheets, or powders, is changed into a molten or liquid state within an evaporation chamber. The phase change results from the input of heat from a source such as a heating coil, electron beam, or laser. The nanoparticles are formed when the evaporated vapor is condensed on a cooling protrusion, and appears as very fine powder, which is collected. Figure 2.6 graphically shows how the technique is used in the creation of nanoparticles.

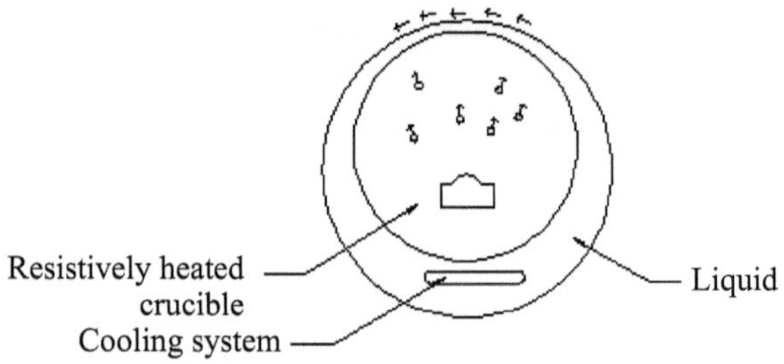

Figure 2.5 Schematic of the single-step, direct evaporation VEROS technique for synthesizing nanofluids.

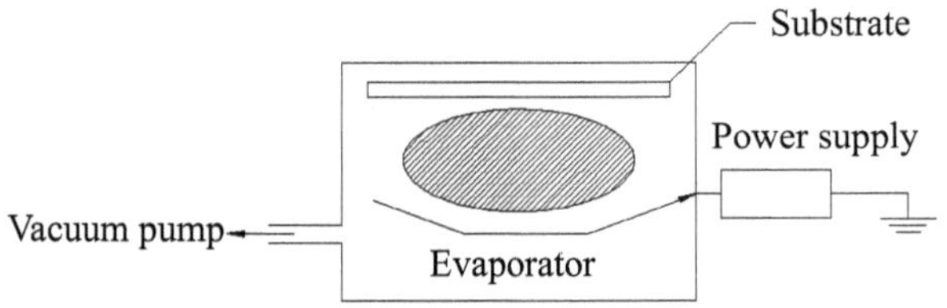

Figure 2.6 Illustration of the direct evaporation technique showing the phase change patch of the parent material.

In order for this technique to be successful, the parent material should have a large mean free path (λ) in its vapor form. The mean free path is the distance between consecutive gas molecules. If this measure is large, then the vapor atoms can travel from the source to the cooling surface without colliding with other atoms. This would protect the structure of the final product [13]. The mean free path is a function of temperature (T), pressure (P), and the collision cross section of the gas (σ), and is given by Eq. (2.9).

$$\lambda = \frac{1}{\sqrt{2}\sigma}\frac{kT}{P} \tag{2.9}$$

Various materials, many of which are oxides, are currently used in the nanoparticle production industry, such as alumina (Al_2O_3) and copper (II) oxide (CuO) [9]. Alumina is a common material used for manufacturing nanoparticles and is used largely in industry in the coating of light bulbs and fluorescent tubes due to its thermal and electrical insulating properties [14].

2.3 Conclusion

When it comes to nanotechnology, many questions are yet to be answered and even more to be formulated. Nanotechnology promises to lead into an entirely new era of advances in different branches of engineering. In order for us to maintain a high status in engineering and technology, engineers have to be exposed to the field at the collegiate level at the latest. There is a direct relationship between the quality of education given to the student engineers and the rate of technological advances made in society. By incorporating these modules into a junior-level thermodynamics/basic energy science course, it is possible to stimulate student engineers so that nanotechnology will become more knowledgeable to them and widespread, which will lead to a more rapid development of the field. The lead author has taught these modules for several semesters at a university. Initial educational evaluation data based on a couple of lectures alone gave an average concept inventory (C.I.) score (geared for the two lectures) of 61.4% for a class of about 20 students (Spring, 2007) in thermodynamics; the corresponding improvement in the average C.I. score for about 10 students (Summer, 2007) was from 43.7% to about 60%. Corresponding thermodynamic C.I. scores for typical thermodynamic courses in U.S. universities at the junior level have been reported to average in a range that included 60%, after the course [1].

Therefore, it is important for all students of engineering to have a solid background in nanotechnology in order to showcase the wider range of design possibilities they will have out in the field and, hence, further integrate nanotechnology into society.

Problems

1. Calculate the density (g/cm^3) and specific gravity of a 150 nm diameter nanoparticle whose weight has been approximated as 3.0×10^{-8} μg.

Solution: Density is defined as the ratio of mass to volume, or

$\rho = m/V$

$\rho = 3.0 \times 10^{-14}$ g/[4π/3*(75 × 10^{-7} cm)^3] = 16.98 g/cm^3

Specific gravity (SG) is the ratio of an object's density to that of water at approximately 4°C.

SG = 16.98 g/cm^3/1.0 g/cm^3 = 16.98

2. A straight brass tube contains a nanofluid made from a dispersion of copper nanoparticles in deionized water. The experimental conditions are such that a constant wall heat flux of 3 kW/m^2 is maintained. If the wall temperature is measured to be 150°C and the fluid temperature is measured to be 107°C, what is the heat transfer coefficient of the nanofluid?

   Solution: The heat transfer coefficient is given by Eq. (2.3):

   $$h_{nf} = q/(T_w - T_f)$$

   or $$h_{nf} = (3000 \text{ W/m}^2)/(43 \text{ K}) = 69.8 \text{ W/m}^2\text{K}$$

3. Given the heat transfer coefficient above, what is the Nusselt number of the nanofluid if the inner diameter of the tube is 10 mm and the thermal conductivity of the nanofluid is 0.11 W/mK?

   Solution: The Nusselt number provides a measure of the enhancement of heat transfer when convection takes place, as compared to simple conduction and is given by Eq. (2.2):

   $$Nu_{nf} = h_{nf}D/k_{nf}$$

   or $$Nu_{nf} = (69.8 \text{ W/m}^2\text{K})(0.01 \text{ m})/(0.11 \text{ W/mK}) = 6.35$$

4. A horizontal tube exchanger contains water as a heating fluid and a nanofluid made of graphite nanoparticles in ethylene glycol. Water flows in at a rate of 8 kg/s and the nanofluid flows in at a rate of 5 kg/s. The initial water temperature is 95°C, and the initial nanofluid temperature is 30°C. The final temperature is 80°C. What is the specific heat of the nanofluid in J/kg°C? Knowing that the constant pressure heat capacity for ethylene glycol is 149.5 J/mol°C and the molecular weight is 62.07 g/mol, find the percent difference in the specific heat of the base fluid and the nanofluid.

   Solution: Assuming that the heat lost to the surroundings is negligible, the heat lost by the water is equal to the heat gained by the nanofluid, or

$$-Q_{\text{nf}} = Q_{\text{hf}}$$

$$m'Cp(T_{\text{in}} - T_{\text{out}}) = m'_{\text{hf}} Cp_{\text{hf}} (T_{\text{out}} - T_{\text{in}})_{\text{hf}}$$

$$Cp = (8\ \text{kg/s})(4184\ \text{J/kg°C})(15°\text{C})/[(5\ \text{kg/s})(50°\text{C})]$$

or $Cp = 2008\ \text{J/kg°C}$

To find the specific heat of ethylene glycol in compatible units, arrange the two known properties so that the unwanted units cancel out:

$$Cp_{\text{ethylene}} = (149.5\ \text{J/mol°C})/(62.07\ \text{g/mol}) = 2.409\ \text{J/g°C}$$
$$= 2409\ \text{J/kg°C}$$

Percent difference = |2008 – 2409|/2409×100% = 16.6%

5. A nanofluid is composed of alumina (Al_2O_3) particles suspended in water. There are 50×10^9 particles per 1 mm^3 of nanofluid, and each particle has a diameter of 80 nm. Given that the thermal conductivity of alumina and water is 36.0 W/mK and 0.610 W/mK, respectively, calculate the ratio of the effective thermal conductivity to that of water alone using the HC model.

   Solution: The Hamilton–Crosser model is given by Eq. (2.7), where

   $$\varepsilon_{\text{p}} = (36.0/0.610) = 59.0$$

   and using the definition of the volume fraction of nanoparticles in the nanofluid by identifying the given information as

   $$n = 50\times10^{9}\ /\text{mm}^3 = 5\times10^{19}\ /\text{m}^3$$
   $$r_{\text{p}} = 40\ \text{nm} = 4\times10^{-8}\ \text{m}$$
   $$\phi = (4)(5\times10^{19})\ \pi\ (4\times10^{-8})\text{\^{}}3/3 = 1.34\times10^{-2}$$
   $$k_{\text{eff}}/k_{\text{f}} = [59.0+2-2(1-59.0)(1.34\times10^{-2})]/[59.0+2+(1-59.0)(1.34\times10^{-2})]$$

   or $k_{\text{eff}}/k_{\text{f}} = 1.0387$

6. If the experimentally determined value of k_{eff} for the nanofluid described in Problem 6 is 0.7 W/mK, determine the percent difference between the model and the experimental (assumed to be accurate) value using your calculated value.

   Solution: Rearranging the ratio to solve for k_{eff}, we have

   $$k_{\text{eff}} = (1.0387)(0.610\ \text{W/mK}) = 0.6336\ \text{W/mK}$$

   Using the equation for percent error, we get

   % error = |actual value – calculated value|/actual value×100%
   = |0.7 – 0.6336|/0.7×100% = 9.5%

References

1. www.foundationcoalition.org/home/keycomponents/concept/index.html, 2007, for concept inventory.
2. Roco, M. C. 2002. Nanotechnology: A frontier for engineering education, *International Journal of Engineering Education*, **18**, 1–14.
3. Choi, S. 1998. Nanofluid technology: Current status and future research, *Korea-U.S. Technical Conference on Strategic Technologies.*
4. Wong, K. V. 2000. *Thermodynamics for Engineers*, CRC Press, 370.
5. Ren, Y., Huaqing, X., and Cai, A. 2005. Effective thermal conductivity of nanofluids containing spherical nanoparticles, *Journal of Physics D: Applied Physics*, **38**, 3958–3961.
6. Liu, M.-S., Lin, M. C.-C., Huang, I.-T., and Wang, C.-C. 2005. Enhancement of thermal conductivity with carbon nanotube for nanofluids, *International Communications in Heat and Mass Transfer*, **32**, 1202–1210.
7. Ma, H. B., Wilson, C., Borgmeyer, B., Park, K., Yu, Q., Choi, S. U. S., and Tirumala, M. 2006. Effect of nanofluid on the heat transport capability in an oscillating heat pipe, *Applied Physics Letter*, **88**, 143116.
8. Daungthongsuk, W. and Wongwises, S. 2007. A critical review of convective heat transfer of nanofluids, *Renewable and Sustainable Energy Reviews*, **11**, 797–817.
9. Wang, X.-Q. and Majumdar, A. S. 2007. Heat transfer characteristics of nanofluids: A review, *International Journal of Thermal Sciences*, **46**, 1–19.
10. Leong, K. C., Yang, C., and Murshed, S. M. S. 2006. A model for the thermal conductivity of nanofluids: The effect of interfacial layer, *Journal of Nanoparticle Research*, **8**, 245–254.
11. Wen, D. and Ding, Y. 2004. Experimental investigation into convective heat transfer of nanofluids at the entrance region under laminar flow conditions, *International Journal of Heat Mass Transfer*, **47**, 5181–5188.
12. Yang, Y., Zhang, Z. G., Grulke, E. A., Anderson, W. B., and Wu, G. 2005. Heat transfer properties of nanoparticle-in-fluid dispersions (nanofluids) in laminar flow, *International Journal of Heat Mass Transfer*, **48**, 1107–1116.
13. Eastman, J. A., Phillpot, S. R., Choi, S. U. S., and Keblinski, P. 2004. Thermal transport in nanofluids, *Annual Review of Materials Research*, **34**, 219–246.

14. Theodore, L. 2006. *Nanotechnology: Basic Calculations for Engineers and Scientists*, John Wiley & Sons, Hoboken.

Chapter 3

Transport Properties of Alumina Nanofluids

Kaufui V. Wong and Tarun Kurma
Department of Mechanical and Aerospace Engineering, University of Miami, Coral Gables, FL 33124, USA
kwong@miami.edu

Recent studies have showed that nanofluids have significantly greater thermal conductivity compared to their base fluids. Large surface-area-to-volume ratio and certain effects of Brownian motion of nanoparticles are believed to be the main factors for the significant increase in the thermal conductivity of nanofluids. In this chapter, all three transport properties, namely, thermal conductivity, electrical conductivity, and viscosity, will be studied for alumina nanofluid (aluminum oxide nanoparticles in water). Experiments were performed both as a function of volumetric concentration (3–8%) and temperature (2–50°C). Alumina nanoparticles with a mean diameter of 36 nm were dispersed in water. The effect of particle

This chapter has been reprinted from the *Nanotechnology*, Vol. 19(34), Kau-Fui Vincent Wong and Tarun Kurma, Transport properties of alumina nanofluids, Copyright 2008, with permission from IOP Publishing.

Nanotechnology and Energy
Edited by Kaufui V. Wong

ISBN 978-981-4774-24-6 (Hardcover), 978-1-315-16357-4 (eBook)
www.panstanford.com

size was not studied. The transient hot wire method as described by Nagaska and Nagashima for electrically conducting fluids was used to test the thermal conductivity [1]. In this work, an insulated platinum wire of 0.003-inch diameter was used. Initial calibration was performed using de-ionized water and the resulting data were within 2.5% of standard thermal conductivity values for water. The thermal conductivity of alumina nanofluid increased with both increase in temperature and concentration. A maximum thermal conductivity of 0.7351 W/mK was recorded for an 8.47% volume concentration of alumina nanoparticles at 46.6°C. The effective thermal conductivity at this concentration and temperature was observed to be 1.1501, which translates to an increase in thermal conductivity by 22% when compared to water at room temperature. Alumina being a good conductor of electricity, alumina nanofluid displays an increasing trend in electrical conductivity as volumetric concentration increases. A microprocessor-based conductivity/TDS meter was used to perform the electrical conductivity experiments. After carefully calibrating the conductivity meter's glass probe with platinum tip, using a standard potassium chloride solution, readings were taken at various volumetric concentrations. A 3457.1% increase in the electrical conductivity was measured for a small 1.44% volumetric concentration of alumina nanoparticles in water. The highest value of electrical conductivity, 314 μS/cm, was recorded for a volumetric concentration of 8.47%. In the determination of the kinematic viscosity of alumina nanofluid, a standard kinematic viscometer with constant temperature bath was used. Calibrated capillary viscometers were used to measure flow under gravity at precisely controlled temperatures. The capillary viscometers were calibrated with de-ionized water at different temperatures, and the resulting kinematic viscosity values were found to be within 3% of the standard published values. An increase of 35.5% in the kinematic viscosity was observed for an 8.47% volumetric concentration of alumina nanoparticles in water. The maximum kinematic viscosity of alumina nanofluid, 2.901 42 mm^2/s, was obtained at 0°C for an 8.47% volumetric concentration of alumina nanoparticles. The experimental results of the present work will help researchers arrive at better theoretical models.

3.1 Background and Motivation

A number of journals, books, and papers were reviewed before they were shortlisted for the few significant references below.

3.1.1 Hamilton and Crosser Model

Lee et al. performed experiments using the hot wire method and discussed the tremendous increase in the thermal conductivity of oxide nanoparticles [2]. They compared their results with the Hamilton and Crosser (HC) model, described below.

Maxwell's model shows that the effective thermal conductivity of suspensions containing spherical particles increases with the volume fraction of the solid particles. For nanospherical particles, the thermal conductivity of suspensions depends on the volume fraction of the particles and also on the shape of the dispersed particles.

Hamilton and Crosser developed an elaborate model for the effective thermal conductivity of two-component mixtures as a function of the conductivity of the pure materials, the composition mixture, and the shape of the dispersed particles [3]. The conductivity of two-component mixtures can be calculated as follows:

$$k = k_o \left[\frac{k + (n-1)k_o - (n-1)\alpha(k_o - k_p)}{k_p + (n-1)k_o + \alpha(k_o - k_p)} \right] \tag{3.1}$$

where k is the thermal conductivity of nanofluid, k_o is the thermal conductivity of the base fluid, k_p is the thermal conductivity of nanoparticle (alumina), α is the volume fraction, $n = 3/\psi$ is the empirical shape factor, where ψ is sphericity, defined as the ratio of the surface area of a sphere (with a volume equal to that of the particle) to the surface area of the particle. For alumina nanofluid, $n = 3$.

3.1.2 Jang and Choi's Model

Jang and Choi described the role of Brownian motion in the enhanced thermal conductivity of a nanofluid [4]. Traditional conductivity theories of solid/liquid suspensions, such as the Maxwell or other macroscale approaches, cannot explain why nanofluids have these intriguing features. The common assumption in the conductivity of solid/fluid suspensions is that the discrete particles are motionless in a continuous material. At very small particle sizes, as in nanofluids, Brownian motion becomes important, requiring a different approach to understand energy transport in nanofluids. Jang and Choi developed a theoretical model based on kinetics, Kapitza resistance, and convection, and derived a general expression for the thermal conductivity of nanofluids [4].

The first mode of heat transfer is energy transport due to collision between base fluid molecules, which physically represents the thermal conductivity of the base fluid.

Assuming that the energy carrier travels freely only over the mean free path l, after which the base fluid molecules collide, the net energy flux J_U across a plane at z is given by

$$J_U = -\frac{1}{3} l_{BF} \hat{C}_{V,BF} \bar{C}_{BF} (1-f) \frac{dt}{dz} = -k_{BF} \frac{dt}{dz} (1-f) \tag{3.2}$$

where $\hat{C}_V$ is the heat capacity per unit volume, $\bar{C}$ is mean speed, T is the temperature of the base fluid molecules, ρ is the volume fraction, k_{BF} is the thermal conductivity of base fluid, l_{BF} is the mean free path of the base fluid molecules.

The second mode of heat transfer is thermal diffusion in nanoparticles in fluids. This can be represented by the equation

$$J_U = -\frac{1}{3} l_{nano} \hat{C}_{V,nano} \bar{v} f \frac{dt}{dz} = -k_{nano} \frac{dt}{dz} f \tag{3.3}$$

where l_{nano} is the mean free path of nanoparticles and k_{nano} is the thermal conductivity of suspended nanoparticles.

The thermal conductivity of suspended nanoparticles involving Kapitza resistance is

$$k_{nano} = \beta\, k_{particle} \tag{3.4}$$

where β is a constant related to Kapitza resistance $\sim \mathbf{O}(10^{-2})$ and $k_{particle}$ is the thermal conductivity of nanoparticles.

The third mode of heat energy transfer is collision between nanoparticles due to Brownian motion. This mode is much smaller than the other modes and can be neglected.

The fourth mode of heat energy transfer is thermal interaction of dynamic or dancing nanoparticles with base fluid molecules.

Therefore, Brownian motion of nanoparticles in nanofluids produces convection-like effects at the nanoscale:

$$J_{\mathrm{U}} = h(T_{\mathrm{nano}} - T_{\mathrm{BF}})f = h\delta_{\mathrm{T}} f \frac{(T_{\mathrm{nano}} - T_{\mathrm{BF}})}{\delta_{\mathrm{T}}} \sim h\delta_{\mathrm{T}} f \frac{dt}{dz} \tag{3.5}$$

where h is the heat transfer coefficient for flow past nanoparticles and δ_{T} is the thermal boundary layer thickness.

Combining the first, second, and fourth modes while neglecting the third, we get

$$k_{\mathrm{eff}} = k_{\mathrm{BF}}(1-f) + k_{\mathrm{nano}} f + fh\delta_{\mathrm{T}} \tag{3.6}$$

For flow past a sphere, the Nusselt number is

$$Nu = 2.0 + 0.5 + RePr + O(Re^2 Pr^2) \tag{3.7}$$

For typical nanofluids, the Reynolds and Prandtl numbers are of the order of 1 and 10, respectively, and Eq. (3.7) can be simplified as

$$Nu \sim Re^2 Pr^2 \tag{3.8}$$

The heat transfer coefficient for flow past nanoparticles is then defined by

$$h \sim \frac{k_{\mathrm{BF}}}{d_{\mathrm{nano}}} Re_{d_{\mathrm{nano}}}^2 Pr^2 \tag{3.9}$$

where d_{nano} is the diameter of a nanoparticle and $Re_{d_{\mathrm{nano}}}$ is the Reynolds number defined by Eq. (3.13). To estimate the thickness of the thermal boundary layer on moving nanoparticles, Jang and Choi postulated that nanolayers of ordered liquid molecules act as a hydrodynamic boundary layer, and that three layers exist at the interface and define the hydrodynamic boundary layer $\boldsymbol{\delta}$ as

$$\delta \sim 3d_{\mathrm{BF}} \tag{3.10}$$

where d_{BF} is the diameter of the base fluid molecule [4]. This idea of a hydrodynamic boundary layer at the nanoscale level was taken into account, considering that base fluid molecules at the interface have much lower velocity than that of free molecules in the bulk fluid. The thermal boundary layer was then estimated from

$$\delta_{\mathrm{T}} \sim \frac{\delta}{Pr} \tag{3.11}$$

By means of Eqs. (3.9)–(3.11), we can rewrite Eq. (3.6) as

$$k_{\text{eff}} = k_{\text{BF}}(1-f) + k_{\text{nano}}f + 3C_1 \frac{d_{\text{BF}}}{d_{\text{nano}}} k_{\text{BF}} Re^2_{d_{\text{nano}}} Prf \tag{3.12}$$

where C_1 is a proportional constant and $Re_{d_{\text{nano}}}$ is the Reynolds number defined by

$$Re_{d_{\text{nano}}} = \frac{\overline{C}_{\text{RM}} d_{\text{nano}}}{v} \tag{3.13}$$

In Eq. (3.13), $\overline{C}_{\text{RM}}$ and v are the random motion velocity of nanoparticles and dynamic viscosity of the base fluid, respectively. Einstein (1956) noted that thermally induced fluctuations would cause a particle to vary in direction many millions of times per second; it was assumed that a nanoparticle freely moves over the distance of the mean free path of a base fluid molecule. Therefore, $\overline{C}_{\text{RM}}$ was defined by

$$\overline{C}_{\text{RM}} = \frac{D_{\text{O}}}{l_{\text{BF}}} \tag{3.14}$$

where the nanoparticle coefficient is given by Einstein as

$$D_{\text{O}} = k_{\text{b}}T/3\pi\mu d_{\text{nano}} \tag{3.15}$$

and $k_{\text{b}} = 1.3807 \times 10^{-23}$ J/K is the Boltzmann constant.

3.1.3 Brownian Model

Prasher et al. described three possible mechanisms for thermal energy transfer in nanofluids: (i) translational Brownian motion, (ii) existence of an interparticle potential, and (iii) convection in the liquid due to the Brownian movement of the particles [5].

1. **Translational Brownian motion.** According to Prasher et al., Brownian motion of the particles is quite possible. The root mean square velocity (v_{N}) of a Brownian particle is defined as

$$v_{\text{N}} = \sqrt{\frac{3k_{\text{b}}T}{m_{\text{N}}}} = \frac{1}{d_{\text{N}}}\sqrt{\frac{18k_{\text{b}}T}{\pi\rho_{\text{N}}d_{\text{N}}}} \tag{3.16}$$

   where k_{b} is the Boltzmann constant, T is temperature, m_{N} is particle mass, ρ_{N} is density, and d_{N} is the diameter of the nanoparticle.

Applying kinetic theory, the enhancement in thermal conductivity due to the Brownian movement of the particles was derived as

$$k_{\text{BROWNIAN}} = \frac{(\phi C_{\text{N}} v_{\text{N}} l)}{3} \tag{3.17}$$

where C_{N} is the heat capacity per unit volume and l is the mean free path (due to the collision of the nanoparticles with each other and with the liquid molecules).

As a first estimate, l was assumed to be the same as d_{N}. Assuming $\phi = 0.05$ and alumina nanoparticles of size 10 nm, the observed enhancement in thermal conductivity for a water-based nanofluid was estimated to be 0.2%, and 0.5% for ethylene glycol, which is too small to explain the observed thermal conductivity data.

2. **Existence of an interparticle potential.** Assuming that interparticle potential can lead to phonon-like energy modes, the highest specific heat per unit volume is $3nk_{\text{b}}$, where n is the number of nanoparticles per unit volume.

   The enhancement in thermal conductivity by applying kinetic theory is

$$k = nk_{\text{b}} v l_{\text{p}} = \left(\frac{6\phi}{\pi d_{\text{N}}^3}\right) k_{\text{b}} v l_{\text{p}} \tag{3.18}$$

   where v is the velocity of the phonon-like mode (= 5000 m/s, which is the speed of sound in typical solids) and l_{p} is the mean free path of these modes (= 100 nm, which is typical for solids at room temperature).

   The estimated enhancement in thermal conductivity for alumina nanoparticles (d_{n} = 10 nm, ϕ = 0.05) is estimated to be 0.1% for water and 0.26% for ethylene glycol. Even for this most optimistic case, the enhancement is negligible due to the very small number of nanoparticles per unit volume in the system.

3. **Convection in the liquid due to the Brownian movement of the particles.** The Reynolds number based on v_{N} given by Eq. (3.16) can be written as

$$Re = \frac{1}{v}\sqrt{\frac{18k_{\text{b}}T}{\pi \rho_{\text{N}} d_{\text{N}}}} \tag{3.19}$$

where v is the kinetic energy of the liquid.

Since the Reynolds number for Brownian motion is inversely proportional to the square root of the diameter of the nanoparticle, *Re* for 10 nm alumina nanoparticles in water was calculated to be 0.0029, which is very small; therefore, the convection flow falls in the Stokes regime.

Consider a sphere embedded in a semi-infinite medium of thermal conductivity k_m. Then Nusselt's number (*Nu*) based on the radius of the particle ($a = d/2$) can be shown to be 1, i.e., $h = k_m/a$ [6].

In the Stokes regime [7], h is given as

$$h = \left(\frac{k_f}{a}\right)[1 + (1/4)\, Re \times Pr] \quad (3.20)$$

These relations are derived analytically from first principles. This means the effective thermal conductivity of the fluid due to the convection caused by the movement of a single sphere is

$$k_m = k_f\,[1 + (1/4)\, Re \times Pr] \quad (3.21)$$

For a 10 nm alumina nanoparticle, the enhancement at room temperature is approximately 4.2% for water and 6.6% for ethylene glycol. Therefore, the enhancement due to convention is one order of magnitude higher than any of the other mechanisms. It should be noted that the analysis was based on a single isolated sphere, whereas there will be interaction in the convection currents from different spheres. Substituting Eq. (3.21) into the Maxwell–Garnett model (MG), we obtain the following:

$$\frac{k}{k_f} = \left(1 + \frac{Re \times Pr}{4}\right)\left[\frac{(1+2\alpha)+2\phi(1-\alpha)}{(1+2\alpha)-\phi(1-\alpha)}\right] \quad (3.22)$$

where the thermal conductivity k of a composite for the Maxwell–Garnett model, for $k_m \ll k_p$, where k_m is the matrix conductivity and k_p is the particle conductivity [8], is given by

$$\frac{k}{k_m} = \left[\frac{(1+2\alpha)+2\phi(1-\alpha)}{(1+2\alpha)-\phi(1-\alpha)}\right] \quad (3.23)$$

where $\alpha = 2R_b k_m/d$ (R_b is the interfacial resistance, which is $\sim 0.77 \times 10^{-8}$ Km^2/W for water), ϕ is the particle volume fraction, and d is particle diameter.

Equation (3.22), together with the definition of *Re* given in Eq. (3.19), has all the necessary ingredients for predicting thermal conductivity of nanofluids, because it includes conduction contribution of the particles, interfacial resistance between the particles and the fluid, and a convection contribution (the effect of smaller d_N). This model was termed a single-sphere Brownian model [5].

The radial velocity field of a fluid due to the Stokes movement of a sphere [9] is given by

$$u_r = u_s - u_s\,[1 - 1.5(a/r) + 0.5(a/r)^3]\cos(\theta) \qquad (3.24)$$

where u_s is the fluid velocity at the surface of the sphere, r is the radial distance from the surface of the sphere, and θ is the polar angle.

Assuming the 99% criteria for the velocity to define the hydrodynamic boundary layer ($u_r = 0.001u_s$), the hydrodynamic boundary layer extends to $r/a \sim 50$. This shows that even at very small ϕ, the flow field due to two particles will interact and Eq. (3.21) thus needs to be modified. Brodkey et al. showed that heat transport in a nanofluid system is in some ways similar to particle-to-fluid heat transfer in fluidized beds, as convective currents from various spheres interact in fluidized beds [10]. They showed that for fluidized beds made of micron-sized particles, *Nu* for particle-to-fluid was 20–100 times more than that of the single-particle *Nu*.

It is also to be noted that, except for very simple conduction cases and geometries, the heat transfer correlations for convection in complicated situations, such as those due to the presence of multiple spheres, are always empirical in nature [6, 10].

Taking the cue from the *Nu* correlations for particle-to-fluid heat transfer in fluidized beds, Prasher et al. [5] proposed a general correlation for h of the form

$$h = \frac{k_f}{a(1 + ARe^m Pr^{0.333}\phi)} \qquad (3.25)$$

for the Brownian-motion-induced convection from multiple nanoparticles, where A and m are constants. This modification leads to

$$\frac{k}{k_f} = (1 + ARe^m Pr^{0.003}\phi)\left[\frac{(1+2\alpha)+2\phi(1-\alpha)}{(1+2\alpha)-\phi(1-\alpha)}\right] \qquad (3.26)$$

where A and m were found to be 4000 and 2.75 for alumina.

3.2 Experimental Results and Discussion

The *thermal conductivity* data presently available for alumina nanofluid are between 1% and 5% volumetric concentration and a temperature range of 21–85°C, whereas the experimental data from the current work expand the range to 8.5% volumetric concentration and the temperature range starting from 2 to 55°C. Table 3.1 shows the thermal conductivity results obtained from the hot wire method experiments.

Table 3.1 Thermal conductivity in W/mK for alumina nanofluid

Temperature (°C)	2.99% alumina nanofluid	4.67% alumina nanofluid	6.49% alumina nanofluid	8.47% alumina nanofluid
1.5	0.5946			
2.6			0.6199	
2.9		0.6111		0.6314
4.3			0.6242	
4.7		0.6168		
5.5	0.6038			
6.1				0.6402
8.0				0.6475
9.0		0.6259		
9.4			0.6376	
10.2	0.6152			
21.0	0.6383	0.6548	0.6680	0.681 88
32.9	0.6649			
34.0		0.6831		
34.4			0.7000	0.7099
34.9				
44.5		0.7042		
44.9	0.6876			
46.6			0.7214	0.7351

In Table 3.1, the highlighted values show the data obtained in the current experiments outside the range of already published data. Figure 3.1 shows the thermal conductivity of alumina nanofluid at different temperatures for various volumetric concentrations.

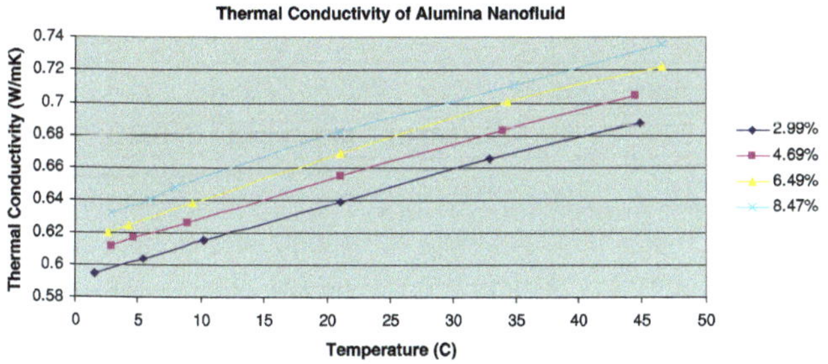

Figure 3.1 Thermal conductivity of alumina nanofluid.

Experiments were performed both as a function of volumetric concentration (3–8%) and temperature (2–50°C). Alumina nanoparticles with a mean diameter of 36 nm were dispersed in water. The transient hot wire method, as described by Nagasaka and Nagashima (1980 and 1981), for electrically conducting fluids was used to test the thermal conductivity [1, 11]. In this work, an insulated platinum wire of 0.003-inch diameter was used as the hot wire for the thermal conductivity experiments. Initial calibration was performed using de-ionized water and the resulting data were within 2.5% of the standard thermal conductivity values for water. The thermal conductivity of alumina nanofluid increased with an increase both in temperature and concentration.

The experimental results were compared to various theoretical and experimental models. The discussion below shows different models and a comparison between the models and the current experimental data.

3.2.1 Model 1: Hamilton and Crosser Model

The thermal conductivity equation developed by Hamilton and Crosser has been cited in Eq. (3.1) [3]. Table 3.2 shows the comparison between the HC model and alumina nanofluid at different volumetric concentrations of alumina nanoparticles; the experimental data were within 6.78% of the predicted model values.

Table 3.2 Comparison of experimental data with HC model for different volumetric concentrations of alumina nanoparticles (Thermal conductivity in W/mK)

Alumina nanofluid (vol%)	k/k_0 (exp)	k/k_0 (HC)	k (exp)	k (HC)
0	1.028	1.004	0.6167	0.6000
2.99	1.104	1.092	0.6625	0.6542
4.67	1.133	1.142	0.6835	0.6833
6.49	1.156	1.200	0.6958	0.7167
8.47	1.179	1.267	0.7083	0.7583

Tables 3.3–3.6 show the comparison with the HC model at different temperatures and concentrations. Alumina nanofluid with 2.99% volume concentration shows a maximum deviation of 3.05% from the HC model. Alumina nanofluid with 4.67% volume concentration shows a maximum deviation of 5.34% from the HC model. Alumina nanofluid with 6.49% volume concentration shows a maximum deviation of 8.55% from the HC model. Alumina nanofluid with 8.47% volume concentration shows a maximum deviation of 11.79% from the HC model.

Table 3.3 Comparison of experimental data with the HC model for 2.99% volumetric concentrations of alumina nanoparticles

Alumina nanofluid (2.99%)		
Temperature (°C)	k (exp)	k (HC)
1.5	0.5946	0.6135
5.5	0.6038	0.6218
10.2	0.6152	0.6306
21.0	0.6383	0.6535
32.9	0.6649	0.6747
44.9	0.6876	0.6935

Table 3.4 Comparison of experimental data with the HC model for 4.67% volumetric concentrations of alumina nanoparticles

Alumina nanofluid (4.67%)		
Temperature (°C)	*k* (exp)	*k* (HC)
2.9	0.6111	0.6450
4.7	0.6168	0.6490
9.0	0.6259	0.6589
21.0	0.6548	0.6835
44.5	0.6831	0.7075
34.0	0.7042	0.7245

Table 3.5 Comparison of experimental data with the HC model for 6.49% volumetric concentrations of alumina nanoparticles

Alumina nanofluid (6.49%)		
Temperature (°C)	*k* (exp)	*k* (HC)
2.6	0.6199	0.6778
4.3	0.6242	0.6826
9.4	0.6376	0.6931
21.0	0.6680	0.7182
34.4	0.7000	0.7446
46.6	0.7214	0.7662

Table 3.6 Comparison of experimental data with the HC model for 8.47% volumetric concentrations of alumina nanoparticles

Alumina nanofluid (8.47%)		
Temperature (°C)	*k* (exp)	*k* (HC)
2.9	0.6314	0.7167
6.1	0.6402	0.7242
8.0	0.6475	0.7273
21.0	0.6819	0.7576
34.4	0.7099	0.7864
46.6	0.7351	0.8076

3.2.2 Model 2: Jang and Choi Model

The Jang and Choi model [4] has been shown in Eq. (3.12). Figure 3.2 shows the comparison between the Jang and Choi model, with their data and the current experimental data, at different volumetric concentrations of alumina nanoparticles.

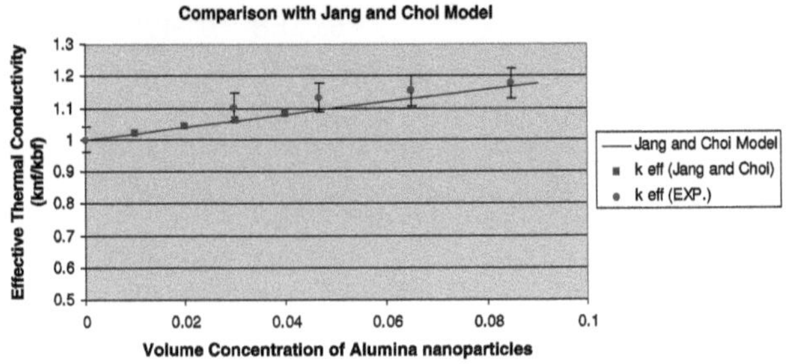

Figure 3.2 Comparison of the current experimental data with Jang and Choi's data [4] at different volumetric concentrations of alumina nanoparticles.

Jang and Choi's data for the thermal conductivity of alumina nanofluid are within 1–4% volumetric concentration of alumina nanoparticles [4]. In the current work, the range was extended to 8.5% volumetric concentration. Figure 3.2 shows the comparison of the current experimental data with Jang and Choi's theoretical model.

The experimentally obtained results were within 4% of the Jang and Choi model. It can be noticed that the experimental data approach the predicted theoretical model as the volumetric concentration of nanoparticles increases.

3.2.3 Model 3: Brownian Model

Prasher et al. in their journal paper "Thermal Conductivity of Nanoscale Colloidal Solutions (Nanofluids)" described the Brownian model for heat transfer as Eq. (3.26) [5].

Figure 3.3 shows the comparison between the Brownian model with Prasher et al.'s experimental data and the current experimental data, at different volumetric concentrations of alumina nanoparticles. The experimentally obtained results were within 2% of the Brownian model. Figure 3.4 shows the comparison with all the different models discussed in this work.

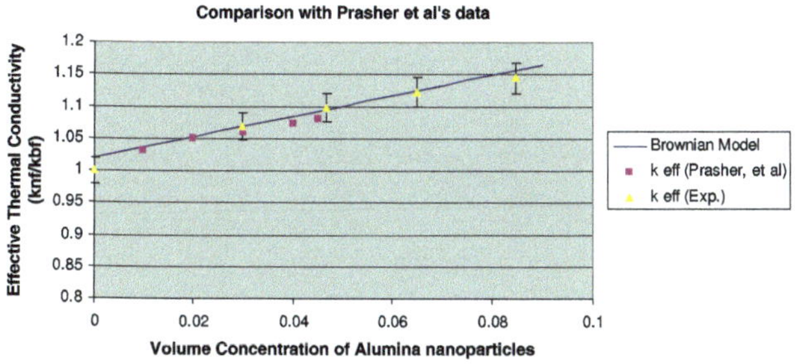

Figure 3.3 Comparison of the current experimental data with the data from Prasher et al. at different volumetric concentrations of alumina nanoparticles [5].

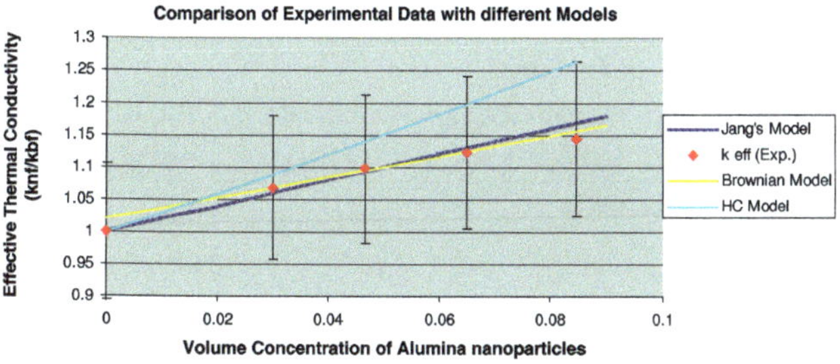

Figure 3.4 Comparison of the experimental data with the Jang and Choi model, the Brownian model, and the HC model.

There is insufficient data published when it comes to the *electrical conductivity* of alumina nanofluid. This has prompted us to perform electrical conductivity experiments on alumina nanofluid; this, along with thermal conductivity and viscosity, would give a better picture about the transport properties of alumina nanofluid.

The results presented in Table 3.7 were obtained by performing electrical conductivity tests using the bench-top microprocessor-based conductivity meter. The results are plotted in Fig. 3.5.

It is observed that the electrical conductivity of alumina nanofluid increases almost linearly with increase in the volumetric concentration of alumina nanoparticles. The increasing trend in the electrical conductivity is due to the high conductivity of the alumina. The base fluid's (de-ionized water) electrical conductivity is negligible. Therefore, it can be safely stated that alumina nanoparticles are the major contributor toward the electrical conductivity of alumina nanofluid.

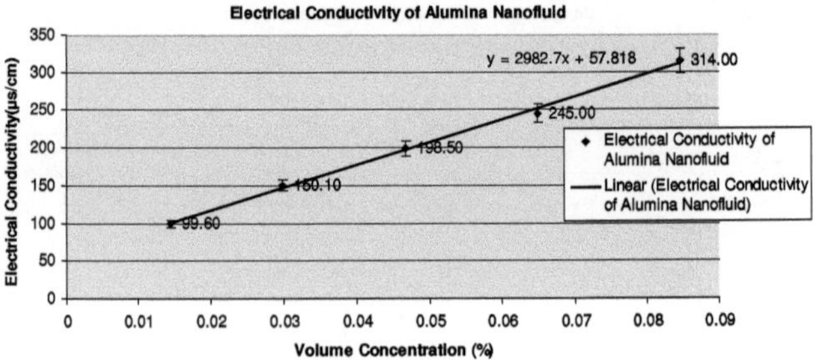

Figure 3.5 Electrical conductivity of alumina nanofluid at different volumetric concentrations of alumina nanoparticles.

Table 3.7 Electrical conductivity data for alumina nanofluid

Alumina nanofluid (vol%)	Electrical conductivity (μS/cm)	Resistivity (MΩ cm)	Total dissolved solids (mg/l)
1.44	99.6	0.010	66
2.99	150.1	0.007	100
4.67	198.5	0.005	131.6
6.49	245	0.004	163.1
8.47	314	0.003	209

Nara et al. described the *kinematic viscosity* using the following equation:

$$\nu = \frac{1}{Re}\sqrt{\frac{18k_{\mathrm{b}}T}{\pi\rho d}} \tag{3.27}$$

where ν is the viscosity of the fluid, Re is the Reynolds number for Brownian motion of nanoparticles (= 0.029 for alumina nanoparticles in water), k_{b} is the Boltzmann constant (= $1.380\,6503 \times 10^{-23}$ m^2 kg/s^2 K), T is temperature, ρ is the density of nanoparticles, and d is the diameter of alumina nanoparticles [12].

According to Ward and Whitmore, any physical interpretation of viscosity/volume concentration measurements made on non-flocculating suspensions of discrete, solid particles is generally based upon comparisons with simple systems of geometrically shaped bodies [13]. The most widely used relationship is that derived by Einstein for smooth, rigid, inertia-less, spherical particles at infinite dilution [14–16], according to which the relative viscosity ν_r is given by

$$\nu_r = \frac{\nu}{\nu_0} = 1 + k\phi \tag{3.28}$$

where ν is the viscosity of the suspension, ν_0 is the viscosity of the base fluid, ϕ is the volume concentration of the suspended particles, and k is constant (= 2.5) [13].

The kinematic viscosity equation stated by the ASTM D446-04 method is as follows:

$$\nu = \left(\frac{10^6 \pi g D^4 H t}{128 V\ L}\right) - \frac{E}{t^2} \tag{3.29}$$

where ν is the kinematic viscosity (in mm^2/s), D is the diameter of the capillary (in m), L is the length of the capillary (in m), H is the average distance between the upper and lower meniscus (in m), V is the timed volume of liquids passing through the capillary (in m^3, approximately the volume of a timing bulb), E is the kinetic energy factor (in mm^2 s), and t is the flow time (in s).

Based on the above relations, an equation for computing the kinematic viscosity was developed. This equation takes into consideration both the kinetic energy factor and the physical dimensions of the viscometer used for the experiment:

$$\nu = Ct - E/t^2 \tag{3.30}$$

where C is calculated to be 0.002 and 0.008 for size 25 and size 75 viscometers, respectively. E is the kinetic energy factor given by the following equation:

$$E = 52.5V^{3/2}/L(Cd)^{1/2} \tag{3.31}$$

where V is the volume of the timing bulb (in ml), L is the capillary working length (in mm), d is the capillary working diameter (in mm), and C is the viscometer constant (in mm^2/s^2).

The kinetic energy factor E is calculated to be 60.246 and 60.552 for viscometers of sizes 25 and 75, respectively. The kinematic viscosity experiments performed by Prasher et al. on alumina nanofluid were within the temperature range of 21–85°C and a volumetric concentration between 1% and 4.5% [5]. In the current work, the temperature range was expanded to 0–55°C, for a volumetric concentration between 2.9% and 8.5%. The results obtained are tabulated in Table 3.8 and plotted in Fig. 3.6.

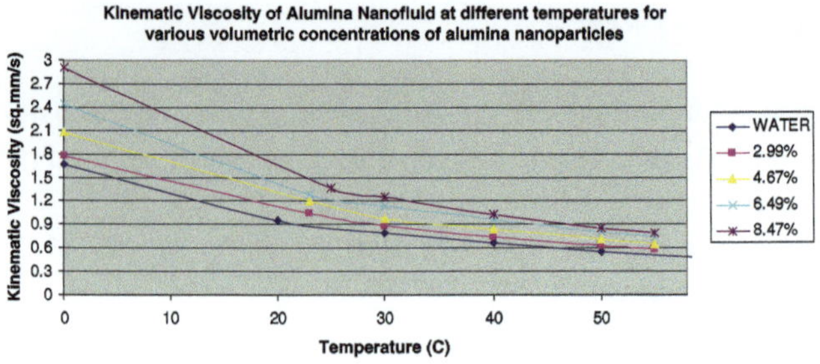

Figure 3.6 Kinematic viscosity of alumina nanofluid at different temperatures and at different volumetric concentrations of alumina nanoparticles.

It can be seen from the results that the viscosity of alumina nanofluid for each concentration decreases with increasing temperature as expected from most liquids. It can also be observed that the viscosity of alumina nanofluid increases with increasing concentration of alumina nanoparticles.

Table 3.8 Kinematic viscosity of alumina nanofluid

Temperature (°C)	2.99% alumina nanofluid	4.67% alumina nanofluid	6.49% alumina nanofluid	8.47% alumina nanofluid
0	1.775 13	2.073 14	2.443 27	2.901 42
23	1.046 13	1.199 89	1.291 59	
25				1.360 67
30	0.888 92	0.989 98	1.147 67	1.255 39
40	0.743 86	0.840 20	0.963 59	1.031 66
50	0.630 36	0.712 62	0.813 10	0.852 41
55	0.583 24	0.649 51	0.738 37	0.788 18

3.3 Conclusion

After performing many thermal conductivity experiments on alumina nanofluid, there is much evidence that the thermal conductivity of water is significantly increased with the addition of alumina nanoparticles. This phenomenon could be due to the large surface-area-to-volume ratio of alumina nanoparticles and to Brownian motion. The effect of Brownian motion on thermal conductivity is due to the collision between the nanoparticles and the base fluid molecules and also causes convection-like effects at the nanoscale due to the thermal interactions of dynamic or dancing nanoparticles with base fluid molecules. This argument can be reinforced by the fact that the kinematic viscosity of alumina nanofluid increased with increasing volume fraction of nanoparticles, which implies that at higher concentrations, the collision between nanoparticles and base fluid molecules should decrease since the nanofluid is more viscous, not allowing the nanoparticles to move freely, and thus decreasing the thermal conductivity. In contrast, the thermal conductivity values increased with increasing volume fraction, therefore showing that the convection-like effects may play a vital role in the alumina nanofluid's thermal conduction. A similar analysis could be applied to the electrical conductivity of alumina nanofluids, as it can be observed from the results that the electrical conductivity increases with increasing volume fraction of nanoparticles even though the electrical conductivity of water, the base fluid in this case, was negligible. The effects of particle size and shape were not studied; the alumina fluid studied was obtained commercially. A maximum thermal conductivity of 0.7351 W/m K was recorded for an 8.47% volume concentration of alumina nanoparticles at 46.6°C. The effective thermal conductivity at this concentration and temperature was observed to be 1.1501, which translates to an increase in thermal conductivity by 22% when compared to water at room temperature. A 3457.1% increase in the electrical conductivity was measured for a meager 1.44% volumetric concentration of alumina nanoparticles in water. The highest value of electrical conductivity, 314 μS/cm, was recorded for a volumetric concentration of 8.47%. An increase of 35.5% in the kinematic viscosity was observed for an 8.47% volumetric concentration of alumina nanoparticles in water. The maximum kinematic viscosity of alumina nanofluid, 2.901 42 mm^2/s, was obtained at 0°C for an 8.47% volumetric concentration of alumina nanoparticles. The experimental results of the present work will aid scientists and engineers gain a better understanding of the transport properties of nanofluids.

References

1. Nagasaka Y. and Nagashima A. 1981. Absolute measurement of the thermal conductivity of electrically conducting liquids by the transient hot-wire method, *J. Phys. E: Sci. Instrum.*, **14**, 1435–1440.
2. Lee S. P., Choi U. S. S., Li S., and Eastman A. J. 1999. Measuring thermal conductivity of fluids containing oxide nanoparticles, *J. Heat Transfer*, **121**, 280–289.
3. Hamilton R. L. and Crosser O. K. 1962. Thermal conductivity of heterogeneous two-component systems, *IEC Fundam.*, **1**, 187–191.
4. Jang P. S. and Choi U. S. S. 2004. Role of Brownian motion in the enhanced thermal conductivity of nanofluids, *Appl. Phys. Lett.*, **84**, 4316–4318.
5. Prasher R., Bhattacharya R., and Phelan E. P. 2005. Thermal conductivity of nanoscale colloidal solutions (nanofluids), *Phys. Rev. Lett.*, **94**, 025901.
6. Incorpera P. F. and Dewitt P. D. 1996. *Fundamentals of Heat and Mass Transfer* (New York: Wiley).
7. Acrivos A. and Tayor D. T. 1962. Conductivity in Stokes regime, *Phys. Fluids*, **5**, 387.
8. Nan C. W. 1997. The effective thermal conductivity or particular composites with interfacial thermal resistance, *J. Appl. Phys.*, **81**, 6692–6699.
9. Lamb H. 1945. *Hydrodynamics* (New York: Dover).
10. Brodkey R. S., Kim D. S., and Sidner W. 1991. Fluid to particle heat transfer in a fluidized bed and to single particles, *Int. J. Heat Mass Transfer*, **34**, 2327.
11. Nagasaka Y. and Nagashima A. 1980. Simultaneous measurement of the thermal conductivity and the thermal diffusivity of liquids by the transient hot-wire method, *Rev. Sci. Instrum.*, **52**, 229–232.
12. Nara S., Bhattacharya P., Vijayan P., Lai W., Rosenthal W., and Phelan E. P. 2005. Experimental determination of the effect of varying the base fluid on static thermal conductivity of nanofluids, *ASME Int. Mechanical Engineering Congr. RD&D Expo*, IMECE2005-81494, 755–758.
13. Ward G. S. and Whitmore L. R. 1950. Studies of the viscosity and sedimentation of suspensions, part 1—The viscosity of suspension of spherical particles, *Br. J. Appl. Phys.*, **1**, 286–290.
14. Einstein A. 1906. A new determination of molecular dimensions, *Ann. d. Phys.*, **19**, 289–306.

15. Einstein A. 1911. A new determination of molecular dimensions, Corrections, *Ann. d. Phys.*, **34**, 591–592.
16. Einstein A. 1956. *Investigation on the Theory of Brownian Movement* (New York: Dover).

Chapter 4

Heat Transfer Mechanisms and Clustering in Nanofluids

Kaufui V. Wong and Michael J. Castillo
Department of Mechanical and Aerospace Engineering, University of Miami, Coral Gables, FL 33124, USA
kwong@miami.edu

This chapter surveys heat transfer in nanofluids. It summarizes and analyzes the theories regarding heat transfer mechanisms in nanofluids, and it discusses the effects of clustering on thermal conductivity. The heat transfer associated with conduction is presented through various experiments followed by a discussion of the theories developed. Relationships between thermal conductivity and various factors such as temperature, concentration, and particle size are also displayed along with a discussion on clustering. There is a brief discussion on convection where the number of studies is limited. There is research currently being performed on the manipulation of the properties governing the thermal conductivity of nanofluids—the particle size, shape, and surface area. Other

---

This chapter has been reprinted from the *Advances in Mechanical Engineering*, Vol. 2010, Kaufui V. Wong and Michael J. Castillo, Heat transfer mechanisms and clustering in nanofluids, SAGE Publications Ltd., Copyright 2010 K. V. Wong and M. J. Castillo.

Nanotechnology and Energy
Edited by Kaufui V. Wong

ISBN 978-981-4774-24-6 (Hardcover), 978-1-315-16357-4 (eBook)
www.panstanford.com

factors that affect heat transfer are the material of the nanoparticle, particle volume concentration, and the fluid used. Although the interest in this relatively new class of fluids has generated many experimental studies, there is still disagreement over several aspects of heat transfer in nanofluids, primarily concerning the mechanisms behind the increased thermal conductivity. Although nanoparticles have greatly decreased the risks, there is still evidence of unwanted agglomeration, which causes erosion and affect the overall conductivity. Research is currently being conducted to determine how to minimize this unwanted clustering.

4.1 Introduction

The growth of technology found in high-tech industries, such as microelectronics, transportation, and manufacturing, has created a cornucopia of ideas that would have wide-ranging effects on many obstacles facing today's scientific world, including energy efficiency, pollution, and reusability. However, there are many factors hindering further development in these industries, one being the ability to rapidly cool the products being used. Cooling is necessary for maintaining the operational performance and reliability of new products, and as a result of increased heat loads and heat fluxes caused by the increase in power and decrease in feature sizes present in new products, the demand for a more efficient cooling process has increased dramatically in the last decade. Consequently, more companies are beginning to invest more capital into the research of more efficient heat transfer processes.

The conventional method for enhancing heat transfer in a thermal system consists of increasing the heat transfer surface area as well as the flow velocity of the working fluid [1]. The dispersion of solid nanoparticles in heat transfer fluids is a relatively new method. Extended surfaces such as fins and microchannels (width <100 μm) have already been used to increase the heat transfer surface area. Their performance in effectively removing as much as 1000 W/cm^2 has shown a great improvement in the area of cooling.

However, further development of this technology is at a standstill because it has already been pushed to its achievable limits. Thus, attention is now turning toward the dispersion of solid particles in fluids.

Since Maxwell's theories in 1873, scientists have attempted to increase the thermal conductivity of a fluid through the combination of solid particles and a heat transfer fluid. Although liquid cooling is prevalent today (i.e., in automobiles and in some microelectronics), it has been severely limited because of the inherent poor thermal conductivity of traditional heat transfer fluids. Efforts to increase this fundamental limit began with the dispersion of millimeter- or micrometer-sized particles in fluids. Although this action increased the thermal conductivity of the heat transfer fluid, it was not practical because the increase in heat transfer required a large number of particles (>10% by volume). This often resulted in a significant pressure drop, thus requiring more pumping power. Furthermore, because of their size, these particles rapidly settled in the liquid, clogged microchannels, and caused wear in pipes, pumps, and bearings.

As technology turned to miniaturization and nanotechnology, the idea of nanofluids was developed at various institutions around the world (initial development was performed at Argonne National Laboratory, and much research is still carried out at this location). Coined by Stephen U. S. Choi in 1995, the term nanofluids (short for nanoparticle fluid suspension) is used to describe stable suspensions of nanoparticles (average size <100 nm) in traditional heat transfer fluids such as water, oil, or ethylene glycol [1]. Experimentation has shown significant improvement in the thermal conductivity of fluids containing oxide and metallic nanoparticles. With volume concentrations between 0.5% and 4%, nanofluids have shown an enhancement of 15%–40% of the thermal conductivity of the base fluid [2]. It has also been observed that nanoparticles stay suspended in the fluid longer than micrometer-sized particles, thus reducing the severity of the obstacles presented by the rapid settling of the particles (such as abrasion and clogging of pipes and microchannels).

Experiments, such as those found in Refs. [3–10], are continuously being conducted to achieve a deeper understanding of the mechanisms behind the increase in thermal conductivity caused by these nanoparticle suspensions. In addition, the paper by Sandhu [11] declared the improved thermal conductivity of magnetic nanofluids. Philip et al. [12] showed that arranging the linear aggregation length from nano- to micron-scales, the thermal conductivity of the nanofluid was enhanced up to 216%, using 4.5 volume percentage of nanoparticles. Repeated magnetic cycling shows that the enhancement is reversible.

There has been a dramatic increase in interest in the field of nanofluids as shown by the exponential increase in the number of publications concerning the subject matter in the Science Citation Index (SCI) journals. Several businesses, corporations, and scientific institutions have begun research toward the common goal of achieving the highest possible thermal properties at the smallest possible concentration (preferably <1% by volume). Nanofluids have the potential of becoming compact, cost-effective liquid cooling systems in high-performance situations.

4.2 Conduction Heat Transfer in Nanofluids

Maxwell first proposed the idea of suspending metallic particles in conventional heat transfer fluids in 1873 [13]. He believed that the metallic particles (which have thermal conductivities that are significantly larger than that of liquids, as shown in Fig. 4.1) would increase the electrical and thermal conductivity of the fluids. This idea was carried on throughout the next century as scientists attempted to create a fluid with millimeter- and micrometer-sized particles that could be used for practical applications. However, even though their efforts showed that particles did increase the heat transfer properties of their base fluids, they could not overcome problems caused by the large size of the particles. It was not until the advent of nanotechnology that the thought of nanoparticles came about.

The idea of nanofluids created an influx of experimentation that has led to the knowledge that we have today. Though experiments have been carried out with nanoparticles formed from a wide array of materials, including aluminum oxide (Al_2O_3), copper (Cu), copper oxide (CuO), gold (Au), silver (Ag), silicon carbide (SiC), titanium carbide (TiC), titanium oxide (TiO_2), and carbon nanotubes, the most common nanoparticles are Al_2O_3 and CuO [14]. The base fluids used most commonly are water, engine oil, and ethylene glycol. The nanofluids used in these experiments are created by mixing the base fluid and the nanoparticles together and then stabilizing the suspension.

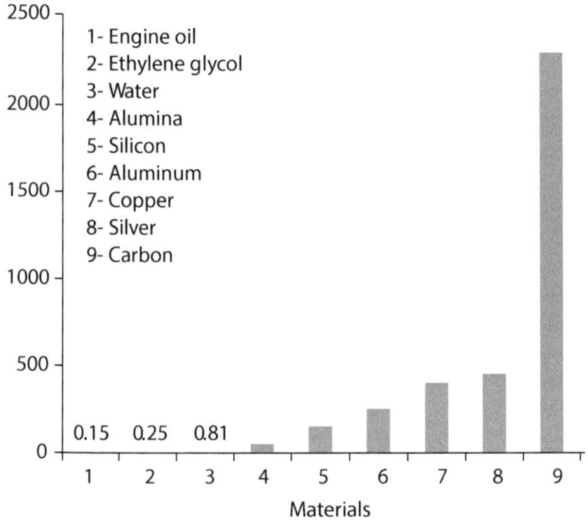

Figure 4.1 Thermal conductivity of typical materials [18].

Nanoparticles can be produced by either physical or chemical means. Current physical processes include mechanical grinding and the inert-gas-condensation technique pioneered by Granqvist and Buhrman. Presented in 1976, the latter technique involves evaporation in a temperature-regulated oven containing an inert gas [15]. Current chemical processes include chemical precipitation, chemical vapor deposition, microemulsions, spray pyrolisis, thermal spraying, and a sonochemical method for the production of iron nanoparticles [16]. The most common processes currently used in the production of metal nanoparticles include mechanical milling, inert-gas-condensation technique, chemical precipitation, spray pyrolisis, and thermal spraying [2]. Although nanoparticles are constantly being produced in small volumes for experimental needs, there is still research being conducted toward achieving more cost-efficient production processes in order to begin the move toward large-scale production.

There are currently two methods used to disperse nanoparticles in the base fluid: the two-step technique and the single-step technique. The *two-step method* involves making the nanoparticles first, by either physical or chemical means, and then dispersing them into the base fluid. In combination with the inert-gas-condensation technique (which has been proven to be a viable process for producing bulk quantities of nanopowders), the two-step method can be used to initiate the move toward commercialization by facilitating the mass production of nanofluids [17]. The *single-step method* involves simultaneously making and dispersing nanoparticles into the base fluid [1]. This method is favorable when using metallic nanoparticles—since the nanoparticles are placed in the base fluid as they are produced, this process helps prevent oxidation of the particles.

Because of the attractive van der Waals forces between the particles, they tend to agglomerate before they are dispersed in the liquid (especially if nanopowders are used); therefore, a means of separating the particles is necessary. Groups of particles will settle out of the liquid and decrease the conductivity of the nanofluid. Only by fully separating all nanoparticle agglomerates into their individual particles in the host liquid will a well-dispersed, stable suspension exist, and only under this condition will the optimum thermal conductivity exist. Xuan and Li [19] proposed different methods for the stabilization of the suspension, including changing the pH value of the nanofluid, using dispersants, and using ultrasonic vibration. The most commonly used method is the ultrasonic vibration, which has been relatively successful in eliminating agglomerated nanoparticles. There is certain hesitation when using dispersants because they can affect the chemical composition of the nanofluid and change the results. If dispersants are used, it cannot be clearly determined if the change in thermal conductivity was affected by the stabilizers placed in the fluid. This problem can be taken into account by comparing the nanofluid conductivity with that of the liquid with dispersant.

Three methods are commonly employed to measure the thermal conductivity of nanofluids: the transient hot wire method, temperature oscillation, and the steady-state parallel plate method [14]. The most commonly used method is the transient hot wire method, which involves the use of small-diameter wires that act as electrical resistance heaters and resistance thermometers [20]. The wire is heated by passing a current through it, and the rise in temperature over the time elapsed is measured. Since the wire is essentially wrapped in the liquid, the heat generated will be diffused into the liquid. The higher the thermal conductivity of the surrounding liquid, the lower the rise in temperature will be.

To calculate the thermal conductivity of the surrounding liquid, a derivation of Fourier's law for radial transient heat conduction is used [1]. This differential equation for the conduction of heat is

$$\frac{\partial^2 T}{\partial x^2}+\frac{\partial^2 T}{\partial y^2}+\frac{\partial^2 T}{\partial z^2}=\frac{1}{\alpha}\cdot\frac{\partial T}{\partial T}. \tag{4.1}$$

Using a solution presented by Carslaw and Jaeger [21], the conductivity of a solution can be expressed as

$$\frac{q}{4\pi(T_2-T_1)}\cdot\ln\frac{t_2}{t_2}, \tag{4.2}$$

where T_1 and T_2 represent the temperature of the heat source at times t_1 and t_2, respectively.

Many experiments have been performed using nanofluids, and although the results vary, they show an amplification of the thermal conductivity of the base fluid. Observations show four important characteristics of nanofluids:

1. Increased thermal conductivity at low nanoparticle concentrations
2. Linear relationship between thermal conductivity and concentration
3. Thermal conductivity being strongly dependent on temperature
4. Thermal conductivity being strongly dependent on the size of the particles used

As a result of debates as to why these characteristics are present in nanofluids, various theories have been developed.

Lee et al. [22] measured the thermal conductivity of nanofluids using CuO and Al_2O_3 nanoparticles and water and ethylene glycol as the base fluids. Results showed an enhancement in the thermal conductivity of ethylene glycol of more than 20% at 4% volume fraction of CuO nanoparticles. Further experimentation revealed that the thermal conductivity increased as the volume concentration of nanoparticles was increased; thus, it was determined that the thermal conductivity of nanofluids was dependent on the thermal conductivity of both the particles and the base fluid in most ranges of k_p/k_0, where k_p is the conductivity of the particle and k_0 is the conductivity of the base fluid. For large values of k_p/k_0, the later statement may not be true. Although the thermal conductivity of the nanofluid is always greater than that of the base fluid without nanoparticles, this increase will be different for each base fluid.

Xie et al. [23] measured the thermal conductivity of nanofluids containing Al_2O_3 nanoparticles. They also investigated the effects of the pH value of the suspension and the specific surface area (SSA) of the dispersed particles. In accordance with the previous results, the thermal conductivity of the fluid was enhanced with the addition of the nanoparticles. However, they also noted that the thermal conductivity increased as the difference between the pH value and the isoelectric point (pH value at which there is no electric charge) of Al_2O_3 increased and that the enhancements were highly dependent on the specific surface area of the nanoparticles. When compared to theoretical models, the measured thermal conductivity was much higher than the calculated values. A relationship between temperature and thermal conductivity was later presented by Das et al. [24]. In this study, not only it was determined that the thermal conductivity increased with an increase in temperature, but also it was shown that nanofluids composed of smaller particles experienced a greater enhancement than with larger particles. One possible explanation for this could be attributed to the Brownian motion of the particles in the fluid. Since temperature represents the overall kinetic energy of the particles, an increase in temperature will cause increased motion in the particles. It is easier for smaller particles to move; therefore, smaller particles will display a higher level of Brownian motion than larger particles. This results in greater heat conduction among smaller particles as the temperature is increased.

Results from these various experiments may have varied, but they all showed that the thermal conductivity of a fluid containing nanoparticles was greater than that of a base fluid with no particle suspension. Moreover, it was shown that the thermal conductivity was affected by factors such as temperature, particle size, and pH level. Several theoretical models have been proposed to explain the behavior of nanoparticles. Many of these models can be categorized as either static or dynamic models [22]. Static models assume that the nanoparticles are stationary in the base fluid, forming a composite material. In these models, the thermal properties of nanofluids are predicted through conduction-based models such as that of Maxwell. One such model is the modified Maxwell theory of Hamilton and Crosser [25], which gives the enhancement of thermal conductivity as

$$k_{\text{eff}} = k_0 \frac{k_p + (n-1)k_0 - (n-1)\varepsilon(k_0 - k_p)}{k_p + (n-1)k_0 + \varepsilon(k_0 - k_p)} \tag{4.3}$$

$$n = \frac{3}{\psi} \tag{4.4}$$

where k_{eff} is the effective conductivity, k_{p} is the conductivity of the particle, k_0 is the conductivity of the base fluid, ε is the particle volume fraction, n is the particle shape factor, and ψ is the sphericity of the particles [1]. This model was in good agreement with experimental data obtained with the use of Al_2O_3 nanoparticles in a nanofluid, but it was not able to accurately predict the thermal conductivity of nanofluids containing CuO nanoparticles. It should be noted that in Eq. (4.3), the effective thermal conductivity is independent of the thermal conductivity of the particle as the ratio of k_{p}/k_0 becomes large.

Dynamic models assume that nanoparticles are in constant random motion in the base fluid (i.e., Brownian motion), as shown in Fig. 4.2. In the dynamic models, it is believed that this random motion may be the main cause of the increased thermal properties associated with nanofluids. Taking Brownian motion to be a key mechanism in the thermal properties of nanofluids, Jang and Choi [26] developed a model that portrayed the relationship between conductivity, temperature, concentration, and particle size. However, there is disagreement with the assumption that random motion plays a key role in the transfer of heat in a nanofluid. Keblinski et al. [27] proposed an explanation of four possible factors for the heat transfer mechanism in nanofluids one of which was Brownian motion. However, the study concluded that the movement of nanoparticles due to Brownian motion was too slow in transporting heat through a fluid. To travel from one point to another, a particle moves a large distance over many different paths in order to reach a destination that may be a short distance from the starting point. Therefore, the random motion of particles, no matter how agitated or energetic they may be, cannot be a key factor in the improvement of heat transfer. Jang and Choi [26] also came to a similar conclusion. Here, it was determined that the collision between nanoparticles due to random motion was a very slow process and could, therefore, be neglected in the calculation of thermal conductivity. Although they contributed to the idea that collisions resulting from Brownian motion did not contribute to the overall conduction of heat, Jang and Choi [26] were able to develop a dynamic model that takes into account convection heat transfer induced by Brownian nanoparticles.

The general expression derived in this study introduced four modes of energy transport in nanofluids:

1. Collision between base fluid molecules (i.e., thermal conductivity of base fluid),
2. Thermal diffusion in nanoparticles in fluids,
3. Collision between nanoparticles due to Brownian motion (neglected because it is a very slow process), and
4. Thermal interactions of dynamic nanoparticles with base fluid molecules (once overlooked, this mode is now considered to be a key factor in the relationship between conductivity, temperature, and particle size).

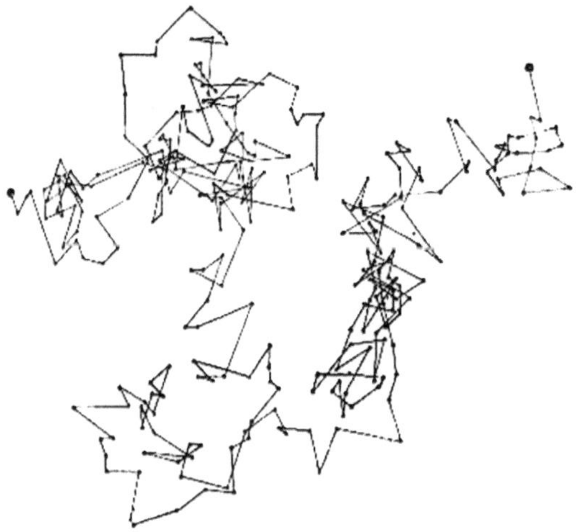

Figure 4.2 Interpretation of Brownian motion [28].

The major aspect of this model was the introduction of the idea that nanoparticles can produce a convection-like effect in a fluid. The thermal conductivity for their model is given by

$$k_{\text{eff}} = k_{\text{Bf}}(1-f) + k_{\text{nano}} f + 3C_1 \frac{d_{\text{Bf}}}{d_{\text{nano}}} k_{\text{Bf}} Re_{d_{\text{nano}}}^2 Prf, \tag{4.5}$$

where k_{eff} is the effective thermal conductivity of the nanofluid, k_{Bf} is the base fluid conductivity, f is the volume fraction of the nanoparticles, k_{nano} is the thermal conductivity of the nanoparticles, C_1 is an empirical constant, d_{Bf} is the diameter of the base fluid molecule, and d_{nano} is the diameter of a nanoparticle [1]. $Re_{d_{\text{nano}}}$ is the Reynolds number defined by

$$Re_{d_{\text{nano}}} = \frac{\bar{C}_{\text{R.M.}} d_{\text{nano}}}{\nu} \tag{4.6}$$

where ν is the dynamic viscosity of the base fluid, and $\bar{C}_{\text{R.M.}}$ is the random motion velocity of nanoparticles defined by

$$\bar{C}_{\text{R.M.}} = \frac{D_0}{l_{\text{Bf}}}, \tag{4.7}$$

where l_{Bf} is the mean-free path of a base fluid molecule and D_0 is the nanoparticle diffusion coefficient given by

$$D_0 = \frac{k_b T}{3\pi\mu d_{\text{nano}}}, \tag{4.8}$$

where μ is the viscosity of the base fluid, T is the temperature of the base fluid, and k_b is the Boltzmann constant. The predictions presented by this model are in excellent agreement with temperature-dependent conductivity data from experiments involving nanofluids containing Al_2O_3 nanoparticles. Models derived from Maxwell's equations fail to correlate with this type of experimental data.

Another model developed by Kumar et al. [29] involves the combination of the stationary particle model and the moving particle model. The stationary particle model looks at the increased surface area as the particle size decreases. By assuming two parallel paths of heat flow (one through base fluid molecules and the other through the nanoparticles), this model shows the linear dependence of thermal conductivity on particle concentration and the inverse dependence of thermal conductivity on the size of the particle. The moving particle model accounts for the temperature effect and is derived from the Stokes Einstein formula. The effective thermal conductivity of the nanofluids, k_{eff}, for this model is given by

$$k_{\mathrm{eff}} = k_{\mathrm{m}}\left[1+\frac{k_{\mathrm{p}}\,\varepsilon r_{\mathrm{m}}}{k_{\mathrm{m}}(1-\varepsilon)r_{\mathrm{p}}}\right], \tag{4.9}$$

where k_{m} is the base fluid conductivity, ε is the nanoparticle volume fraction, r_{m} is the radius of the base fluid molecules, and r_{p} is the radius of the nanoparticles. The nanoparticle's thermal conductivity k_{p} is defined as

$$k_{\mathrm{p}} = c\bar{u}_{\mathrm{p}} \tag{4.10}$$

The mean velocity of the nanoparticle, $\bar{u}_{\mathrm{p}}$, is derived from the Stokes–Einstein formula

$$\bar{u}_{\mathrm{p}} = \frac{2k_{\mathrm{p}}T}{\left(\pi\eta d_{\mathrm{p}}^{2}\right)}, \tag{4.11}$$

where T is the fluid temperature, η is the dynamic viscosity of the fluid, and d_{p} is the diameter of the nanoparticle. The combination of the stationary particle model and the moving particle model shows the dependence of thermal conductivity on particle size, concentration, and temperature. Although this model correlates with the experimental data for nanofluids containing gold at small concentrations, there is a discrepancy in the stationary particle model. It has been pointed out that if the radius of the nanoparticles is larger than that of the liquid molecules, then the calculated thermal conductivity for the nanofluid will equal that of the base fluid, an unrealistic situation. Therefore, when using this model, an unphysical assumption that the mean-free path of a nanoparticle is on the order of 1 cm is made in the moving particle model [29]. There are still some issues over the benefits of this model, and as a result, it is the source of some debates today concerning the dynamic model.

The influence of particle anisotropy on the effective thermal conductivity of a suspension was experimentally studied by Cherkasova and Shan [30]. Suspensions of micron-sized, silicon-carbide particles with varying aspect-ratio distributions were prepared and measured. It was shown that the conductivity of the silicon-carbide suspensions can be quantitatively predicted by the effective medium theory presented by Nan et al. [31], as long as the volume-weighted aspect ratio of the particles is used.

The effect of Kapitza resistance between the particle and fluid can significantly impact the effective thermal conductivity of nanofluids. This effect was included into the model by Nan et al. [31], which predicted the experimental data of carbon nanotube suspensions reasonably well. Ju and Li [32] and Xue [33] also presented models for the effective thermal conductivities of carbon nanotube-based mixtures with an interfacial thermal resistance effect.

References [34, 35] discuss experiments where the interfacial layers of nanoparticles were examined. Document [36] discusses the relationship between temperature, particle size, and the thermal conductivity of the nanofluids examined. The survey done here does not include any work involving synthetic nanofluids, which form a class by themselves [37–39].

4.3 Convective Heat Transfer in Nanofluids

If nanofluids can improve the heat transfer coefficient of heat exchangers and energy systems, then they can aid in reducing the size of such systems while leading to increased energy and fuel efficiencies. Nanofluid convective heat transfer research may be classified by fluid conditions of laminar flow, turbulent flow, and pool boiling. The number of studies in these areas is limited, with the smallest number of studies having been reported in the last class of pool boiling.

Works in laminar flow include Refs. [40–44]. Some of the works show heat transfer enhancements for laminar flow with different particle types (alumina and copper oxide) and sizes. Nanofluid heat transfer results for multiwalled carbon nanotubes (MWCNTs) show excellent heat transfer and thermal conductivity enhancement [42]. The results published by Yang et al. [44] displayed a heat transfer enhancement trend that is opposite to that for thermal conductivity—there is a drop in enhancement with increased temperature. But the temperature range studied is small, and the temperature dependency is not strong. More test data need to be obtained before any definite conclusions can be drawn.

Heat transfer enhancements in turbulent flow of nanofluids include Refs. [45–48]. The studies were performed on water-based nanofluids containing alumina, titanium oxide, copper particles, and amorphous carbonic nanoparticles. There seems to be a consensus that the Reynolds number has little or no effect on the enhancement of heat transfer. Furthermore, the heat transfer improvement increased as the particle volume concentration was increased.

Putra et al. [49] investigated the convection of Al_2O_3 and CuO in water-based nanofluids. It was discovered that the natural convection of nanofluids was less intense than that of the base fluid. Moreover, it deteriorated as the particle density and concentration were increased. Smaller particles experienced even worse convective heat transfer because the particle density increases as the particle size decreases. However, Khanafer et al. [50] developed a model for the convective heat transfer of nanofluids that produced different results. With the assumption that the base fluid and the nanoparticles were in thermal equilibrium and flowed at the same velocity (i.e., nanofluid is in single phase), it was shown that the heat transfer rate increased as the particle volume fraction increased. Another model proposed by Kim et al. [51] introduced a factor f to include the effect of the thermal conductivity ratio between the nanoparticles and the base fluid, the shape factor and volume fraction of particles, and the ratio of density and heat capacity of nanoparticles to the base fluid. Results from this model showed that the amount of heat transfer in the nanofluid increased as the particle volume fraction was increased.

It is evident that the results of Kim and Khanafer contradict the results of Putra. Possible explanations for these differences could be dependent on the assumptions made in each model. Before nanofluids can be introduced to industrialized applications, there must be a better understanding of convection heat transfer and its effects on the overall thermal conductivity of the nanoparticle suspension.

4.4 Clustering in Nanofluids

One of the main obstacles encountered in microfluid experiments was the agglomeration of particles. Even though research, such as that documented in Refs. [22, 23], shows a substantial increase in the thermal conductivity of the base fluid with the addition of nanoparticles, the movement toward practical applications has been hampered by the rapid settling of the nanoparticles. The settling of particles not only decreased the overall heat transfer of the fluid (by decreasing the effective surface area used for heat transfer), but also led to the abrasion of surfaces, clogging of microchannels, and a decrease in pressure, which resulted in an increase in pumping power [14].

Although nanosized particles have greatly reduced the problem of agglomerated particles, it still occurs and can hinder the thermal conductivity of the nanofluid, especially at concentrations over 5%; agglomeration is more apparent when using oxide nanoparticles because they require a higher volume concentration compared to metallic nanoparticles in order to achieve the same thermal conductivity enhancement [2]. The tendency of particles to group together before they are dispersed in the fluid is due to the van der Waals forces. This is particularly seen in metallic particles since dipoles can occur easily in the molecules of these particles. The creation of dipoles prompts the attraction of other dipoles in the vicinity. The van der Waals forces stem from the attraction of these dipoles, which can be induced even in neutral particles. This attractive force is considered to be the main culprit behind the agglomeration of particles, especially in nanopowders.

To alleviate this problem, there have been various proposals for the manufacture and dispersion of nanoparticles in fluids. One proposal involves adding surface treatments to the nanoparticles. It was seen that when copper nanoparticles were coated with a 2–10 nm thick organic layer, a stable suspension would be achieved in ethylene glycol [16]. There is research currently being conducted toward improving the two-step process to produce well-dispersed nanofluids. Moreover, there exist a few one-step processes that result in nanoparticles being uniformly dispersed and stably suspended in the base fluid. One such method involves condensing copper nanopowders directly from the vapor phase into flowing ethylene glycol in a vacuum chamber [52]. Documents [53–55] also show stable, well-dispersed suspensions in nanofluids containing TiO_2, CuO, and Cu. In these experiments, a one-step process called submerged arc nanoparticle synthesis was used to create the nanoparticles.

Various techniques have been implemented to reduce the clustering of particles once they are in the fluid [56, 57]. Usually, they involve some sort of agitation within the nanofluid to separate the clusters into individual particles and keep them from settling. These methods include the use of dispersants, changing the pH value of the base fluid, and using ultrasonic vibration to excite the particles [6]. Among these methods, the most commonly used ones are ultrasonic vibration and the use of dispersants. Both techniques are relatively effective, but when using dispersants, the amount added to the fluid must be a very low percentage (usually 1% or less). This is done to minimize its effects on the thermal conductivity of the nanofluid.

However, it should be noted that loose particle chains may be responsible for some of the high thermal conductivities of nanofluids; see Prasher et al. [58].

The Argonne National Laboratory also developed the single-step and two-step processes for the dispersion of nanoparticles in a fluid [1]. The single-step process consists of simultaneously making and dispersing the particles in the fluid. The two-step method separates the manufacture and dispersion of particles into two steps (particles are manufactured first and then dispersed into the base fluid). The two-step process is the more commonly used method and is usually used in conjunction with ultrasonic vibration to reduce the amount of clustered particles in the fluid.

Analysis of the reviewed literature shows that there is still no conclusive theory concerning the prevention of clustering in the nanoparticle suspensions. Before using nanofluids in practical applications, the problem of clustering must be consistently kept to a minimum. When looking at long-term effects, clustering of the particles will eventually cause a decrease in the thermal conductivity of the nanofluid and may also cause wear in the pipes or pumps through which it is flowing. Therefore, nanofluids cannot be used in systems designed for long-term use until this problem is solved. Otherwise, the use of nanofluids may decrease the life expectancy of a system, even if it improves the overall efficiency. In the meantime, an optimization and design problem persists when nanofluids are used in the field.

4.5 Conclusion

Even though nanofluids are still relatively new, they have caused a dramatic increase in the interest of ultra-high-performance cooling. Experiments are continuously being conducted to achieve a deeper understanding of the mechanisms behind the increase in thermal conductivity caused by these nanoparticle suspensions. As a result, various conclusions have been drawn regarding the characteristics of nanofluids, including the relationship between thermal conductivity, particle size, nanoparticle concentration, and temperature.

Heat can be transferred through conduction, convection, and radiation. There is a more thorough understanding of conduction in nanofluids than the other two, and various models have been formulated to predict the thermal behavior of nanofluids. Static models investigate the thermal conductivity of a nanofluid assuming a stationary suspension of particles in the base fluid. This allows for the use of derived forms of Maxwell's equation. Dynamic models assume that the particles are in constant, random motion while dispersed in the fluid. These models are the source of much debate today over the involvement of Brownian motion in the thermal conductivity of nanofluids. While some scientists believe that this random motion is the source of conduction in nanofluids, studies show that it is insignificant compared to other factors because the transfer of heat is a very slow process in Brownian motion. However, some researchers were able to develop a dynamic model that portrayed four modes of energy transport. The major aspect of this model was the convection heat transfer induced by Brownian motion in particles. Results from this model correlated excellently with experimental data of nanofluids containing aluminum oxide particles.

Convection heat transfer is recognized within nanofluids, but there is not enough research results published to develop a model that fully explains this behavior in nanofluids. Furthermore, several of the research papers available seem to contradict each other as some data show an increase in convection as the particle volume fraction is increased, while other data show deterioration in convective heat transfer as the particle density and concentration were increased.

Clustering still poses a problem in nanofluids even though the occurrence of agglomeration has decreased from the previous micrometer-sized particles suspensions. Various methods are currently used to keep particles from clustering together, but in the long run, it is inevitable. Clustering is a problem that must be solved before nanofluids can be considered for long-term practical uses. Although the increase in thermal conductivity would increase the efficiency of the systems where nanofluids are used, the life of the system may be decreased over time if particles begin to form clusters.

Nanofluids have the potential to open the doors to major advancements in many high-tech industries where limits on cooling have posed limits on innovation. Since all other cooling options have been exhausted, nanofluids are the only option left with the possibility of increasing heat transfer capabilities of current systems. However, a full understanding of the mechanisms behind the enhancement of thermal conductivity in nanofluids has not been reached and there is still disagreement between some of the experimental results. This lack of agreement has led to the generation of various models. Once a general model that fully explains the behavior of nanoparticle suspensions is developed, steps can be taken toward practical uses. Moreover, better techniques for the dispersion of particles in fluids must be created to minimize clustering. When these objectives have been reached, nanofluids will enter the practical arenas of science in a more meaningful way. At the present time, there is quite an amount of work going on to create synthetic nanofluids for various applications. This is evidenced by the number of patented nanofluids. However, the literatures on these are not generally available.

References

1. S. K. Das, S. U. S. Choi, W. Yu, and T. Pradeep, *Nanofluids: Science and Technology*, John Wiley & Sons, Hoboken, NJ, USA, 2008.
2. W. Yu, D. M. France, J. L. Routbort, and S. U. S. Choi, "Review and comparison of nanofluid thermal conductivity and heat transfer enhancements," *Heat Transfer Engineering*, **29**(5), pp. 432–460, 2008.
3. D. J. Jeffrey, "Conduction through a random suspension of spheres," in *Proceedings of the Royal Society of London. Series A*, **335**, pp. 355–367, 1973.

4. J. Koo and C. Kleinstreuer, "A new thermal conductivity model for nanofluids," *Journal of Nanoparticle Research*, **6**(6), pp. 577–588, 2004.
5. B.-X. Wang, L.-P. Zhou, and X.-F. Peng, "A fractal model for predicting the effective thermal conductivity of liquid with suspension of nanoparticles," *International Journal of Heat and Mass Transfer*, **46**(14), pp. 2665–2672, 2003.
6. Y. Xuan and Q. Li, "Heat transfer enhancement of nanofluids," *International Journal of Heat and Fluid Flow*, **21**(1), pp. 58–64, 2000.
7. T.-K. Hong, H.-S. Yang, and C. J. Choi, "Study of the enhanced thermal conductivity of Fe nanofluids," *Journal of Applied Physics*, **97**(6), Article ID 064311, 4 pages, 2005.
8. S. A. Putnam, D. G. Cahill, P. V. Braun, Z. Ge, and R. G. Shimmin, "Thermal conductivity of nanoparticle suspensions," *Journal of Applied Physics*, **99**(8), Article ID 084308, 6 pages, 2006.
9. S. M. S. Murshed, K. C. Leong, and C. Yang, "Enhanced thermal conductivity of TiO_2—water based nanofluids," *International Journal of Thermal Sciences*, **44**(4), pp. 367–373, 2005.
10. S. M. S. Murshed, K. C. Leong, and C. Yang, "A combined model for the effective thermal conductivity of nanofluids," *Applied Thermal Engineering*, **29**(11–12), pp. 2477–2483, 2009.
11. A. Sandhu, "Magnetic nanofluids: chain reaction," *Nature Nanotechnology*. doi:10.1038/nnano.2008.38, 2008.
12. J. Philip, P. D. Shima, and B. Raj, "Nanofluid with tunable thermal properties," *Applied Physics Letters*, **92**(4), Article ID 043108, 2008.
13. J. C. Maxwell, *Treatise on Electricity and Magnetism*, Clarendon Press, Oxford, UK, 1873.
14. V. Trisaksri and S. Wongwises, "Critical review of heat transfer characteristics of nanofluids," National Research Council of Thailand (NRCT), 2006, http://www.energy-based.nrct.go.th/Article/Ts-3%20critical%20review%20of%20heat%20transfer%20characteristics%20of%20nanofluids.pdf.
15. C. G. Granqvist and R. A. Burhman, "Ultrafine metal particles," *Journal of Applied Physics*, **47**(5), pp. 2200–2219, 1976.
16. K. S. Suslick, M. Fang, and T. Hyeon, "Sonochemical synthesis of iron colloids," *Journal of the American Chemical Society*, **118**(47), pp. 11960–11961, 1996.
17. J. M. Romano, J. C. Parker, and Q. B. Ford, "Application opportunities for nanoparticles made from the condensation of physical vapors," in

Proceedings of the International Conference on Powder Metallurgy and Particulate Materials, **2**, pp. 12–13, Chicago, Ill, USA, June 1997.

18. M. Kostic, "Nanofluids: advanced flow and heat transfer fluids," Northern Illinois University, 2004, http://www.kostic.niu.edu/DRnanofluids.
19. Y. Xuan and Q. Li, "Heat transfer enhancement of nanofluids," *International Journal of Heat and Fluid Flow*, **21**(1), pp. 58–64, 2000.
20. "Thermal Conductivity of Liquids and Gases," National Institute of Standards and Technology (NIST), http:// fluidproperties.nist.gov/thermal.html.
21. H. S. Carslaw and J. C. Jaeger, *Conduction of Heat in Solids*, Oxford University Press, New York, NY, USA, 1967.
22. S. Lee, S. U.-S. Choi, S. Li, and J. A. Eastman, "Measuring thermal conductivity of fluids containing oxide nanoparticles," *Journal of Heat Transfer*, **121**(2), pp. 280–289, 1999.
23. H. Xie, J. Wang, T. Xi, Y. Liu, F. Ai, and Q. Wu, "Thermal conductivity enhancement of suspensions containing nanosized alumina particles," *Journal of Applied Physics*, **91**(7), pp. 4568–4572, 2002.
24. S. K. Das, N. Putra, P. Thiesen, and W. Roetzel, "Temperature dependence of thermal conductivity enhancement for nanofluids," *Journal of Heat Transfer*, **125**(4), pp. 567–574, 2003.
25. R. L. Hamilton and O. K. Crosser, "Thermal conductivity of heterogeneous two-component systems," *Industrial and Engineering Chemistry Fundamentals*, 1(3), pp. 187–191, 1962.
26. S. P. Jang and S. U. S. Choi, "Role of Brownian motion in the enhanced thermal conductivity of nanofluids," *Applied Physics Letters*, **84**(21), pp. 4316–4318, 2004.
27. P. Keblinski, S. R. Phillpot, S. U. S. Choi, and J. A. Eastman, "Mechanics of heat flow in suspensions of nano-sized particles (nanofluids)," *International Journal of Heat and Mass Transfer*, **307**, pp. 313–317, 2003.
28. K. Hoon and Y. K. Lee, "Brownian motion of a microscopic particle," http://www.hasdeu.bz.edu.ro/softuri/fizica/mariana/Termodinamica/Brownian 1/files/report.br1.
29. D. H. Kumar, H. E. Patel, V. R. R. Kumar, T. Sundararajan, T. Pradeep, and S. K. Das, "Model for heat conduction in nanofluids," *Physical Review Letters*, **93**(14), Article ID 144301, 2004.

30. A. S. Cherkasova and J. W. Shan, "Particle aspect-ratio effects on the thermal conductivity of micro- and nanoparticle suspensions," *Journal of Heat Transfer*, 130(8), Article ID 082406, 7 pages, 2008.

31. C.-W. Nan, R. Birringer, D. R. Clarke, and H. Gleiter, "Effective thermal conductivity of particulate composites with interfacial thermal resistance," *Journal of Applied Physics*, **81**(10), pp. 6692–6699, 1997.

32. S. Ju and Z. Y. Li, "Theory of thermal conductance in carbon nanotube composites," *Physics Letters A*, **353**(2–3), pp. 194–197, 2006.

33. Q. Z. Xue, "Model for the effective thermal conductivity of carbon nanotube composites," *Nanotechnology*, **17**(6), pp. 1655–1660, 2006.

34. W. Yu and S. U. S. Choi, "The role of interfacial layers in the enhanced thermal conductivity of nanofluids: a renovated Maxwell model," *Journal of Nanoparticle Research*, **5**(1–2), pp. 167–171, 2003.

35. W. Yu and S. U. S. Choi, "The role of interfacial layers in the enhanced thermal conductivity of nanofluids: a renovated Hamilton-Crosser model," *Journal of Nanoparticle Research*, **6**(4), pp. 355–361, 2004.

36. C. H. Chon, K. D. Kihm, S. P. Lee, and S.U. S. Choi, "Empirical correlation finding the role of temperature and particle size for nanofluid (Al_2O_3) thermal conductivity enhancement," *Applied Physics Letters*, **87**(15), Article ID 153107, 3 pages, 2005.

37. M.-S. Liu, M. C. C. Lin, I.-T. Huang, and C.-C. Wang, "Enhancement of thermal conductivity with carbon nanotube for nanofluids," *International Communications in Heat and Mass Transfer*, **32**(9), pp. 1202–1210, 2005.

38. F. D. S. Marquis and L. P. F. Chibante, "Improving the heat transfer of nanofluids and nanolubricants with carbon nanotubes," *JOM*, **57**(12), pp. 32–43, 2005.

39. L. Vekas and B. Doina, "Magnetic nanofluids, preparation, properties and some applications," in *Proceedings of the 1st Nanoforum Workshop*, Sinaia, Romania, October 2003.

40. D. Wen and Y. Ding, "Experimental investigation into convective heat transfer of nanofluids at the entrance region under laminar flow conditions," *International Journal of Heat and Mass Transfer*, **47**(24), pp. 5181–5188, 2004.

41. S. Z. Heris, S. Gh. Etemad, and M. N. Esfahany, "Experimental investigation of oxide nanofluids laminar flow convective heat transfer," *International Communications in Heat and Mass Transfer*, **33**(4), pp. 529–535, 2006.

42. Y. Ding, H. Alias, D. Wen, and R. A. Williams, "Heat transfer of aqueous suspensions of carbon nanotubes (CNT nanofluids)," *International Journal of Heat and Mass Transfer*, **49**(1–2), pp. 240–250, 2006.
43. D. J. Faulkner, D. R. Rector, J. J. Davidson, and R. Shekarriz, "Enhanced heat transfer through the use of nanofluids in forced convection," in *Proceedings of the ASME International Mechanical Engineering Congress and Exposition (IMECE '04)*, pp. 219–224, Anaheim, Calif, USA, November 2004.
44. Y. Yang, Z. G. Zhang, E. A. Grulke, W. B. Anderson, and G. Wu, "Heat transfer properties of nanoparticle-in-fluid dispersions (nanofluids) in laminar flow," *International Journal of Heat and Mass Transfer*, **48**(6), pp. 1107–1116, 2005.
45. B. C. Pak and Y. I. Cho, "Hydrodynamic and heat transfer study of dispersed fluids with submicron metallic oxide particles," *Experimental Heat Transfer*, **11**(2), pp. 151–170, 1998.
46. Y. Xuan and Q. Li, "Investigation on convective heat transfer and flow features of nanofluids," *Journal of Heat Transfer*, **125**(1), pp. 151–155, 2003.
47. D. Kim, Y. Kwon, Y. Cho, C. Li, S. Cheong, Y. Hwang, J. Lee, D. Hong, and S. Moon, "Convective heat transfer characteristics of nanofluids under laminar and turbulent flow conditions," *Current Applied Physics*, **9**(2), supplement 1, pp. e119–e123, 2009.
48. J. Buongiorno, "Convective transport in nanofluids," *Journal of Heat Transfer*, **128**(3), pp. 240–250, 2006.
49. N. Putra, W. Roetzel, and S. K. Das, "Natural convection of nano-fluids," *Heat and Mass Transfer*, **39**(8–9), pp. 775–784, 2003.
50. K. Khanafer, K. Vafai, and M. Lightstone, "Buoyancy-driven heat transfer enhancement in a two-dimensional enclosure utilizing nanofluids," *International Journal of Heat and Mass Transfer*, **46**(19), pp. 3639–3653, 2003.
51. J. Kim, Y. T. Kang, and C.-K. Choi, "Analysis of convective instability and heat transfer characteristics of nanofluids," *Physics of Fluids*, **16**(7), pp. 2395–2401, 2004.
52. J. A. Eastman, S. U. S. Choi, S. Li, W. Yu, and L. J. Thompson, "Anomalously increased effective thermal conductivities of ethylene glycol-based nanofluids containing copper nanoparticles," *Applied Physics Letters*, **78**(6), pp. 718–720, 2001.
53. H. Chang, T. T. Tsung, Y. C. Yang, L. C. Chen, H. M. Lin, C. K. Lin, and C. S. Jwo, "Nanoparticle suspension preparation using the arc spray

nanoparticle synthesis system combined with ultrasonic vibration and rotating electrode," *International Journal of Advanced Manufacturing Technology*, **26**(5–6), pp. 552–558, 2005.

54. C.-H. Lo, T.-T. Tsung, L.-C. Chen, C.-H. Su, and H.-M. Lin, "Fabrication of copper oxide nanofluid using submerged arc nanoparticle synthesis system (SANSS)," *Journal of Nanoparticle Research*, **7**(2–3), pp. 313–320, 2005.

55. C.-H. Lo, T.-T. Tsung, and L.-C. Chen, "Shape-controlled synthesis of Cu-based nanofluid using submerged arc nanoparticle synthesis system (SANSS)," *Journal of Crystal Growth*, **277**(1–4), pp. 636–642, 2005.

56. Q. Cao and J. Tavares, "Dual-plasma synthesis of coated nanoparticles and nanofluids," November 2006, http://aiche.confex.com/aiche/2006/techprogram/P65561.HTM.

57. S. M. S. Murshed, K. C. Leong, and C. Yang, "Thermophysical and electrokinetic properties of nanofluids—a critical review," *Applied Thermal Engineering*, **28**(17–18), pp. 2109–2125, 2008.

58. R. Prasher, P. Bhattacharya, and P. E. Phelan, "Brownian motion-based convective-conductive model for the effective thermal conductivity of nanofluids," *Journal of Heat Transfer*, **128**(6), pp. 588–595, 2006.

Chapter 5

Nanoscience and Technology in Solar Cells

Kaufui V. Wong, Nicholas Perilla, and Patrick Andrew Paddon
Department of Mechanical and Aerospace Engineering, University of Miami, Coral Gables, FL 33124, USA
kwong@miami.edu

Energy will be a big challenge in the coming decades. The global population is increasing steadily. Not only are there more people in the world, but the worldwide drive to increase living standards has increased individual energy demands. Growing energy needs were typically met by finding new sources of fossil fuels. People have fortunately begun to realize the adverse environmental impacts of burning fossil fuels and that this practice cannot be maintained indefinitely, leading to renewed interest in photovoltaic technologies. The discovery of the photoelectric effect brought hope to the objective of helping to fill the world energy needs with an already continuously delivered source. This discovery was the birth of the

This chapter has been reprinted from the *Proceedings of IMECE2011*, 2011 ASME International Mechanical Engineering Congress and Exposition, Kaufui V. Wong, Nicholas Perilla, and Andrew Paddon, Nanoscience and nanotechnology in solar cells, Copyright 2011, with permission from ASME.

Nanotechnology and Energy
Edited by Kaufui V. Wong

ISBN 978-981-4774-24-6 (Hardcover), 978-1-315-16357-4 (eBook)
www.panstanford.com

idea, but it was the development of the crystalline silicon cell that marked the birth of the industry. The cost and inefficiency of these solar panels have prevented them from becoming an economically competitive form of everyday power generation. Cost was reduced with the introduction of amorphous silicon thin-film cells despite slightly lower efficiencies. Their lower manufacturing costs have allowed solar energy to be included in more applications; the costs have not been reduced enough to compete with current grid rates. The current trend in research suggests that the application of nanotechnology may be the awaited break needed to crack this cost barrier. Nanotechnology promises to reduce cost because they require less controlled conditions, which will greatly reduce the cost per cell, and the initial cost of a new cell type. Nanoscience and nanotechnology are being researched and developed to help solve problems that have prevented the use of other promising technologies, and improving efficiencies of those technologies that have been developed. The addition of nanoparticles to the matrix is a possible way to improve electron transport, and nanotubes could be used in conjunction with nanoparticles.

The science of interactions and addition of nanoparticles and their function in solar photovoltaic cells is known, but still developing. Nanoscience has produced proof-of-concept photovoltaic cells made of small perfect crystals, rather than large, perfect silicon crystals that are more expensive to produce. However, larger scale, practical photovoltaic cells have not been reproduced efficiently and at a reasonable cost for small, perfect crystals not made of silicon.

Nanowhiskers are being experimented as new antireflective coating. Sensitizing dyes are being used to increase the range and location of the wavelengths that can be absorbed to be more favorable to sunlight, allowing the use of materials that lack this key characteristic. Quantum dots (QDs) could be an improvement to these dyes, as the smaller particles will have the added benefit of having multiple electrons created per photon without impeding electron transfer. Recent research has also shown a method to transform optical radiation into electrical current that could lead to self-powering molecular circuits and efficient data storage. The many possible applications of nanotechnology make photovoltaic cells a promising pursuit.

5.1 Introduction

In the current age of energy awareness, nations are looking for solutions. There are many solutions that are promising. There is one in particular that stands out. This is electricity produced from photovoltaic (PV) cells. These cells allow us to tap into the vast amounts of energy that the sun already delivers to earth. While this green energy solution is a benign one when installed, it cannot change the economics that drive the market. In the past 50 years, the efficiency of solar cells has been steadily increasing and the prices have been dropping. Much progress has been made especially when there have been economic incentives from various governments, and solar cells are now approaching the prices that will make them competitive. These advances have led to three distinct generations of solar cells. The first generation was the crystal silicon that still dominates the market. The second is the amorphous silicon thin-film cells, which have reduced prices greatly and show potential of reducing them even further. The third generation is being developed right now and consists of nanotechnology, which uses flexible, printable substrates and mixes them with electronically conducting nanomaterial [1]. These new advancements seem to be on the verge of making an idea from many years ago into a reality.

The basic concept of photovoltaics is the opposite of electroluminescence. An amazing characteristic of illumination is that even if the p and n terminals become shorted, there still exists a current due to the photogenerated carriers. Still, however, the maximum current, or I_{SC}, that a photovoltaic cell can deliver is dependent on the illumination's intensity. The issue, however, is that at maximum current, voltage is equal to zero, and at maximum voltage, current is equal to zero. Both these phenomena result in zero power output, thus sacrifices must be made to operate the cell at a point between 0 and V_{OC}. The maximum power output, I_{SC}, and V_{OC} all determine the fill factor of a photovoltaic cell, which is also hoped to be maximized. Some issues that the nanoparticles face include loss of long-term efficiency due to oxygen. Also the materials so far discovered absorb in the mid-visible range, rather than the near-infrared range where the most photons can be harvested [2].

These p–n junctions, which are a vital feature of solar cells, are silicon doped with impurities, including boron for the p-type and phosphorus for the n-type, with other elements substituted for experimental purposes as well [3].

The science behind these technological advances has undergone changes over the past 40 years, with the move from aluminum to highly doped polysilicon gate electrodes in the 1970s, inclusion of lightly doped drain extensions in the 1980s, and the inclusion of nitrogen to the SiO_2 gate dielectric to form silicon oxynitride in the 1990s. Not to mention that the MOSFET, or planar silicon metal oxide semiconductor field-effect transistor, has gone from dimensions of micrometers to nanometers. Today insulators made with high dielectric constants, such as HfO_2, are replacing silicon oxynitride because they withstand a higher capacitance, reducing the amount of current that leaks. Also a large group of metals are being looked at to replace the highly doped polysilicon. More non-classical designs are being approached in search of further enhancing the performance of electrical devices such as a silicon on insulator. This new technique involves a layer of silicon dioxide covered by a thinner layer of active silicon, increasing the velocity of field-effect transistors and exhibiting almost perfect characteristics of a drain current gate voltage [4].

The many structures of nanoparticles determine what range of frequencies they can resonate at, or accept plasmon energy levels from. A few of the structure resonances are roughly between 575 and 9000 nm or 2.25 and 0 eV for nanoshells, between 475 and 1400 nm or 2.6 and 1.0 eV, and between 600 and 1200 nm or 2.2 and 1.25 eV for cubes. These can stretch from collecting UV light to mid-infrared. The wavelength ranges and corresponding nanostructures are shown in Fig. 5.1 [5].

The two types of solar cells reported recently are the dye-sensitized solar cell (or DSSC) and CdSe QD solar cell. Using DSSC, electrons pass through a TiO_2 layer and gather on a layer of conductive fluorine-doped SnO_2, which is on top of a glass surface. Opposed to the CdSe dot system, the split and transfer process occurs between a polymer and the CdSe dots, which provide tunnels to the electrodes. However, TiO_2 only collects 5% of the solar spectrum with a bandgap of 3.2 eV. Doping the TiO_2 with nitrogen can make improvements to the photo activity of the system as well.

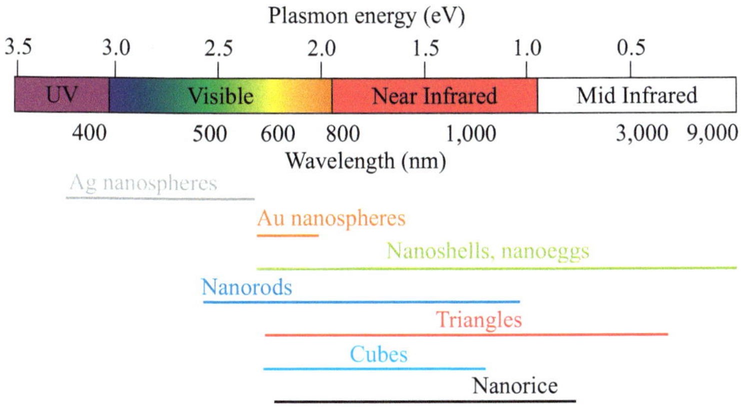

Figure 5.1 Nanoparticle resonances for an array of morphologies [5].

Another improvement found was antireflection (AR) coating, which increased the efficiency of both solar collection and the conversion of solar to electricity [6]. This efficiency of solar collection is extremely important because any photons with energy higher than the bandgap are lost to thermalization of the charge carriers and those with lower energy are simply lost. The benefit of a multi-junction cell (MJC) is that it allows for a much larger range of wavelengths in the solar spectrum that can be absorbed. This is done by the method discussed of stacking different bandgap-valued solar cells in series. The current world record for a 3-junction device is about 40.7% efficiency under 240-sun illumination [7]. Tunneling-assisted photoconductivity devices have also proved to be more superior to that of the typical bulk devices that have flooded the markets over the years. This improvement has been found from the integration of quantum wells into the p-i-n photovoltaic devices [8].

5.1.1 History

The idea of powering the earth with the sunlight it already receives began when the transoceanic telegraph cables were being laid and it was discovered that selenium produces electricity when exposed to light. Research continued on selenium, but the conversion efficiency was unacceptably low. Work on selenium was abandoned when it was discovered by Bell Labs that purposefully contaminated silicon was five times more efficient at converting light into electricity.

However, silicon is not the ideal semiconducting material, yet is most used because of the fact that it is the second most copious element in the crust [7].

The two main preparation techniques used for thin films are applicable based on the type of material being used. For example, the evaporation technique requires thermal stability, such as small molecules, which are less soluble. On the other hand, polymers will decay from extreme heat and are oversized for evaporation, yet have greater solubility. Semiconductor nanocrystals are currently a center of focus for solar photoconversion, such as QDs, wires, and rods. Certain properties of the material must be taken into account. For instance, solubilizing side chains may not maximize in harvesting light in order to convert to energy, but this passiveness as well as the possibility of softening materials could instigate better charge transport [9–11].

5.1.2 Applications

With this performance enhancement, Bell Labs started marketing this new technology. The costs were too high, and the efficiencies were too low, which limited the technology to novelty items, such as toys. The Pentagon created the first real need for photovoltaic cells when they decided to test powering satellites with them in 1955. These initial tests proved the solar cells to be reliable and created a large market for solar cells. This market allowed photovoltaic companies to exist and do research and sell their products. As a result, these companies kept developing the technology. After the success of photovoltaic, they began to be applied in other remote locations where power requirements were traditionally met with batteries. The oil companies first used this to provide electrical power to the oil platforms in the oceans, since the oil companies had the resources and the alternative was expensive. Other industries, including Coast Guard buoys, telecommunications companies, and water pumps, adopted this use of solar cells.

Photovoltaic electrification of homes began in French Polynesia by the French Atomic Energy Commission, and by 1987, it powered half of the homes. This has allowed children to study with better lighting and no smoke, increasing scholastic performance.

In developing nations, solar panels are used to power small electronics and are the only source of electricity in rural areas. In the developed world, photovoltaic cells are used where the cost of installing new lines is greater than the investment in the panels and power requirements are low, including bus shelters, and emergency phones.

When photovoltaic cells were first considered for providing grid electricity, plants were modeled after conventional power plants. These plants used vast amounts of space and were located far away from where the electricity was going to be used. This has overlooked important advantages of using photovoltaic cells. Solar cells can easily be placed in the area where the energy is needed such as the top of a building or house. This prevents transmission losses since the electricity does not have to be sent across long distances. As the trend moves to installing photovoltaic on location, it has prompted the use of building materials that double as photovoltaic, giving more benefit for the consumer. Photovoltaic cells also have the advantage that they produce the most electricity in summer, when consumption is the highest [12].

5.1.3 Advancements

The first photovoltaic modules had conversion efficiencies of about 2%, which was mainly due to their low packing densities. Sharp Corporation was then able to make modules that had denser packing resulting in a conversion efficiency of about 4.5%. These modules were used to provide electricity to 256 lighthouses off the coast of Japan between 1961 and 1972. The oil embargo of 1973 and the growing applications of photovoltaic technologies helped stimulate interest in the technology. Government funding for photovoltaic projects soon began.

Most of the advances in photovoltaic panels occurred when there was government backing. When the government began funding these projects, the Block I purchase, cells had about 4.8–6.5% efficiency and cost about $20–39 per watt in 1980 US dollars. By the Block III purchase, efficiencies were in the 6.5–8.4% range and costs were down to about $12–18 per Watt US 1980 and by Block V in 1980, they were above 10% [13–16].

There were two important technological advancements that resulted in these improvements. The first of these was glass printing and glass structural design. The second was using multicrystalline designs. Multicrystalline cells have slightly lower efficiencies because the grain boundaries obstruct the process by which the photons produce electricity. They do, however, have much lower costs since the process to produce one large crystal is more difficult than producing a multicrystalline layer. Many more of the improvements in the field came in the form of improvements to the multicrystalline design to improve the refinement of the materials, improve processes, and automate them [13].

Despite the advancements, solar cells still have many issues and challenges that must be overcome. The fundamental issue is that there is no set stone design rules for nanostructures, resulting in little being known about certain parameters of elemental choices or configuration. Another issue is that of transport for all uses of nanostructures, in terms of cost and efficiency. Solutions for the problem of transport can be mitigated by the use of tightly packed nanostructured arrays, or excite the carriers [17].

5.2 Nanoscience

The macro-properties of materials at the nanoscale depend on the boundary conditions and, hence, the size of the nanoparticles. If a lump of a pure substance, say gold, is divided into two, the macro-properties of the two parts will be identical, e.g., melting point, hardness, and density. If one continued the division, the parts will have identical macro-properties until a critical size when this no longer holds true. It is more difficult to create perfect large crystals than perfect smaller crystals. Thus, the nanoscience of photovoltaic cells teaches us to reproduce these small crystals (just above this critical size) so that the efficiency of reproducing these particles is optimized. The defects in smaller crystals are more easily removed by heating (e.g., annealing) than the defects in larger crystals such as silicon. The procedure of creating smaller organic perfect crystals by boiling is a lot simpler than that of creating large silicon perfect crystals. Quantum detainment of electronic particles (positive holes and negative electrons) in nanocrystals creates unique optical and electronic properties that have the promise to increase the power conversion efficiency of solar cells for photovoltaic and solar fuels production at a cheaper price. Chief among these special properties is the efficient creation of more than one electron–hole pair (named excitons in nanocrystals) from a single absorbed photon. In separated nanocrystals that have three-dimensional detention of charge carriers (QDs) or two-dimensional detention (quantum wires and rods), this process is called multiple exciton generation (MEG). These excitons also have a strong correlation with open-circuit voltage as well as bandgap energy, as shown in Fig. 5.2a. The figure shows the changes witnessed in work function, or electron volts, in relation to the open-circuit voltage due to the type of material used.

An example of this increase is using all-inorganic materials that produced a large short circuit, in fact the largest of any nanostructured solar cell to date. It was created by a Schottky junction at the negative electrode, producing a photocurrent greater than 21 mA/cm^2. The cell consisted of a PbSe NC film yielding a power conversion efficiency of 2.1% in the visible and up to 25% infrared regions of the solar spectrum. All of this was done without the use of sintering, super lattice order, or separate phases for electron and hole transport [18, 19].

The bulk inorganic semiconductors have high dielectric constants. At room temperature, the photogenerated electronic particles are uncorrelated and move unrestrained in conduction and valence bands (free carriers). An internal electric field is needed to separate free electrons and holes so they can be gathered at opposite electrodes and used in PV cell. A p–n junction in the device typically creates the field, but Schottky junctions between semiconductor and metal or liquid contacts with the correct work function differences with respect to the semiconductor material can also be used [20].

In organic photovoltaic cells, excitons must be isolated for free carriers to form, then gathered at opposite electrodes. This can be achieved from a heterojunction having enough band-offset that drives electrons and holes apart. "Hot carriers/excitons", photons having energies greater than the semiconductor bandgap, create these free carriers or excitons with excess energy of the bandgap. Kinetic energy can be dissipated quickly (picoseconds or lower timescale) by electron–phonon scattering, turning it into heat. Then they occupy the lowest energy levels, where they are able to do electrical or chemical work or lost through radiative or nonradiative recombination [9].

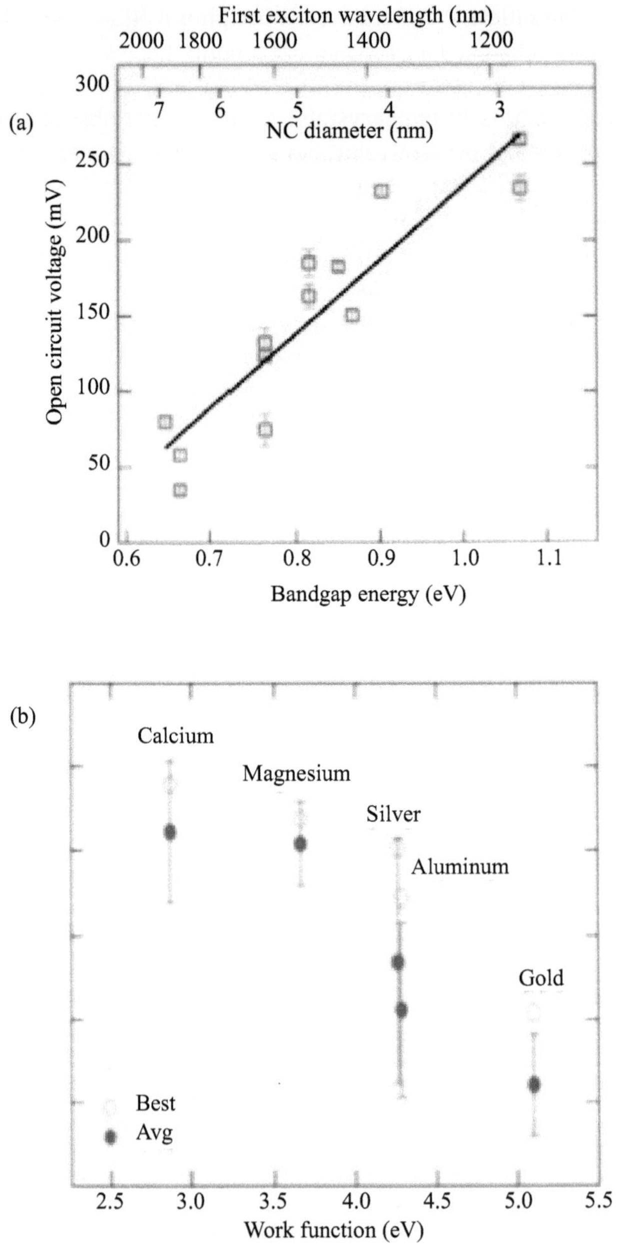

Figure 5.2 (a) The effect that exciton wavelength has on the open-circuit voltage as well as bandgap energy. (b) The differences in work function and voltage (which is the left-hand axis) caused by alterations in materials used [18].

Shockley and Queisser [21] calculated in 1961 that the maximum possible thermodynamic efficiency of converting solar irradiance into electrical free energy in a photovoltaic cell was 31–33%, with optimum bandgaps between 1.1 and 1.4eV (gaps of Si and GaAs are near to these).

5.2.1 Energy Loss Reduction

Energy loss to carrier cooling can be decreased by stacking a series of semiconductors with dissimilar bandgaps in tandem with the largest gap first followed by smaller gaps. With this arrangement, conversion efficiency can reach 67% at 1-sun intensity.

In practice, only two to three gaps are used because the most gain in efficiency is obtained with three; diminishing returns follow with more gaps.

With 2 suns, the maximum efficiency can be calculated as 43%; 48% with 3 suns, 52% with 4 suns, and 55% with 5 suns. The present record in the laboratory of multi-junction solar cell based, with three junctions (GaInAs), is 41% under solar concentrations of 140–240 suns [22]. The highest 1 sun of a single-junction Si photovoltaic cell is 25% [7, 14, 15, 22]. Most of today's Si solar cells are roughly between 14% and 17% efficient, for commercial use [7].

5.2.2 Use of Hot Carriers to Enhance Conversion Efficiency

One way of enhancing conversion efficiency is to transport hot carriers to collecting contacts before the carriers cool, called hot carrier solar cells. The second approach is to use excess kinetic energy to produce additional electron–hole pairs, called impact ionization, in bulk semiconductors. However, the maximum power conversion efficiency cannot be improved in Si, CdTe, $CuIn_xGa_{1-x}Se_2$ semiconductors because the maximum quantum yield for impact ionization does not produce extra carriers until photon energies reach the ultraviolet band of the sun's radiation.

The rate of impact ionization must vie with the rate of energy relaxation by phonon emission through electron–phonon scattering. It has been shown that the rate of impact ionization becomes competitive with phonon scattering rates only when the kinetic energy of the electron is many multiples of the bandgap energy [23–25].

5.2.3 Quantum Dot Solar Cells

Quantum dots have been used in three principal ways for solar cells. Photoelectrodes containing QD arrays have been employed, QD-sensitized nanocrystalline titanium oxide solar cells have been made, and QDs dispersed in organic semiconductor polymer matrices have been successfully manufactured. These configurations for QD solar cells are being studied in different laboratories throughout the world [10]. Quite high and reliable MEG efficiencies have now been claimed, including photocurrent quantum yields of 200% from QDs held to single crystal TiO_2 surfaces at 77 K [26]; no QD-based photovoltaic device has displayed improved conversion efficiency owing to MEG effects.

Conversion efficiency of the QD-based solar cells is about three times that of bulk semiconductors with low bandgaps (<0.5 eV) of the same material even without MEG. If the bandgap were 0.9–1.3 eV, it would be about four to five times more efficient.

5.3 Current Conditions of Nanotechnology

Current energy prices are rising, and the biggest rise is in the form of electricity. This sector is also the one with the highest carbon dioxide emission [27]. This is encouraging to the photovoltaic field because it becomes easier to become economically competitive with grid prices. Currently, the photovoltaic field is dominated by crystalline silicon solar cells. The current trend in the market is slow improvement in median efficiency as proven technologies are refined and rapid improvement as new bolder techniques are tried [13]. The increase in efficiency of solar cells has a direct relation with the reduction of cost. The original solar cells were not efficient enough to be able to payback in their lifetimes. In 2011, cells were sold at a price of about $2.50/W and paid back in 3–5 years depending on sunshine [27]. This price continued to drop and in 2012, it reached $1.82/W [13].

The latest research in solar structures consists of n-type nanocones surrounded by p-type semiconductors. These nanocones are made of zinc oxide and work as a junction framework and the electron conductor. On the other side, the p-type matrix is made of polycrystalline cadmium telluride and serves as the primary photon absorber medium and hole conductor. Using this structure, Xu et al. were able to increase the light-to-power conversion efficiency of 1.8% of conventional planar structures of the same materials to 3.2%, an increase of 80%. The three-dimensional structure creates an intrinsic electric field distribution that increases the efficiency of charge transportation and conversion of energy from sunlight into electricity. The key features are the unique electric field distribution, the synthesis of nanocones from inexpensive methods, and the minimal defects and voids in semiconductors, enhancing electric and optical properties for converting photons to electricity. The cone shape creates an electric field near the top junction, effectively separating, injecting, and collecting minority carriers, which creates a higher efficiency than of conventional planar cells made of same materials. These solar cells can tolerate defective materials as well as be generated at reduced costs due to more efficient charge transport [28].

Another recent discovery was by colleagues of Washington University in St. Louis and the Department of Energy's Oak Ridge National Laboratory as the use of chlorosomes for converting sunlight to energy. These have been analyzed in green photosynthetic bacteria through the use of small-angle neutron scattering and have proven efficient even in low-light and extreme environments. Researchers referred to it as one of the most efficient light-harvesting complexes in nature. It was observed under a spectrum of thermal and ionic conditions, in which they altered very little structurally ensuring their stability. Certain factors, such as the size, shape, and organization, are important to light-harvesting complexes such as chlorosomes and affect electron transfer to semiconductor electrodes. The complicated complex is also able to capture light efficiently over vast areas and focus the light to the reaction center without losing large amounts of it [29].

5.3.1 Crystalline Silicon

Crystalline silicone currently dominates the photovoltaic market and appears that it will continue to do so for the next 5–10 years. In the 1950s, during the early years of development and testing, silicon solar cells had efficiencies of approximately 6%, with levels now reaching near 25% [3]. This predominance of silicon-based cells is because the raw material is relatively available, has low ecological impact, and in its pure form has very low degradation. Silicon-based cells also have a dominating role in application because they have been around the longest. Most of silicon cells are in the monocrystalline and multicrystalline forms [30]. As of 2008, the efficiency for monocrystalline silicon solar cells has reached 24.7% at the research level and 20.3% at the production level. For polysilicon, the efficiencies are 20.3% at the research level and 17.7% at the production level. A survey of monocrystalline and multicrystalline modules in 2004 showed that the efficiency ranged between 9.1% and 16.1% with the median close to 12.5% [31].

Current research in crystalline silicon photovoltaic cells is focused on a few things. The first is the reduction of degradation. Currently most silicon crystals are made using the Czochralski method. Degradation of silicon cells is caused by oxygen that contaminates the production of the Czochralski crystals. The float zone method produces purer crystals and reduces contamination and is used for research and development (R & D) but is too expensive for production cells. A new method of production, the Magnetically Grown Czochralski Method, has been shown to reduce the contamination and evenly distributes the contamination that does form. Another one of interest is the development of other methods of crystal formation. Tricrystals have been developed that provide better mechanical stability and can be grown faster. They have been shown to provide up to a 40% cost reduction.

Silicon still has many disadvantages though. From a solid state physics point of view, silicon is not an ideal material because of its very limited bandgap. The ideal material should have a bandgap of about 1.1–1.7 eV, have a direct band structure, and have materials that are readily available, non-toxic, and consists of processes that are easily producible and can be applied to large area production, have good conversion efficiency, and have long stability.

A great disadvantage of silicon is that it is an indirect semiconductor, so instead of requiring 1 μm thickness for 90% absorption, as is the case for a direct semiconductor, it would require 100 μm.

Another factor disfavoring silicon cells is that after the crystal has been made, it must be cut into thin wafers. This has high costs because only diamond-tipped saws are capable of cutting through the crystals and material loss can be as high as 50%. Recently, a new method of using abrasive wires has been capable of cutting thinner wafers while reducing waste by up to 30%.

New techniques have been researched for new architecture to enhance harvesting of solar energy. It has been discovered that wire arrays have less than 5% of the areal fraction of wires, yet capture up to 96% peak absorption as well as enhanced near-infrared absorption. Also using a radial semiconductor/liquid junction at the surface, they are able to obtain near-unity quantum efficiency, approximately 0.89. This breakthrough means that efficiency may be increased, while using a hundredth of the material previously used, resulting in a much lower production cost [32].

Silicon also has high reflectivity, which prevents all the light that is available from being captured by the cell. Reducing the reflectivity of the cell can improve the efficiency of the overall cell. This can be accomplished by applying an antireflective coating, or a more effective option is to add surface texture. The most common texture is isotropic etching on monocrystalline cells. This technique, however, can only be applied to these cells, but research is being done on finding methods of applying it to other types of cells.

Even though one of the reasons that silicon is currently used is its relative availability, its price has not been stable. This is because they are most commonly made from rejected semiconductor material. If both industries begin to grow at different rates, getting enough material for the photovoltaic industry can be difficult. An extreme example is during a tech boom in which even silicon that would have normally been rejected was being bought by the semiconductor industry. This caused the photovoltaic industry to compete with the semiconductor industry for material resources. Since almost half of the cost of the cells is the silicon, the unit price is greatly dependent on the price of silicon.

The main advantage of monocrystalline silicon cells lies in the fact that larger crystals have been found to have greater conversion efficiencies. Multicrystalline cells tend to have lower efficiencies at the crystal level but utilize the module area better, resulting in comparable modular conversion efficiencies. The greatest advantage of multicrystalline cells is that they can be produced with processes that are less expensive and easier to reproduce. The cost reduction works to nullify the reduced efficiency. Efficiency may be sacrificed if it can be made cheaply enough to justify the increase in the number of cells needed. There is a limit to how low efficiency should go (about 10%) before the increase in area raises the costs of the modules [30].

Another interesting feature of the crystalline structure is that the results with the maximum doping effect and smallest grains also presented a decrease in the activation energy. Based on this study, the room-temperature conductivity jumped to over 50 times the regular amount [33]. On top of this 50-fold increase in conductivity, absorption can be increased by another 50 times normal when at resonance frequency through the help of spherical silver nanoparticles from a field enhancement produced by surface plasmons. The maximum enhancement of this trait was found to be occurring with a shell thickness of 0.26 times the radius of the particle [34].

5.3.2 Thin Films

Another type of photovoltaic cells is a thin-film cell. The main advantage of thin films is that since they are much thinner, the material requirements are much less. The added advantage of thin cells is that their theoretical efficiencies should also increase. There are a variety of thin-film cells. These include ribbon silicon, amorphous silicon, heterostructures, CISG cells, cadmium telluride cells, as well as much structural advancement that can be made. There has been discoveries of strong correlations between the rate of conductivity increase and the structure of films. However, with increasing film grain sizes, there is a diminishing return effect to the rate of conductivity increase [33].

5.3.2.1 Ribbon silicon

Ribbon silicon cells have many advantages over the crystalline silicon cells. The first advantage is that since the ribbon silicon is already very thin, it does not have to be sawed. This eliminates the cost of the sawing equipment as well as the cost of the discarded material. Production costs are also much lower than for crystalline silicon because there are a variety of methods for producing the ribbon.

These methods result in efficiencies ranging between 11.1% and 17.3%. These production methods also allow for much higher production rates that reduce cost by allowing higher yield with the same equipment. High-temperature deposition methods have been successful in creating large crystals that have good conversion efficiencies. These methods have the ability to be expanded to high continuous throughput. The fact that they require high temperature prevents the methods from being implemented. The high temperature causes impurities in the substrate to migrate to the active region, thereby increasing cost unnecessarily. A new method is being researched that could solve this problem. Seeding allows these high-temperature methods to be used on high-purity substrates and then be transferred to less expensive substrates, while allowing the pure substrate to be reused [30].

5.3.2.2 Amorphous silicon

The newest type of silicon cells is amorphous cells. They have the advantage of larger bandgaps; they can be produced at lower temperatures, thereby using less expensive substrates, and can be made to be only a few nanometers thick. Their disadvantages are that they have lower efficiencies and suffer from high degradation rates, especially during initial exposure. Since these films can be made so thin, they can be assembled in tandem. Tandem cells have the highest theoretical efficiencies with 41.9% for two cells combined in series, or more than 50% for combined multiple cells. The reason for this improved efficiency is because they allow maximum efficiency conversion at different parts of the solar spectrum, thereby increasing the overall bandgap, and also reducing the amount of degradation. Even with multiple layers, the total thickness would still be about 1 μm.

Current improvements with amorphous cells are with understanding the material and the stability, improve the process of hydrogen dilution, understand the properties between amorphous and crystalline cells, increase overall efficiencies, and optimize stacking and tandem cells for the best prevention of degradation [30].

Another application of amorphous silicon is in the heterostructure design, which is a combination of amorphous and crystalline layers in a cell. This combination has shown to produce efficiencies of up to 20.7%. This efficiency increase is accompanied by a reduction in cost. Since all processes are done at temperatures of less than 200°C, the manufacturing process is also improved.

5.3.2.3 Ternary compounds

This class of ternary compound semiconductors includes copper indium diselenide ($CuInSe_2$), copper gallium diselenide ($CuGaSe_2$), copper indium disulfide ($CuInS_2$), and the multinary CIGS ($Cu(In,Ga)(S,Se)_2$). As of 2003, the efficiencies for CIGS cell modules ranged from 5.6% to 16.8%, but given the flexibility in material and process properties that can achieve big changes in properties, it is expected the more rapid progress can be made with copper indium diselenide and related component cells [30].

5.3.2.4 Cadmium telluride

Cadmium telluride has many advantageous properties, including a bandgap close to ideal, and relatively easy production techniques with low temperatures. An important feature of the production techniques is that they allow the material to be deposited on the substrate at relatively high rates. These cells have been found to have greater efficiencies with polycrystals rather than monocrystals, facilitating production without reducing efficiency since polycrystals are easier to grow than monocrystals.

When research began on this type of cell, it had the highest initial efficiencies of any type. This cell type looks very promising since the highest quality crystals come from one of the easiest deposition processes. The highest efficiencies for CdTe have been in the form of CdTe–CdS cells. These have faced several difficulties, the main ones being formation of the junction, crystallizations of the films, and the formation of stable back contacts. Even though these cells are fairly new, the highest wattage achieved by a monolithic module has been 70 W made from CdTe.

There have been some difficulties with their production nonetheless. It has been difficult to dope p-Type CdTe, find low resistance contacts, and they have high recombination losses associated with the junction interphase. One of the biggest issues the cells might face is a nontechnical issue of public acceptance. Cd and Te are both toxic chemicals even though the CdTe compound itself is stable and harmless [30].

5.3.2.5 Organic cells

Another thin-film technology is organic solar cells. These cells have very high optical absorption coefficients that allow for cells less than 1 nm thick. This means the cells can be made with much less material. One of the disadvantages, however, is that the width of the charge generation layer is much smaller for organic cells. One possible solution is by adding an interpenetrating network of fullerene [30, 33] or nanoparticles. Despite this drawback, organic and polymer solar cells are preferred to silicon cells because they have lower costs, have better ease of fabrication and lower weight [35]. The thin layers of organic materials result in poor absorbance, hence poor efficiency. However, the thicker ones suffer bulk recombination losses. In an attempt to counterbalance this issue, plasmonic structures have been placed in, or around, the organic cell layer. Thus, enhancing the electromagnetic field and improving the light absorption of the active layer without having to bulk up the cell layer. Through this technique, there have been observations of a 70% enhancement in efficiency [36].

New strategies have also been discovered using the synthesis of C_{60} derivatives. The first uses highly soluble fullerene derivatives and showed an increase of approximately 100 mV when mixed with DPM, diphenylmethanofullerene, opposed to the usual PCBM, phenyl C61 butyric acid methyl ester. This resulted in over 800 mV for the open-circuit voltage. The second contained oligomers and dendrimers covalently bonded to the C_{60} core, which created a control over the energy and electron transduction, as well as a white light efficiency of 0.2% [37]. Also using organic liquid crystals, which show self-organized molecular columns, it was revealed that changes occurred in the electronic properties when changing from a disordered orientation to columnar phases. The significant change was a decrease in ionization potential by 1.8 eV after the appearance of new electronic states at low binding energy. Also witnessed was energy dispersion on upper bands of the columns by 1.1 eV [38, 39].

5.4 Nanotech Application

Numerous applications of nanotechnology have surfaced as it continues its development in all fields. One of the most promising aspects of nanotechnology is the ease with which it could be manufactured. It is predicted that nano-semiconductors will be made in simple laboratory vats. This reduces the costs involved with using clean rooms and vacuum chambers. These fabrication methods will also allow for rapid roll-to-roll and low temperature production [1]. Several new materials are being developed in Taiwan for application in photovoltaic cells. These include replacing indium tin oxide (ITO), and the development of nanomaterials; low-cost, high-efficiency high-temperature crystalline silicon thin-film technologies; and high-density thin-film manufacturing technology and processes [31]. Most of the applications of nanotechnology are in an effort to increase the performance of existing technologies.

These methods include sensitizing the active layer, combining them with conducting polymers, and reduction of reflection and QDs. The biggest hurdle this technology is experiencing is being able to prove the longevity that consumers have become accustomed to. Since the technology has not been around for very long, even accelerated experiments have not been completed to prove that the technology can last 25 years like other solar cells [1].

5.4.1 Dye-Sensitized Solar Cells

Dye-sensitized solar cells work by sensitizing materials to a different part of the electromagnetic spectrum so that more of the energy being received by the sun may be used by the cell. Nanocrystalline dye-sensitized solar cells are a completely new type of cells. The new development with these cells is that the absorption material is completely separate from the charge carrier transport. This method seems to be effective only in transferring the electrons to the charge carrier transport material, when the two are in contact. An attempt to minimize this problem is to have a nonporous microlayer. This method is based on an increased contact between the dye and the charge carrier transport material [30]. Current research is using this design to make the absorption range of titania closer to the visible region of the electromagnetic spectrum, increasing the percentage of the energy from the sun. Since the advent of the dye-sensitized solar cell architecture, cells have now reached efficiencies of up to 11% [14, 40].

5.4.2 Conducting Polymers

Another method that employs nanotechnology in the use of solar cells is the combination of nanocrystalline titanium dioxide and conducting polymers, as well as Ti–Au electrodes on silicon oxide. Notable progress has been made by Siemens in this area by reaching 10% efficiency by combining buckyballs and electrically conducting polymers. Los Alamos National Lab has also found that the energy output can be doubled with the use of lead selenium nanocrystals [1, 41].

When working with polymers, it has been found that layered donor–acceptor heterojunction is an appealing architecture for obtaining a larger efficiency for solar cells, however, depending on if appropriate donor and acceptor semiconductors with the optimum properties can be created. Using these polymers, records show power conversion efficiency for photovoltaic cells as high as 2–3% from white light alone, as well as a 2–3 times larger photovoltages to that of electrodes. The disorder acts as both a limiting and a defining characteristic for polymers [42, 43].

5.4.3 Titania Nanotubes

Another new application of this technology is in titania, TiO_2, nanotubes. These tubes are proposed to replace titania nanoparticles. While the nanoparticles have much higher efficiencies of up to 11%, the electron transport that they utilize is a slow mechanism. It is expected that the nanotubes will greatly increase the electron transport speed, and as the efficiencies improve, they should prove better than the nanoparticles.

The disadvantage of the current method of production for the titania nanotubes is that they have the working electrode on the back. As a result, some of the photon energy is decreased as the electrons pass through the glass and electrolyte. By developing a method of growing the nanotubes on the front, this can be avoided and more photon energy can be captured, thereby increasing the efficiency of the cell [9, 44].

Currently titania arrays are made by electrochemical annodization methods from titanium foil. There are several difficulties with the nanotubes grown in this manner. The first is that the nanotubes have a closed tip, which has the potential for trapping air, increasing surface tension, and it makes electron transport difficult. Work is being done to develop a method of growing the nanotubes without the closed tip. Unfortunately, most of these processes are slow, difficult, and use toxic materials, making their use in large-scale production not feasible.

Current research is focused on finding new production methods that can be adapted to large-scale production. There is a relatively new process currently being used for the production of titania nanotube arrays that can easily be adapted to larger area productions by using a larger size titanium foil up to a size of 16.5 cm^2 [45].

Dispersing nanoparticles raise the conversion efficiency by a factor of about 3 as compared to omitting the particles [37]. It has been found that using carbon nanotube arrays, or CdTe-based solar cells, a theoretical efficiency of 29% can be obtained, but the maximum reported has only been 16% [46]. In an attempt to increase efficiency, studies focus on reducing the crystal defects and maximizing interface characteristics between the CdTe and its complementary cell material.

5.4.4 Antireflectivity

Reducing the reflectivity of the solar cell is an important step in increasing the efficiency of the cells. Nanotechnology offers several solutions to this problem, through processing for texture or surface coating.

5.4.4.1 Nano-imprint lithography

One of the new methods for reducing reflectivity is the use of nano-scaled cathodes. Nano-scaled cathodes can reduce reflection losses by 20–40% in organic photovoltaic cells. One method being used to develop this size cathode is nano-imprint lithography. It permits the printing of a variety of different shapes. Trying non-textured, grating, and discrete diamonds has shown that grating provides the best overall performance [47].

5.4.4.2 Coatings

Using nanotechnology to develop better coatings has proven promising. In a recent experiment, it was found that they had lower reflectivity and higher transmission. When ITO nanowhiskers are deposited on the surface, they provide an antireflective layer with a reflectivity of less than 5%, better than the current Si_3N_4 antireflective coating used. With improved optical transmission in the 700–1000 nm range, the conversion efficiencies of solar cells are improved from 13.93% to 14.37% [46]. It has also been found that these coatings could be used alone. A recent experiment tested the deposition of nanowhiskers on both a bare cell and one coated with Si_3N_4. The results of the experiment showed that the nanowhiskers provided the largest effect. In addition to increasing efficiency by reducing reflectivity, if the open-circuit voltage is assumed to remain constant with the nanowhiskers, then this method should also increase short-circuit current from 33.27 to 34.30.

Other nanotechnologies are also being investigated to be used as antireflective coatings. Nanopyramids and nanorods show promise because they have excellent antireflective properties through a wide range of wavelengths and incident angles. As with most new technologies, the high cost of manufacturing and depositing them on solar cells has thus far kept them from being used in production cells [48].

5.4.5 Electron Transfer

Research in nanotechnology is also attempting to increase electron transfer. There is a better transfer of electrons in well-separated layers, but these tend to limit electron transfer to the interphase between the two materials. There is an attempt to solve this problem by having two interpenetrating layers of electron emitters and acceptors. This concept has been tried by a variety of methods, including two-phase blends of polymer/inorganic nanocrystals, dye/dye, polymer/fullerene, polymer/polymer, but the most exploited are polymer matrix with physically dispersed dye. Another possible method of efficiency enhancement has undergone extensive research and that is using dyes that are capable of singlet fission into two triplets, meaning that two electron–hole pairs could be produced from only one photon [11].

One of the methods being employed to accomplish this task is improving holes transfer to the ITO electrode. A method that has proven successful is to cover the ITO with a layer of polyaniline (PANI). Current improvements being researched are production. Current methods include deposition, spinning, or applying it in the form of nanotubes, all of which are very difficult to do. An added difficulty is that the PANI absorbs sunlight, thereby reducing the amount that reaches the active layer. To resolve this issue, the solar cell's architecture has been modified. The PANI has been incorporated in nano-sized grains directly into the active, which still provides the necessary interpenetrating network, but since it is not a continuous layer, it still allows all the light to shine through. The nanoparticles were made from pure SiC. A portion was made conductive by using camphorsulfonic acid to dope them chemically. The conducting form of PANI greatly increased the energy conversion efficiency. This method has the important benefit that it is easily mixed into the matrix, making it a realistic candidate for large-scale production [16, 25].

5.4.6 Quantum Dots

Many materials have favorable properties, but they may at times lack the right absorption wavelength to make them effective with sunlight. This is the case with titania. While it has attracted a lot of attention in the photovoltaic community, its large bandgap puts its absorption region in the ultraviolet range, which only compromises at most 5% of the solar radiation that reaches earth. As described earlier, DSSC is one method of solving this issue. The use of DSSCs is limited, however, because they become unstable at high temperatures.

Another option being explored is modifying the oxide with QDs. One of the great promises of QDs is that they may theoretically be able to generate multiple electrons per photon [40]. Another recent proposal also makes QDs very favorable. It proposes that a metallic band in the mid gap would allow photons that have less energy than the gap energy, to be utilized by the cell. This would allow for higher efficiencies up to a theoretical value of 46%. Recently, QDs have been proposed as a way to manufacture this material [49].

One design that is being experimented is combining titania sensitized with nanocrystalline CdS QDs. The idea behind this design is that CdS has a bandgap that matches favorably with that of the sun. This would tune the film of titania to the visible region and speed up separation, thereby improving the overall efficiency of the cell.

With the addition of the QDs, the absorbance of titania is greatly changed. The titania film alone has an absorption peak at about 380 nm and has no noticeable absorption past approximately 410 nm. After the addition of 11 cycles of QDs, the absorption peak moves to about 475 nm and is about 50% greater and had appreciable absorption to about 580 nm and had measurable absorption well past 600 nm. Most importantly, the absorbance after 11 cycles included most of the region of absorbance for no cycles, so there was virtually no loss of the original region. The use of the QDs had a very dramatic increase in efficiency. The conversion efficiency of the cell with zero cycles was 0.1% and hit a maximum at 1.91% for 7 cycles. It is believed that the efficiency began dropping for more than 7 cycles because the QDs began inhibiting the electron transfer rate, but this is believed to be avoidable as smaller QDs are used [40]. Analyzing research, it has been hypothesized that in order to obtain a strong reaction between CdSe and thiol functional groups may need a lower amount of QDs in the mixture to induce an efficient charge transfer. Two weight ratios were used, 5:1 and 1.8:1, producing power conversion efficiencies of 0.60% and 0.25%, respectively [50].

One of the latest technological advancement using QDs has been made by scientists at Brookhaven National Laboratory, pairing them with fullerenes to increase the efficiency of solar panels. It has been the first instance of development of a hybrid inorganic/organic photovoltaic cell, working as an electron donor–bridge–acceptor system in the conversion of light to current [51].

5.5 Discussion and Conclusion

Since the discovery of the photovoltaic effect, the aim of fulfilling the earth's energy needs, with the energy of the sun that already reaches earth every day, has seemed a possible but elusive goal. Many improvements in efficiencies, materials, and processes have led to a huge decrease in cost. As prices for photovoltaic cell decrease, prices for electricity from the grid are increasing. This coupled with promising technologies being developed make it seem possible that this technology will be economically competitive with traditional energy sources. The science of this nanotechnology is developed, but not complete.

Current research in the field is focused on continued improvement in the crystalline silicon cells, as well as new materials and processes. Thin films are showing promise as ribbon silicon, amorphous silicon, ternary compounds, cadmium telluride, and organic cells. These thin cells generally have lower efficiencies but have other advantages that make up for this deficiency. The biggest advantage they have is that they are easier to manufacture and use much less material.

The other set of improvements are due to the incorporation of new nanotechnologies into existing designs. This has been accomplished through the use of DSSCs, mixing them with conducting polymers, titania nanotubes, reducing the reflectivity of the cell through the use of nano-imprinted lithography and coatings, improving electron transfer, and QDs. These advancements improve efficiencies by allowing the cells to receive more light, absorb energy from a larger, more ideal spectrum, and improving the ease with which photons are converted to electrons and transferred from the active region to be used to power something. In addition to improving efficiencies, these technologies can typically be produced in less controlled environments. These processes are also typically capable of being expanded to large production runs, without greatly decreasing efficiency and quality or increasing cost.

The environmental damage caused by the use of fossil fuels is motivation for a growing number of people to use renewable energy. People are more willing to pay to safeguard the environment as the price approaches that of conventional energy sources. For this technology to become mainstream and an important percentage of world energy production, it has to become economically competitive when compared to conventional energy sources. The future looks bright for photovoltaic technology as these technologies continue to be developed. They have the potential to greatly increase efficiency while greatly reducing cost. It is possible that in the foreseeable future, prices for energy from photovoltaic sources will be at least as low as those from conventional sources and will start replacing them, allowing energy independence from fossil fuels.

References

1. Trombly, J., "Technology solutions: Nano-PV set to accelerate solar-energy use," *Environ. Sci. Technol.*, **38**, pp. 376–376A, 2004.
2. Kelsall, R. W., Hamley, I. W., and Geoghegan, M., *Nanoscale Science and Technology*, John Wiley & Sons, Ltd., England, Chapter 6, 2005.
3. Catchpole, K. R., "Nanostructures in photovoltaics," *Phil. Trans. Math. Phys. Eng. Sci.*, **364**, pp. 3493–3503, 2006.
4. Vogel, E. M., "Technology and metrology of new electronic materials and devices," *Nature Nanotechnol.*, **2**, pp. 25–30, 2007.
5. Lal, S., Link, S., and Halas, N. J., "Nano-optics from sensing to waveguiding," *Natrue Photonics*, **1**, pp. 641–642, 2007.
6. Schaefer, H. E., *Nanoscience: Nanoscience and Technology Series*, Springer, USA, Chapter 10, 2010.
7. Tsakalakos, L., "Nanostructures for photovoltaics," *Mater. Sci. Eng.*, **62**(6), pp. 175–189, 2008.
8. Varonides, A. C., "Tunneling photoconductivity computations of multi-quantum well p-I (nano)-n photovoltaic nanostructures by mean of the causal Green's function," *Thin Solid Films*, **451–452**, pp. 393–396, 2004.
9. Hoppe, H. and Sariciftci, N. S., "Organic solar cells: An overview," *J. Mater. Res.*, **19**(7), pp. 1924–1945, 2004.
10. Nozik, A. J., Beard, M. C., Luther, J. M., Law, M., Ellingson, R. J., and Johnson, J. C., "Semiconductor quantum dots and quantum dot arrays and applications of multiple exciton generation to third-generation photovoltaic solar cells," *Chem., Rev.*, **110**, pp. 6873–6890, 2009.
11. Wang, Z., Wang, Y., Hu, N., Wei, L., Chen, S., and Zhang, Y., "Polythiophene microspheres synthesized by transition metal mediated oxidative dispersion polymerization," *J. Polym. Sci. A Polym. Chem.*, **48**, pp. 5265–5269, doi: 10.1002/pola.24337, 2010.

12. Perlin, J., "Solar energy, history of," in *Encyclopedia of Energy*, C. J. Cleveland, Ed., Elsevier, New York, pp. 607–622, 2004.
13. Green, M., "Silicon photovoltaic modules: A brief history of the first 50 years," *Prog. Photovolt.*, **13**, pp. 447–455, 2005.
14. Jadhav, M. V., Todkar, A. S., Gambhire, V. R., and Sawant, S. Y., "Nanotechnology for powerful solar energy," *Int. J. Adv. Biotechnol. Res.*, **2**, pp. 208–212, 2011.
15. Kapur, V. K., *Photovoltaics for the 21st Century*, The Electrochemical Society, USA, pp. 99–101, 1999.
16. Hagfeldt, A. and Gratzel, M., "Molecular photovoltaics," *Acc. Chem. Res.*, **33**, pp. 269–277, 2000.
17. Honsberg, C. B., Barnett, A. M., and Kirkpatrick, D., "Nanostructured solar cells for high efficiency photovoltaics," in *Photovoltaic Energy Conversion, IEEE 4th World Conference*, Hawaii, USA, 2006.
18. Luther, J. M., Law, M., Beard, M. C., Song, Q., Reese, M. O., Ellingson, R. J., and Nozik, A. J., "Schottky solar cells based on colloidal nanocrystal films," *Nano Lett.*, **8**(10), pp. 3488–3492, 2008.
19. Xia, Y., Yang, P., Sun, Y., Wu, Y., Mayers, B., Gates, B., Yin, Y., Kim, F., and Yan, H., "One-dimensional nanostructures: Synthesis, characterization, and applications," *Adv. Mater.*, **15**, pp. 353–389, 2003.
20. Nozik, A. J., "Nanoscience and nanostructures for photovoltaics and solar fuels," *Nano Lett.*, **10**(8), pp. 2735–2741, 2010.
21. Shockley, W. and Queisser, H. J., "Detailed balance limit of efficiency of p-n junction solar cells," *J. Appl. Phys.*, **32**, 510, 1961.
22. Green, M. A., Emery, K., Hishikawa, Y., and Warta, W., "Solar cell efficiency tables (Version 34)," *Prog. Photovolt. Res. Appl.*, **17**, pp. 320–326, 2009.
23. Bude, J. and Hess, K. J., "Thresholds of impact ionization in semiconductors," *J. Appl. Phys.*, **72**, 3554, 1992.
24. Jung, H. K., Taniguchi, K., and Hamaguchi, C., "Impact ionization model for full band Monte Carlo simulation in GaAs," *J. Appl. Phys.*, **79**, 2473, 1996.
25. Harrison, D., Abram, R. A., and Brand, S., "Characteristics of impact ionization rates in direct and indirect gap semiconductors," *J. Appl. Phys.*, **85**, 8186, 1999.
26. Sambur, J. B., Novet, T., and Parkinson, B. A., "Perspective of the prospects of carrier multiplication nanocrystal solar cell," *Science*, **330**, pp. 63–66, 2010.

27. Dincer, F., "The analysis on photovoltaic electricity generation status, potential and policies of the leading countries in solar energy," *Renew. Sustainable Energy Rev.*, **15**, pp. 713–720, 2011.

28. Xu, J., "New ORNL solar cell technology cranks up efficiency," Oak Ridge National Laboratory, April 29, 2011.

29. Tang, K. H, Zhu, L., Urban, V. S., Collins, A. M., Biswas, P., and Blankenship, R. E., "Temperature and ionic strength effects on the chlorosome light-harvesting antenna complex," *Langmuir*, **27**, pp. 4816–4828, 2011.

30. Goetzberger, A., Hebling, C., and Schock, H., "Photovoltaic materials, history, status and outlook," *Mater. Sci. Eng. R Rep.*, **40**, pp. 1–46, 2003.

31. Liou, H. M., "Overview of the photovoltaic technology status and perspective in Taiwan," *Renew. Sustainable Energy Rev.*, **14**, pp. 1202–1215, 2010.

32. Kelzenberg, M. D., Boettcher, S. W., Petykiewicz, J. A., Turner-Evans, D. B., Putnam, M. C., Warren, E. L., Spurgeon, J. M., Briggs, R. M., Lewis, N. S., and Atwater, H. A., "Enhanced absorption and carrier collection in Si wire arrays for photovoltaic applications," *Nature Mater.*, **9**, pp. 239–244, 2010.

33. Katz, E. A., Faiman, D., Tuladhar, S. M., Shtutina, S., Froumin, N., Polak, M., and Strzhemechny, Y., "Diffusion process for doping of C_{60} (fullerene) thin films," *Sol. Energ. Mat. Sol. Cells*, 75, pp. 421–426, 2003.

34. Hu, L., Chen, X., and Chen, G., "Surface-plasmon enhanced near-bandgap light absorption in silicon photovoltaics," *J. Comput. Theor. Nanosci.*, **5**, pp. 2096–2101, 2008.

35. Niziol, J., Gondek, E., Danel, A., and Sanetra, J., "Influence of dispersed core-shell nano-sized particles on P3OT based photovoltaic device," in *ICTON Mediterranean Winter Conference, 2009*, pp. 1–4, 2009.

36. Kochergin, V., Neely, L., Jao, C. Y., and Robinson, H. D., "Aluminum plasmonic nanostructures for improved absorption in organic photovoltaic devices," *Organic Electron. Photonics*, **98**, 3, 2011.

37. Segura, J. L., Giacalone, F., Gomez, R., Martin, N., Guildi, D. M., Luo, C., Swartz, A., Riedel, I., Chirvase, D., Parisi, J., Dyakonov, V., Sariciftci, N. S., and Padinger, F., "Design, synthesis and photovoltaic properties of [60] fullerene based molecular materials," *Mater. Sci. Eng. C*, **25**, pp. 835–842, 2005.

38. Crispin, X., Comil, J., Friedlein, R., Okudaira, K. K., Lemaur, V., Crispin, A., Kestemont, G., Lehmann, M., Fahlman, M., Lazzaroni, R., Geerts, Y., Wendin, G., Ueno, N., Bredas, J. L., and Salaneck, W. R., "Electronic

delocalization in discotic liquid crystals: A joint experimental and theoretical study," *J. Am. Chem. Soc.*, **126**, pp. 11889–11899, 2004.

39. Fan, Z., Ruebusch, D. J., Rathore, A. A., Kapadia, R., Ergen, O., Leu, P. W., and Javey, A., "Challenges and prospects of nanopillar-based solar cells," *Nano Res.*, **2**, pp. 829–843, 2009.
40. Chen, H., Fu, W., Yang, H., Sun, P., Zhang, Y., Wang, L., Zhao, W., Zhou, X., Zhao, H., Jing, Q., Qi, X., and Li, Y., "Photosensitization of TiO_2 nanorods with CdS quantum dots for photovoltaic devices," *Electrochim. Acta*, **56**, pp. 919–924, 2010.
41. Akbayir, C., Bulut, F., Farrell, T., Goldschmidt, A., Guntner, R., Kam, A. P., Miclea, P., Scherf, U., Seekamp, J., Solovyev, V. G., and Torres, C. M. S., "Nanostructured conjugated polymeric systems for photovoltaic applications," *Rev. Adv. Mater. Sci.*, **5**, pp. 205–210, 2003.
42. Alam, M. M. and Jenekhe, S. A., "Efficient solar cell from layered nanostructures of donor and acceptor conjugated polymers," *Chem. Mater.*, 16, pp. 4647–4656, 2004.
43. Dykstra, T., Hennebicq, E., Beljonne, D., Gierschner, J., Claudio, G., Bittner, E. R., Knoester, J., and Scholes, G. D., "Conformational disorder and ultrafast exciton relaxation in PPV-family conjugated polymers," *J. Phys. Chem. B*, **113**, pp. 656–667, 2009.
44. Banerjee, S., Misra, M., Mohapatra, S. K., Howard, C., Mohapatra, S. K., and Kamilla, S. K., "Formation of chelating agent driven anodized TiO_2 nanotubular membrane and its photovoltaic application," *Nanotechnology*, **21**, pp. 145201, 2010.
45. Jadhav, M. V., Todkar, A. S., Gambhire, V. R., and Sawant, S. Y., "Nanotechnology for powerful solar energy," *Adv. Biotech. Res.*, **2**(1), pp. 208–212, 2011.
46. Camaco, R. E., Morgan, A. R., Flores, M. C., McLeod, T. A., Kumsomboone, V. S., Mordecai, B. J., Bhattacharjea, R., Tong, W., Wagner, B. K., Flicker, J. D., Turano, S. P., and Ready, W. J., "Carbon nanotube arrays for photovoltaic applications," *JOM*, **59**, pp. 39–42, 2007.
47. Yin, L., Liu, H., Ding, Y., Lu, B., and Fan, D., "Fabrication of 3D micro-structures in polymer photovoltaic devices based on soft nanoimprint lithography technology," in *2007 International Nano-Optoelectronics Workshop*, Beijing, pp. 300–301, 2007.
48. Chang, C. H., Hsu, M. H., Yu, P., Kuo, H. C., Chang, W. L., and Sun, W. C., "Novel indium tin oxide nano-whiskers for enhanced transmission of surface-textured silicon photovoltaic cells," in *34th IEEE Photovoltaic Specialists Conference (PVSC), 2009*, pp. 000540–000543, 2009.

49. Palaniappan, K., Murphy, J. W., Khanam, N., Horvath, J., Alshareef, H., Quevedo-Lopez, M., Biewer, M. C., Park, S. Y., Kim, M. J., Gnade, B. E., and Stefan, M. C., "Poly(3-hexylthiophene)-CdSe quantum dot bulk heterojunction solar cells: Influence of the functional end-group of the polymer," *Macromolecules*, **42**, pp. 3845–3848, 2009.

50. Paci, I., Johnson, J. C., Chen, X., Rana, G., Popovic, D., David, D. E., Nozik, A. J., Ratner, M. A., and Michi, J., "Singlet fission for dye-sensitized solar cells: Can a suitable sensitizer be found?" *J. Am. Chem. Soc.*, **128**, pp. 16546–16553, 2006.

51. Xu, Z. and Cotlet, M., "Quantum dot-bridge-fullerene heterodimers with controlled photoinduced electron transfer," *Angewandte Chemie, Int. Ed.*, **50**, pp. 1–6, 2011.

Chapter 6

Applications of Nanofluids: Current and Future

Kaufui V. Wong and Omar De Leon
Department of Mechanical and Aerospace Engineering, University of Miami, Coral Gables, FL 33124, USA
kwong@miami.edu

Nanofluids are suspensions of nanoparticles in fluids that show significant enhancement of their properties at modest nanoparticle concentrations. Most of the publications on nanofluids are about understanding their behavior so that they can be utilized where straight heat transfer enhancement is paramount as in many industrial applications, nuclear reactors, transportation, electronics as well as biomedicine and food. Nanofluid has also been reported as a smart fluid, where heat transfer can be reduced or enhanced at will. This chapter focuses on presenting the broad range of current and future applications that involve nanofluids, emphasizing their improved heat transfer properties that are controllable and the

This chapter has been reprinted from the *Advances in Mechanical Engineering*, Vol. 2010, Kaufui V. Wong and Omar De Leon, Applications of nanofluids: current and future, SAGE Publications Ltd., Copyright 2010 K. V. Wong and O. De Leon.

Nanotechnology and Energy
Edited by Kaufui V. Wong

ISBN 978-981-4774-24-6 (Hardcover), 978-1-315-16357-4 (eBook)
www.panstanford.com

specific characteristics that these nanofluids possess, which make them suitable for such applications.

6.1 Introduction

Nanofluids are dilute liquid suspensions of nanoparticles with at least one of their principal dimensions smaller than 100 nm. From previous investigations, nanofluids have been found to possess enhanced thermophysical properties such as thermal conductivity, thermal diffusivity, viscosity, and convective heat transfer coefficients compared to those of base fluids such as oil or water [1–6].

From the current review, it can be seen that nanofluids clearly exhibit enhanced thermal conductivity, which goes up with increasing volumetric fraction of nanoparticles. The current review does concentrate on this relatively new class of fluids and not on colloids, which are nanofluids because the latter have been used for a long time. Review of experimental studies clearly showed a lack of consistency in the reported results of different research groups regarding thermal properties [7, 8]. The effects of several important factors such as particle size and shapes, clustering of particles, temperature of the fluid, and dissociation of surfactant on the effective thermal conductivity of nanofluids have not been studied adequately. It is important to do more research so as to ascertain the effects of these factors on the thermal conductivity of a wide range of nanofluids.

Classical models cannot be used to explain adequately the observed enhanced thermal conductivity of nanofluids. Recently, the most developed models only include one or two postulated mechanisms of nanofluid heat transfer. For instance, there has not been much fundamental work reported on the determination of the effective thermal diffusivity of nanofluids or on heat transfer coefficients for nanofluids in natural convection [9].

There is a growth in the use of colloids, which are nanofluids in the biomedical industry for sensing and imaging purposes. This is directly related to the ability to design novel materials at the nanoscale level alongside recent innovations in analytical and imaging technologies for measuring and manipulating nanomaterials. This has led to the fast development of commercial applications that use a wide variety of manufactured nanoparticles. The production, use, and disposal of manufactured nanoparticles will lead to discharges to air, soils, and water systems. Negative effects are likely, and quantification and minimization of these effects on environmental health are necessary. True knowledge of concentration and physicochemical properties of manufactured nanoparticles under realistic conditions is important for predicting their fate, behavior, and toxicity in the natural aquatic environment. The aquatic colloid and atmospheric ultrafine particle literature both offer evidence of the likely behavior and impacts of manufactured nanoparticles [10], and there is no pretense that a review duplicating similar literature about the use of colloids that are also nanofluids is attempted in the current review.

Owing to their enhanced properties as thermal transfer fluids, for instance, nanofluids can be used in a plethora of engineering applications ranging from use in the automotive industry to the medical arena to use in power plant cooling systems as well as computers.

6.2 Heat Transfer Applications

6.2.1 Industrial Cooling Applications

Routbort et al. [11] started a project in 2008 employing nanofluids for industrial cooling that could result in great energy savings and resulting emissions reductions. For the US industry, the replacement of cooling and heating water with nanofluids has the potential to conserve 1 trillion Btu of energy. For the US electric power industry, using nanofluids in closed-loop cooling cycles could save about 10–30 trillion Btu per year (equivalent to the annual energy consumption of about 50,000–150,000 households). The associated emissions reductions would be approximately 5.6 million metric tonnes of carbon dioxide, 8600 metric tonnes of nitrogen oxides, and 21,000 metric tonnes of sulfur dioxide.

For Michelin North America tire plants, the productivity of numerous industrial processes is constrained by the lack of facility to cool the rubber efficiently as it is being processed. This requires the use of more than 2 million gallons of heat transfer fluids for Michelin's North American plants. Michelin has set a goal to obtain a 10% productivity increase in its rubber-processing plants if suitable water-based nanofluids can be developed and commercially produced in a cost-effective manner.

Han et al. [12] have used phase change materials as nanoparticles in nanofluids to simultaneously enhance the effective thermal conductivity and specific heat of the fluids. As an example, a suspension of indium nanoparticles (melting temperature, 157°C) in polyalphaolefin has been synthesized using a one-step, nanoemulsification method. The fluid's thermophysical properties, i.e., thermal conductivity, viscosity, and specific heat, and their temperature dependence were measured experimentally. The observed melting–freezing phase transition of the indium nanoparticles significantly augmented the fluid's effective specific heat.

This work is one of the few to address thermal diffusivity; similar studies allow industrial cooling applications to continue without thorough understanding of all the heat transfer mechanisms in nanofluids.

6.2.2 Smart Fluids

In this new age of energy awareness, our lack of abundant sources of clean energy and the widespread dissemination of battery-operated devices, such as cell phones and laptops, have accented the necessity for a smart technological handling of energetic resources. Nanofluids have been demonstrated to be able to handle this role in some instances as a smart fluid.

In a recent paper published in the March 2009 issue of *Physical Review Letters*, Donzelli et al. [13] showed that a particular class of nanofluids can be used as a smart material working as a heat valve to control the flow of heat. The nanofluid can be readily configured either in a "low" state, where it conducts heat poorly, or in a "high" state, where the dissipation is more efficient. To leap the chasm to heating and cooling technologies, the researchers will have to show more evidence of a stable operating system that responds to a larger range of heat flux inputs.

6.2.3 Nuclear Reactors

Kim et al. [14, 15], at the Nuclear Science and Engineering Department of the Massachusetts Institute of Technology (MIT), performed a study to assess the feasibility of nanofluids in nuclear applications by improving the performance of any water-cooled nuclear system that is heat removal limited. Possible applications include pressurized water reactor (PWR) primary coolant, standby safety systems, accelerator targets, plasma divertors, etc. [16].

In a PWR nuclear power plant system, the limiting process in the generation of steam is the critical heat flux (CHF) between the fuel rods and the water, when vapor bubbles that end up covering the surface of the fuel rods conduct very little heat as opposed to liquid water. Using nanofluids instead of water, the fuel rods become coated with nanoparticles such as alumina, which actually push newly formed bubbles away, preventing the formation of a layer of vapor around the rod and subsequently increasing the CHF significantly.

After testing in MIT's Nuclear Research Reactor, preliminary experiments have shown promising success, where it is seen that PWR is significantly more productive. The use of nanofluids as a coolant could also be used in emergency cooling systems, where they could cool down overheated surfaces more quickly, leading to an improvement in power plant safety.

Some issues regarding the use of nanofluids in a power plant system include the unpredictability of the amount of nanoparticles that are carried away by the boiling vapor. Another concern is what extra safety measures that have to be taken in the disposal of the nanofluid. The application of nanofluid coolant to boiling water reactors (BWRs) is predicted to be minimal because nanoparticle carryover to the turbine and condenser would raise erosion and fouling concerns.

From Jackson's study [17], it was observed that considerable enhancement in the CHF can be achieved by creating a structured surface from the deposition of nanofluids. If the deposition film characteristics such as the structure and thickness can be controlled, it may be possible to increase the CHF with little decrease in the heat transfer. Whereas the nanoparticles themselves cause no significant difference in the pool-boiling characteristics of water, the boiling of nanofluids shows promise as a simple way to create an enhanced surface.

The use of nanofluids in nuclear power plants seems like a potential future application [16]. Several significant gaps in knowledge are evident at this time, including demonstration of the nanofluid thermal–hydraulic performance at prototypical reactor conditions and the compatibility of the nanofluid chemistry with the reactor materials.

Another possible application of nanofluids in nuclear systems is the alleviation of postulated severe accidents during which the core melts and relocates to the bottom of the reactor vessel. If such accidents were to occur, it is desirable to retain the molten fuel within the vessel by removing the decay heat through the vessel wall. This process is limited by the occurrence of CHF on the vessel outer surface, but analysis indicates that the use of nanofluid can increase the in-vessel retention capabilities of nuclear reactors by as much as 40% [18].

Many water-cooled nuclear power systems are CHF-limited, but the application of nanofluid can greatly improve the CHF of the coolant so that there is a bottom-line economic benefit while also raising the safety standard of the power plant system.

6.2.4 Extraction of Geothermal Power and Other Energy Sources

The world's total geothermal energy resources were calculated to be more than 13,000 ZJ, according to an MIT report (2007) [19]. Currently, only 200 ZJ would be extractable; however, with technological improvements, more than 2000 ZJ could be extracted, which would meet the world's energy needs for several millennia. When extracting energy from the earth's crust, which varies in length between 5 and 10 km and temperature between 500 and 1000°C, nanofluids can be employed to cool the pipes exposed to such high temperatures. When drilling, nanofluids can serve in cooling the machinery and equipment working in high-friction and high-temperature environment. As a "fluid superconductor," nanofluids can be used as a working fluid to extract energy from the earth's core and processed in a PWR power plant system, producing large amounts of work energy.

In the sub-area of drilling technology, so fundamental to geothermal power, improved sensors and electronics cooled by nanofluids capable of operating at higher temperature in downhole tools, and revolutionary improvements utilizing new methods of rock penetration cooled and lubricated by nanofluids, will lower production costs. Such improvements will enable access to deeper, hotter regions in high-grade formations or to economically acceptable temperatures in lower-grade formations.

In the sub-area of power conversion technology, improving heat transfer performance for lower-temperature nanofluids and developing plant designs for higher resource temperatures to the supercritical water region would lead to an order of magnitude (or more) gain in both reservoir performance and heat-to-power conversion efficiency.

Tran et al. [20], funded by the United States Department of Energy (USDOE), performed research targeted at developing a new class of highly specialized drilling fluids that may have superior performance in high-temperature drilling. This research is applicable to high-pressure, high-temperature drilling, which may be pivotal in opening up large quantities of previously unrecoverable domestic fuel resources. Commercialization would be the bottleneck of progress in this sub-area.

6.3 Automotive Applications

Engine oils, automatic transmission fluids, coolants, lubricants, and other synthetic high-temperature heat transfer fluids found in conventional truck thermal systems—radiators, engines, and heating, ventilation, and air conditioning (HVAC)—have inherently poor heat transfer properties. These could benefit from the high thermal conductivity offered by nanofluids that resulted from the addition of nanoparticles [3, 21].

6.3.1 Nanofluid Coolant

In looking for ways to improve the aerodynamic designs of vehicles, and subsequently the fuel economy, manufacturers must reduce the amount of energy needed to overcome wind resistance on the road.

At high speeds, approximately 65% of the total energy output from a truck is expended in overcoming the aerodynamic drag. This fact is partly due to the large radiator in front of the engine positioned to maximize the cooling effect of oncoming air.

The use of nanofluids as coolants would allow for smaller size and better positioning of the radiators. Owing to the fact that there would be less fluid due to the higher efficiency, coolant pumps could be shrunk and truck engines could be operated at higher temperatures allowing for more horsepower while still meeting stringent emission standards.

Argonne researchers, Singh et al. [22], have determined that the use of high thermal conductive nanofluids in radiators can lead to a reduction in the frontal area of the radiator by up to 10%. This reduction in aerodynamic drag can lead to a fuel savings of up to 5%. The application of nanofluid also contributed to a reduction in friction and wear, reducing parasitic losses, operation of components such as pumps and compressors, and subsequently leading to more than 6% fuel savings. It is conceivable that greater improvement of savings could be obtained in the future.

In order to determine whether nanofluids degrade radiator material, they have built and calibrated an apparatus that can emulate the coolant flow in a radiator and are currently testing and measuring material loss of typical radiator materials by various nanofluids. Erosion of radiator material is determined by weight loss measurements as a function of fluid velocity and impact angle.

In their tests, they observed no erosion using nanofluids made from base fluids ethylene and tri-cloroethylene gycols with velocities as high as 9 m/s and at 90°–30° impact angles. There was erosion observed with copper nanofluid at a velocity of 9.6 m/s and impact angle of 90°. The corresponding recession rate was calculated to be 0.065 mils/year of vehicle operation.

Through preliminary investigation, it was determined that copper nanofluid produces a higher wear rate than the base fluid and this is possibly due to oxidation of copper nanoparticles. A lower wear and friction rate was seen for alumina nanofluids in comparison to the base fluid. Some interesting erosion test results from Singh et al. [22] are shown in Tables 6.1 and 6.2.

Table 6.1 Erosion test results for 50% ethlyene glycol, 50% H_2O aluminum 3003–50°C Rig

Impact Angle (°)	Velocity (m/s)	Time (h)	Weight Loss (mg)
90	8.0	236	0±0.2
90	10.5	211	0±0.2
50	6.0	264	0±0.2
50	10.0	244	0±0.2
30	8.0	283	0±0.2
30	10.5	293	0±0.2

Source: Ref. [22].

Table 6.2 Erosion test results for Cu nanoparticles in trichloroethylene glycol on Al 3003–50°C Rig

Impact Angle (°)	Velocity (m/s)	Time (h)	Weight Loss (mg)
90	4.0	217	0±0.2
30	4.0	311	0±0.2
90	7.6	341	0±0.2
30	7.6	335	0±0.2
30	9.6	336	0±0.2

Source: Ref. [22].

Shen et al. [23] researched the wheel wear and tribological characteristics in wet, dry, and minimum quantity lubrication (MQL) grinding of cast iron. Water-based alumina and diamond nanofluids were applied in the MQL grinding process, and the grinding results were compared with those of pure water. Nanofluids demonstrated the benefits of reducing grinding forces, improving surface roughness and preventing burning of the workpiece. Contrasted to dry grinding, MQL grinding could considerably lower the grinding temperature.

More research must be conducted on the tribological properties using nanofluids of a wider range of particle loadings as well as on the erosion rate of radiator material in order to help develop predictive models for nanofluid wear and erosion in engine systems. Future research initiatives involve nanoparticle materials containing aluminum and oxide-coated metal nanoparticles. Additional research and testing in this area will assist in the design of engine cooling and other thermal management systems that involve nanofluids.

Future engines that are designed using the cooling properties of nanofluids would be able to run at more optimal temperatures allowing for increased power output. With a nanofluid engine, components would be smaller and weigh less, allowing for better gas mileage, saving consumers money and resulting in fewer emissions for a cleaner environment.

6.3.2 Nanofluid in Fuel

Aluminum nanoparticles, produced using a plasma arc system, are covered with thin layers of aluminum oxide, owing to the high oxidation activity of pure aluminum, thus creating a larger contact surface area with water and allowing for increased decomposition of hydrogen from water during the combustion process. During this combustion process, the alumina acts as a catalyst and the aluminum nanoparticles then serve to decompose the water to yield more hydrogen. It was shown that the combustion of diesel fuel mixed with aqueous aluminum nanofluid increased the total combustion heat while decreasing the concentration of smoke and nitrous oxide in the exhaust emission from the diesel engine [24, 25].

6.3.3 Brake and Other Vehicular Nanofluids

As vehicle aerodynamics is improved and drag forces are reduced, there is a higher demand for braking systems with higher and more efficient heat dissipation mechanisms and properties such as brake nanofluid.

A vehicle's kinetic energy is dispersed through the heat produced during the process of braking and this is transmitted throughout the brake fluid in the hydraulic braking system. If the heat causes the brake fluid to reach its boiling point, a vapor-lock is created, which retards the hydraulic system from dispersing the heat caused from braking. Such an occurrence will, in turn, cause a brake malfunction and poses a safety hazard in vehicles. Since brake oil is easily affected by the heat generated from braking, nanofluids with enhanced characteristics maximize performance in heat transfer as well as remove any safety concerns.

Copper oxide brake nanofluid (CBN) is manufactured using the method of arc-submerged nanoparticle synthesis system (ASNSS). Essentially, this is done by melting bulk copper metal used as the electrode, which is submerged in dielectric liquid within a vacuum-operating environment and the vaporized metals are condensed in the dielectric liquid [24, 25].

Aluminum oxide brake nanofluid (AOBN) is made using the plasma charging arc system. This is performed in a very similar fashion to that of the ASNSS method. The aluminum metal is vaporized by the plasma electric arc at a high temperature and mixed thoroughly with the dielectric liquid [24, 25].

CBN has a thermal conductivity 1.6 times higher than that of the brake fluid designated DOT3, while AOBN's thermal conductivity is only 1.5 times higher than DOT3. This enhanced thermal conductivity optimizes heat transmission and lubrication.

CBN and AOBN both have enhanced properties such as a higher boiling point, higher viscosity, and a higher conductivity than those of traditional brake fluids (DOT3). By yielding a higher boiling point, conductivity, and viscosity, CBN and AOBN reduce the occurrence of vapor-lock and offer increased safety while driving. Important findings of Kao et al. [24, 25] are shown in Fig. 6.1 and Table 6.3.

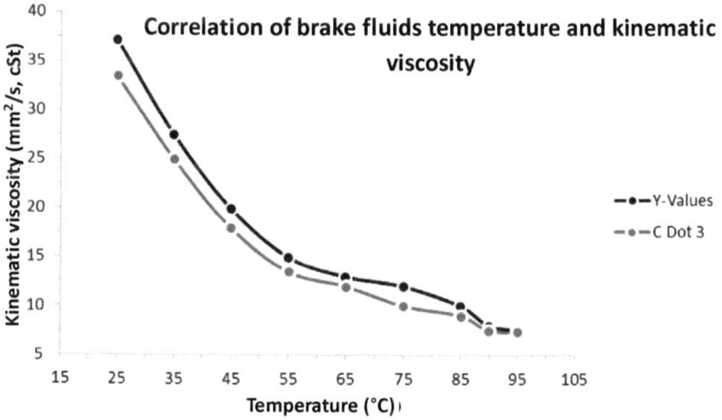

Figure 6.1 CBN temperature and viscosity fluctuations [24, 25].

Table 6.3 CBN and AOBN boiling point and thermal conductivity fluctuations

	DOT3*	CBN 2 wt% (CuO + DOT3)	DOT3*	AOBN 2 wt% (Al_2O_3 + DOT3)
Boiling point	270°C	278°C	240°C	248°C
Conductivity (25°C)	0.03 W/m°C	0.05 W/m°C	0.13 W/m°C	0.19 W/m°C

Source: Refs. [24, 25].
*Different DOT3 brake fluids were used.

In the nanofluid research applied to the cooling of automatic transmissions, Tzeng et al. [26] dispersed CuO and Al_2O_3 nanoparticles into engine transmission oil. The experimental setup was the transmission of a four-wheel-drive vehicle. The transmission had an advanced rotary–blade coupling, where high local temperatures occurred at high rotating speeds. Temperature measurements were taken on the exterior of the rotary–blade coupling transmission at four-engine operating speeds (range from 400 to 1600 rpm), and the optimum composition of nanofluids with regard to heat transfer performance was studied. The results indicated that CuO nanofluids resulted in the lowest transmission temperatures both at high and low rotating speeds. Therefore, the use of nanofluid in the transmission has a clear advantage from the thermal performance viewpoint. As in all nanofluid applications, however, consideration must be given to such factors as particle settling, particle agglomeration, and surface erosion.

In automotive lubrication applications, Zhang [27] reported that surface-modified nanoparticles stably dispersed in mineral oils are effective in reducing wear and enhancing load-carrying capacity. Results from a research project involving industry and academia point to the use of nanoparticles in lubricants to enhance tribological properties such as load-carrying capacity, wear resistance, and friction reduction between moving mechanical components. Such results are promising for enhancing heat transfer rates in automotive systems through the use of nanofluids.

6.4 Electronic Applications

Nanofluids are used for cooling of microchips in computers and elsewhere. They are also used in other electronic applications that use microfluidic applications.

6.4.1 Cooling of Microchips

A principal limitation on developing smaller microchips is the rapid heat dissipation. However, nanofluids can be used for liquid cooling of computer processors due to their high thermal conductivity. It is predicted that the next generation of computer chips will produce localized heat flux over 10 MW/m^2, with the total power exceeding 300 W. In combination with thin film evaporation, the nanofluid oscillating heat pipe (OHP) cooling system will be able to remove heat fluxes greater than 10 MW/m^2 and serve as the next-generation cooling device that will be able to handle the heat dissipation coming from new technology [28, 29].

In order to observe the oscillation, researchers had to modify the metal pipe system of the OHP to use glass or plastic for visibility. However, since OHP systems are usually made of copper, the use of glass or plastic changes the thermal transfer properties of the system and subsequently altering the performance of the system and the legitimacy of the experimental data [28, 29].

So as to obtain experimental data while maintaining the integrity of the OHP system, Arif [30] employed neutron imaging to study the liquid flow in a 12-turn nanofluid OHP. As a consequence of the high-intensity neutron beam from an amorphous silicon imaging system, they were able to capture dynamic images at 1/30th of a second. The nanofluid used was composed of diamond nanoparticles suspended in water.

Even though nanofluids and OHPs are not new discoveries, combining their unique features allows for the nanoparticles to be completely suspended in the base liquid, increasing their heat transport capability. Since nanofluids have a strong temperature-dependent thermal conductivity and they show a nonlinear relationship between thermal conductivity and concentration, they are high-performance conductors with an increased CHF. The OHP takes intense heat from a high-power device and converts it into kinetic energy of fluids while not allowing the liquid and vapor phases to interfere with each other since they flow in the same direction.

In their experiment, Ma et al. [28, 29] introduced diamond nanoparticles into high-performance liquid chromatography (HPLC) water. The movement of the OHP keeps the nanoparticles from settling and thus improving the efficiency of the cooling device. At an input power of 80 W, the diamond nanofluid decreased the temperature difference between the evaporator and the condenser from 40.9°C to 24.3°C (Fig. 6.2).

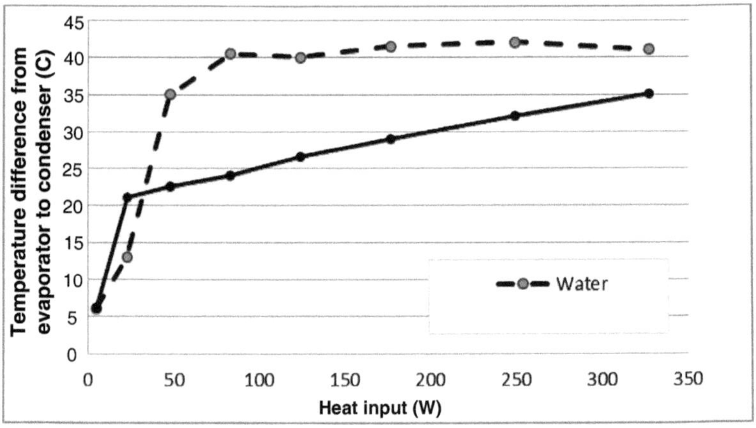

Figure 6.2 Effect of nanofluid on heat transport capability in an OHP [28, 29].

However, as the heat input increases, the oscillating motion increases and the resultant temperature difference between the evaporator and condenser does not continue to increase after a certain power input. This phenomenon inhibits the effective thermal conductivity of the nanofluid from continuously increasing. However, at its maximum power level of 336 W, the temperature difference for the nanofluid OHP was still less than that for the OHP with pure water. Hence, it has been shown that the nanofluid can significantly increase the heat transport capability of the OHP.

Lin et al. [31] investigated nanofluids in pulsating heat pipes by using silver nanoparticles and discovered encouraging results. The silver nanofluid improved heat transfer characteristics of the heat pipes.

Nguyen et al. [32] investigated the heat transfer enhancement and behavior of Al_2O_3–water nanofluid with the intention of using it in a closed cooling system designed for microprocessors or other electronic devices. The experimental data support that the inclusion of nanoparticles into distilled water produces a significant increase in the cooling convective heat transfer coefficient. At a given particle concentration of 6.8%, the heat transfer coefficient increased as much as 40% compared to the base fluid of water. Smaller Al_2O_3 nanoparticles also showed higher convective heat transfer coefficients than the larger ones.

Further research of nanofluids in electronic cooling applications will lead to the development of the next generation of cooling devices that incorporate nanofluids for ultrahigh-heat-flux electronic systems.

6.4.2 Microscale Fluidic Applications

The manipulation of small volumes of liquid is necessary in fluidic digital display devices, optical devices, and microelectromechanical systems (MEMS) such as lab-on-chip analysis systems. This can be done by electrowetting, or reducing the contact angle by an applied voltage, the small volumes of liquid. Electrowetting on dielectric (EWOD) actuation is a very useful method of microscale liquid manipulation.

Vafaei et al. [33] discovered that nanofluids are effective in engineering the wettability of the surface and possibly of surface tension. Using a goniometer, it was observed that even the addition of a very low concentration of bismuth telluride nanofluid dramatically changed the wetting characteristics of the surface. Concentrations as low as 3×10^{-6} increased the contact angle to greater than 40°, distinctly indicating that the nanoparticles change the force balance in the vicinity of the triple line. The contact angle θ_o (Fig. 6.3) rises with the concentration of the nanofluid, reaches a maximum, and then decreases, as seen in Fig. 6.4 [34].

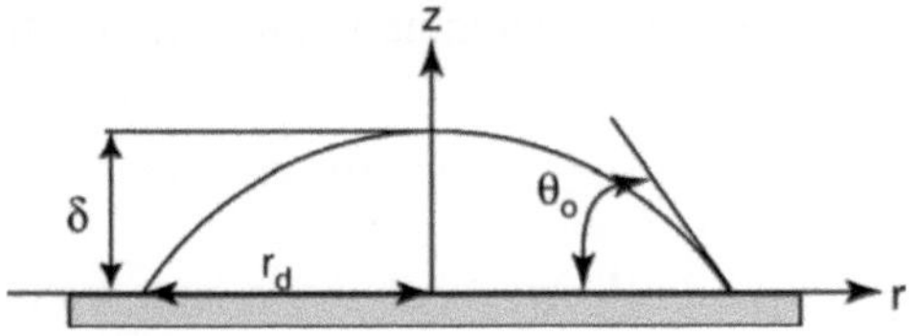

Figure 6.3 Schematic diagram of droplet shape.

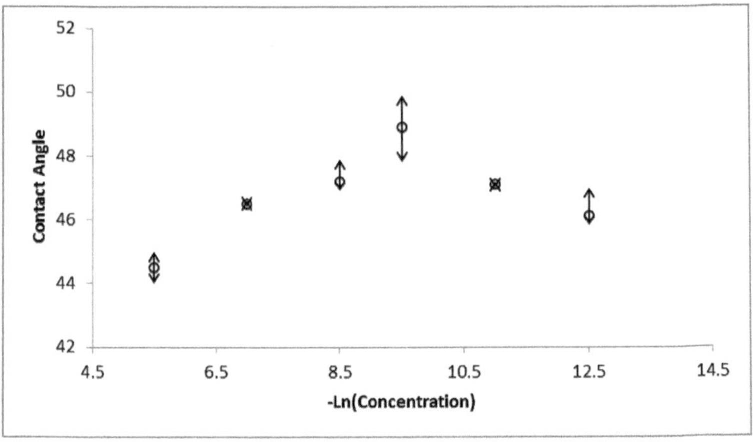

Figure 6.4 Variation of contact angle for 10 nm bismuth telluride nanoparticle concentration on a glass substrate [34].

The droplet contact angle was observed to change depending on the size of the nanoparticles as well. Smaller nanoparticles are more effective in increasing the contact angle. The reason for this effect is that smaller particles would provide more surface-to-volume area for the same concentration.

Dash et al. used the EWOD effect to demonstrate that nanofluids display increased performance and stability when exposed to electric fields. The experiment consisted of placing droplets of water-based solutions containing bismuth telluride nanoparticles onto a Teflon-coated silicon wafer. A strong change in the angle at which the droplet contacted the wafer was observed when an electric field was applied to the droplet. The change noticed with the nanofluids was significantly greater than when not using nanofluids. The bismuth telluride nanofluid also displayed enhanced droplet stability and the absence of the contact angle saturation effect compared to solutions of 0.01 N Na_2SO_4 and thioglycolic acid in deionized water.

That the contact angle of droplets of nanofluids can be changed has potential applications for efficiently moving liquids in microsystems, allowing for new methods for focusing lenses in miniature cameras as well as for cooling computer chips.

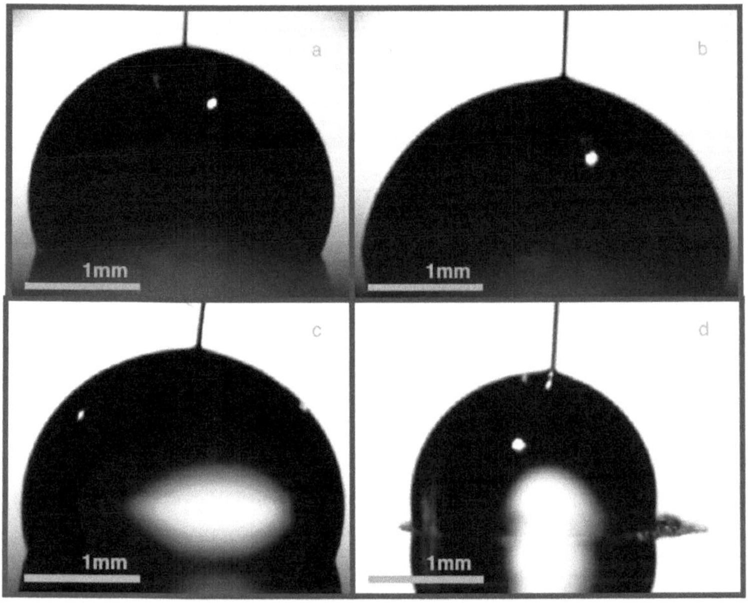

Figure 6.5 Nanofluid droplet with applied voltage (a) 0 V and (b) 64.5 V 0.01 M Na_2SO_4 droplet under (c) 0 V and (d) 40 V. The Na_2SO_4 droplet becomes unstable at approximately 40 V seen by the gas bubbles forming in the contact line [34].

6.5 Biomedical Applications

6.5.1 Nano-drug Delivery

Most bio-MEMS studies were done in academia in the 1990s, while recently commercialization of such devices has started. Examples include an electronically activated drug delivery microchip [35]; a controlled delivery system via integration of silicon and electroactive polymer technologies; an MEMS-based DNA sequencer developed by Cepheid [36]; and arrays of in-plane and out-of-plane hollow micro-needles for dermal/transdermal drug delivery (Ovsianikov et al. and Kim and Lee [37, 38], among others) as well as nanomedicine applications of nanogels or gold-coated nanoparticles [39]. An objective of the advanced endeavors in developing integrated micro- or nano-drug delivery (ND) systems is the interest in easily monitoring and controlling target-cell responses to pharmaceutical stimuli, to understand biological cell activities, or to enable drug development processes.

While conventional drug delivery is characterized by the "high-and-low" phenomenon, micro-devices facilitate precise drug delivery by both implanted and transdermal techniques. This means that when a drug is dispensed conventionally, drug concentration in the blood will increase, peak, and then drop as the drug is metabolized, and the cycle is repeated for each drug dose. Employing ND systems, controlled drug release takes place over an extended period of time. Thus, the desired drug concentration will be sustained within the therapeutic window as required.

A nano-drug supply system, i.e., a bio-MEMS, was introduced by Kleinstreuer et al. [40]. Their principal concerns were the conditions for delivering uniform concentrations at the microchannel exit of the supplied nano-drugs. A heat flux that depends on the levels of nanofluid and purging fluid velocity was added to ascertain that drug delivery to the living cells occurs at an optimal temperature, i.e., 37°C. The added wall heat flux had also a positive influence on drug concentration uniformity. In general, the nano-drug concentration uniformity is affected by channel length, particle diameter, and the Reynolds number of both the nanofluid supply and main microchannels. Since the transport mechanisms are dependent on convection–diffusion, longer channels, smaller particle diameters as well as lower Reynolds numbers are desirable for best, i.e., uniform drug delivery.

6.5.2 Cancer Therapeutics

There is a new initiative that takes advantage of several properties of certain nanofluids to use in cancer imaging and drug delivery.

This initiative involves the use of iron-based nanoparticles as delivery vehicles for drugs or radiation in cancer patients. Magnetic nanofluids are to be used to guide the particles up the bloodstream to a tumor with magnets. It will allow doctors to deliver high local doses of drugs or radiation without damaging nearby healthy tissue, which is a significant side effect of traditional cancer treatment methods. In addition, magnetic nanoparticles are more adhesive to tumor cells than non-malignant cells and they absorb much more power than microparticles in alternating current magnetic fields tolerable in humans; they make excellent candidates for cancer therapy.

Magnetic nanoparticles are used because as compared to other metal-type nanoparticles, these provide a characteristic for handling and manipulation of the nanofluid by magnetic force [41]. This combination of targeted delivery and controlled release will also decrease the likelihood of systemic toxicity since the drug is encapsulated and biologically unavailable during transit in systemic circulation. The nanofluid containing magnetic nanoparticles also acts as a super-paramagnetic fluid, which absorbs energy in an alternating electromagnetic field, producing a controllable hyperthermia. By enhancing the chemotherapeutic efficacy, the hyperthermia is able to produce a preferential radiation effect on malignant cells [42].

There are numerous biomedical applications that involve nanofluids such as magnetic cell separation, drug delivery, hyperthermia, and contrast enhancement in magnetic resonance imaging. Depending on the specific application, there are different chemical syntheses developed for various types of magnetic nanofluids that allow for the careful tailoring of their properties for different requirements in applications. Surface coating of nanoparticles and the colloidal stability of biocompatible water-based magnetic fluids are the two particularly important factors that affect successful application [43, 44].

Nanofluids could be applied to almost any disease treatment techniques by re-engineering the nanoparticles' properties. In their study, the nanoparticles were laced with the drug docetaxel to be dissolved in the cells' internal fluids, releasing the anti-cancer drug at a predetermined rate. The nanoparticles contain targeting molecules called aptamers, which recognize the surface molecules on cancer cells preventing the nanoparticles from attacking other cells. In order to prevent the nanoparticles from being destroyed by macrophages (cells that guard against foreign substances entering our bodies), the nanoparticles also have polyethylene glycol molecules. The nanoparticles are excellent drug delivery vehicles because they are so small that living cells absorb them when they arrive at the cells' surface.

For most biomedical uses, the magnetic nanoparticles should be below 15 nm in size and stably dispersed in water. A potential magnetic nanofluid that could be used for biomedical applications is one composed of FePt nanoparticles. This FePt nanofluid possesses an intrinsic chemical stability and a higher saturation magnetization making it ideal for biomedical applications. However, before magnetic nanofluids can be used as drug delivery systems, more research must be conducted on the nanoparticles containing the actual drugs and the release mechanism.

6.5.3 Cryopreservation

Conventional cryopreservation protocols for slow-freezing or vitrification involve cell injury due to ice formation/cell dehydration or toxicity of high cryoprotectant (CPA) concentrations, respectively. In their study, He et al. [45] developed a novel cryopreservation technique to achieve ultra-fast cooling rates using a quartz micro-capillary (QMC). The QMC-enabled vitrification of murine embryonic stem (ES) cells using an intracellular cryoprotectant concentration in the range used for slowing freezing (1–2 M). More than 70% of the murine ES cells post-vitrification attached with respect to non-frozen control cells, and the proliferation rates of the two groups were alike. Preservation of undifferentiated properties of the pluripotent murine ES cells post-vitrification cryopreservation was verified using three different types of assays. These results indicate that vitrification at a low concentration (2 M) of intracellular cryoprotectants is a viable and effective approach for the cryopreservation of murine ES cells.

6.5.4 Nanocryosurgery

Cryosurgery is a procedure that uses freezing to destroy undesired tissues. This therapy is becoming popular because of its important clinical advantages. Although it still cannot be regarded as a routine method of cancer treatment, cryosurgery is quickly becoming an alternative to traditional therapies.

Simulations were performed by Yan and Liu [46] on the combined phase change bioheat transfer problems in a single cell level and its surrounding tissues, to explicate the difference of transient temperature response between conventional cryosurgery and nanocryosurgery. According to theoretical interpretation and existing experimental measurements, intentional loading of nanoparticles with high thermal conductivity into the target tissues can reduce the final temperature, increase the maximum freezing rate, and enlarge the ice volume obtained in the absence of nanoparticles. Additionally, introduction of nanoparticle-enhanced freezing could also make conventional cryosurgery more flexible in many aspects such as artificially interfering in the size, shape, image, and direction of iceball formation. The concepts of nanocryosurgery may offer new opportunities for future tumor treatment.

With respect to the choice of particles for enhancing freezing, magnetite (Fe_3O_4) and diamond are perhaps the most popular and appropriate because of their good biological compatibility. Particle sizes less than 10 μm are sufficiently small to start permitting effective delivery to the site of the tumor, either via encapsulation in a larger moiety or suspension in a carrier fluid. Introduction of nanoparticles into the target via a nanofluid would effectively increase the nucleation rate at a high-temperature threshold.

6.5.5 Sensing and Imaging

Colloidal gold has been used for several centuries now, be it as colorant of glass ("Purple of Cassius") and silk, in medieval medicine for the diagnosis of syphilis or, more recently, in chemical catalysis, nonlinear optics, supramolecular chemistry, molecular recognition, and the biosciences. Colloidal gold is often referred to as the most stable of all colloids. Its history, properties, and applications have been reviewed extensively. For a thorough and up-to-date overview, the paper by Daniel and Astruc [47] and the references cited therein may be consulted. As stated in the introduction, no attempt is made here to review the use of colloids that are also nanofluids. An increase in colloids that are nanofluids is expected in this category.

6.6 Other Applications

6.6.1 Nanofluid Detergent

Nanofluids do not behave in the same manner as simple liquids with classical concepts of spreading and adhesion on solid surfaces [7, 48, 49]. This fact opens up the possibility of nanofluids being excellent candidates in the processes of soil remediation, lubrication, oil recovery, and detergency. Future engineering applications could abound in such processes.

Wasan and Nikolov [50] of the Illinois Institute of Technology in Chicago were able to use reflected-light digital video microscopy to determine the mechanism of spreading dynamics in liquid-containing nanosized polystyrene particles. They were able to demonstrate the two-dimensional crystal-like formation of the polystyrene spheres in water and how this enhances the spreading dynamics of a micellar fluid at the three-phase region [50].

When encountering an oil drop, the polystyrene nanoparticles concentrate and rearrange around the drop creating a wedge-like region between the surface and the oil drop. The nanoparticles then diffuse into the wedge film and cause an increase in concentration and subsequently an increase in disjoining pressure around the film region. Owing to the increase in pressure, the oil–solution interface moves forward, allowing the polystyrene nanoparticles to spread along the surface. It is this mechanism that causes the oil drop to detach completely from the surface.

Wasan and Nikolov [50] performed an additional experiment where they introduced an electrolyte into the process in order to decrease the interfacial tension at the interface of the oil and the nanofluid, but found that the drop did not become detached from the surface. They actually observed a diminished disjoining pressure contrary to the logical prediction. Additional work must be done in this area to determine such behavior of the nanofluid.

Overall, this phenomenon involving the increased spreading of the detergent surfactants, which are not only limited to polystyrene nanoparticles and enhanced oil removal process, offers a new way of removing stains and grease from surfaces. This type of nanofluid also has potential in the commercial extraction of oil from the ground as well as the remediation of oil spills.

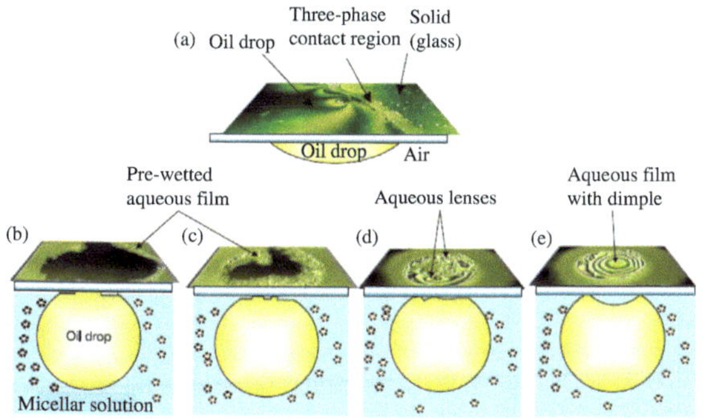

Figure 6.6 (a) Photomicrograph showing the oil drop placed on a glass surface and differential interference patterns formed at the three-phase contact region [50]. (b) Photomicrographs taken after addition of the nanofluid at b, 30 s; 2 min; d, 4 min; e, 6 min. region [50].

6.7 Conclusion

Nanofluids are important because they can be used in numerous applications involving heat transfer, and other applications such as in detergency. Colloids that are also nanofluids have been used in the biomedical field for a long time, and their use will continue to grow. Nanofluids have also been demonstrated for use as smart fluids. Problems of nanoparticle agglomeration, settling, and erosion potential all need to be examined in detail in the applications. Nanofluids employed in experimental research have to be well characterized with respect to particle size, size distribution, shape, and clustering so as to render the results most widely applicable. Once the science and engineering of nanofluids are fully understood and their full potential is researched, they can be reproduced on a large scale and used in many applications. Colloids that are also nanofluids will see an increase in use in biomedical engineering and biosciences.

Further research still has to be done on the synthesis and applications of nanofluids so that they may be applied as predicted. Nevertheless, there have been many discoveries and improvements identified about the characteristics of nanofluids in the surveyed applications and we are a step closer to developing systems that are more efficient and smaller, thus rendering the environment cleaner and healthier.

References

1. S. U. S. Choi. (2009). Nanofluids: From vision to reality through research, *J. Heat Transfer*, **131**(3), 033106.
2. S. U. S. Choi, Z. G. Zhang, P. Keblinski, and H. S. Nalwa. (eds.) (2004). Nanofluids. In: *Encyclopedia of Nanoscience and Nanotechnology*, vol. 6, pp. 757–737, American Scientific Publishers, Los Angeles, California, USA.
3. W. Yu, D. M. France, J. L. Routbort, and S. U. S. Choi. (2008). Review and comparison of nanofluid thermal conductivity and heat transfer enhancements, *Heat Transfer Eng.*, **29**(5), pp. 432–460.
4. T. Tyler, O. Shenderova, G. Cunningham, J. Walsh, J. Drobnik, and G. McGuire. (2006). Thermal transport properties of diamond-based nanofluids and nanocomposites, *Diamond Related Mater.*, **15**, pp. 2078–2081.
5. S. K. Das, S. U. S. Choi, and H. E. Patel. (2006). Heat transfer in nanofluids: A Review, *Heat Transf. Eng.*, **27**(10), pp. 3–19.
6. M. S. Liu, M. C. C. Lin, I. T. Huang, and C. C. Wang. (2005). Enhancement of thermal conductivity with carbon nanotube for nanofluids, *Int. Commun. Heat Mass*, **32**, pp. 1202–1210.
7. S. M. S. Murshed, S. H. Tan, and N. T. Nguyen. (2008). Temperature dependence of interfacial properties and viscosity of nanofluids for droplet-based microfluidics, *J. Phys. D Appl. Phys.*, **41**, 085502.
8. K. V. Wong and T. Kurma. (2008). Transport properties of alumina nanofluids, *Nanotechnology*, **19**, 345702.
9. K. V. Wong, B. Bonn, S. Vu, and S. Samedi. (2007). Study of nanofluid natural convection phenomena in rectangular enclosures, *Proc. IMECE 2007*, November 2007, Seattle, WA.

10. Y. Ju-Nam and J. R. Lead. (2008). Manufactured nanoparticles: An overview of their chemistry, interactions and potential environmental implications, *Sci. Total Environ.*, **400**, pp. 396–414.

11. J. Routbort et al. (2009). Argonne National Lab, Michellin North America, St. Gobain Corp., http://www1.eere.energy.gov/industry/nanomanufacturing/pdfs/nanofluids_industrial_cooling.pdf.

12. Z. H. Han, F. Y. Cao, and B. Yang. (2008). Synthesis and thermal characterization of phase-changeable indium/polyalphaolefin nanofluids, *Appl. Phys. Lett.*, **92**(24), 243104.

13. G. Donzelli, R. Cerbino, and A. Vailati. (2009). Bistable heat transfer in a nanofluid, *Phys. Rev. Lett.*, **102**, 104503.

14. S. J. Kim, I. C. Bang, J. Buongiorno, and L. W. Hu. (2007). Study of pool boiling and critical heat flux enhancement in nanofluids, *Bull. Polish Acad. Sci. Tech. Sci.*, **55**(2), pp. 211–216.

15. S. J. Kim, I. C. Bang, J. Buongiorno, and L. W. Hu. (2007). Surface wettability change during pool boiling of nanofluids and its effect on critical heat flux, *Int. J. Heat Mass Transf.*, **50**, pp. 4105–4116.

16. J. Buongiorno, L. Hu, S. J. Kim, R. Hannink, B. Truong, and E. Forrest. (2008). Nanofluids for enhanced economics and safety of nuclear reactors: An evaluation of the potential features, issues and research gaps, *Nucl. Technol.*, **162**(1), pp. 80–91.

17. J. E. Jackson. (2007). Investigation into the pool-boiling characteristics of gold nanofluids, M.S. Thesis, University of Missouri, Columbia.

18. J. Buongiorno, L. W. Hu, G. Apostolakis, R. Hannink, T. Lucas, and A. Chupin. (2009). A feasibility assessment of the use of nanofluids to enhance the in-vessel retention capability in light-water reactors, *Nucl. Eng. Design*, **239**, pp. 941–948.

19. M.I.T. (2007). *The Future of Geothermal Energy*, Sponsored by U.S.D.O.E., 372 pp.

20. P. X. Tran, D. K. Lyons., et al. (2007). Nanofluids for use as ultra-deep drilling fluids, USDepartment of Energy, http://www.netl.doe.gov/publications/factsheets/rd/R&D108.pdf.

21. M. Chopkar, P. K. Das, and I. Manna. (2006). Synthesis and characterization of nanofluid for advanced heat transfer applications, *Scripta Materialia*, **55**, pp. 549–552.

22. D. Singh, J. Toutbort, G. Chen, J. Hull, R. Smith, O. Ajayi, and W. Yu. (2006). Heavy vehicle systems optimization merit review and peer evaluation, *Argonne Nat. Lab. Annual Rep.*

23. B. Shen, A. J. Shih, S. C. Tung, and M. Hunter. (2009). Application of nanofluids in minimum quantity lubrication grinding, *Tribol. Lubr. Technol.*, **51**(6), pp. 730–737.

24. M. J. Kao, C. H. Lo, T. T. Tsung, Y. Y. Wu, C. S. Jow, and H. M. Lin. (2007). Copper-oxide brake nanofluid manufactured using arc-submerged nanoparticle synthesis system, *J. Alloys Compd.*, **434–435**, pp. 672–674.

25. M. J. Kao, H. Chang, Y. Y. Wu, T. T. Tsung, and H. M. Lin. (2007). Producing aluminum-oxide brake nanofluids using plasma charging system, *J. Chin. Soc. Mech. Eng.*, **28**(2), pp. 123–131.

25. S. C. Tzeng, C. W. Lin, and K. D. Huang. (2005). Heat transfer enhancement of nanofluids in rotary blade coupling of four-wheel-drive vehicles, *Acta Mechanica*, **179**, pp. 11–23.

27. Z. Zhang and Q. Que. (1997). Synthesis, structure and lubricating properties of dialkyl-dithiophosphate-modified Mo-S compound nanoclusters, *Wear*, **209**, pp. 8–12.

28. H. B. Ma, C. Wilson, B. Borgmeyer, K. Park, Q. Yu, U. S. Choi, and M. Tirumala. (2006). Effect of nanofluid on the heat transport capability in an oscillationg heat pipe, *Appl. Phys. Lett.*, **88**, pp. 1–3.

29. H. B. Ma, C. Wilson, Q. Yu, K. Park, U. S. Choi, and M. Tirumala. (2006). An experimental investigation of heat transport capability in a nanofluid oscillating heat pipe, *J. Heat Transfer*, **128**, pp. 1213–1216.

30. M. Arif. (2006). Neutron imaging for fuel cell research, *Imaging Neutron Workshop*, Oak Ridge, TN., October. 2006.

31. Y. H. Lin, S. W. Kang, and H. L. Chen. (2008). Effect of silver nanofluid on pulsating heat pipe thermal performance, *Appl. Therm. Eng.*, **28**(11–12), pp. 1312–1317.

32. C. T. Nguyen, G. Roy, C. Gautheir, and N. Galanis. (2007). Heat transfer enhancement using Al_2O_3–water nanofluid for an electronic liquid cooling system, *Appl. Therm. Eng.*, **27**, pp. 1501–1506.

33. S. Vafaei, T. B. Tasciuc, M. Z. Podowski, A. Purkayastha, G. Ramanath, and P. M. Ajayan. (2006). Effect of nanoparticles on sessile droplet contact angle, *Nanotechnology*, **17**, pp. 2523–2527.

34. R. K. Dash, T. B. Tasciuc, A. Purkayastha, and G. Ramanath. (2007). Electrowetting on dielectric-actuation of microdroplets of aqueous bismuth telluride nanoparticles suspensions, *Nanotechnology*, **18**, 475711.

35. R. S. Shawgo, A. C. R. Grayson, Y. Li, and M. J. Cima. (2002). Bio-MEMS for drug delivery, *Curr. Opin. Solid State Mater. Sci.* **6**, pp. 329–334.

36. Cepheid. (2009). <http://www.Cepheid.Com>

37. A. Ovsianikov, B. Chichkov, P. Mente, N. A. Monteiro-Riviere, A. Doraiswamy, and R. J. Narayan. (2007). Two photon polymerization of polymer–ceramic hybrid materials for transdermal drug delivery, *Int. J. Ceramic Technol.*, **4**, pp. 22–29.

38. K. Kim and J. B. Lee. (2007). High aspect ratio tapered hollow metallic microneedle arrays with microfluidic interconnector, *Microsyst. Technol.*, **13**, pp. 231–235.

39. V. Labhasetwar and D. L. Leslie-Pelecky. (2007). *Biomedical Applications of Nanotechnology*, Wiley-Interscience, John Wiley & Son, Inc.

40. C. Kleinstreuer, J. Li and J. Koo. (2008). Microfluidics of nano-drug delivery, *Int. J. Heat Mass Transfer*, **51**, pp. 5590–5597.

41. D. Bica, L. Vekas, M. V. Avdeev, O. Marinica, V. Socoliuc, M. Balasoiu, and V. M. Garamus. (2007). Sterically stable water-based magnetic fluids: Synthesis, structure and properties, *J. Magn. Magn. Mater.*, **311**, pp. 17–21.

42. P. C. Chiang, D. S. Hung, J. W. Wang, C. S. Ho, and Y. D. Yao. (2007). Engineering water-dispersible FePt nanoparticles for biomedical applications, *IEEE Trans. Magn.*, **43**(6), pp. 2445–2447.

43. L. Vekas, D. Bica, and O. Marinica. (2006). Magnetic nanofluids stabilized with various chain length surfactants, *Romanian Rep. Phys.*, **58**, 3, pp. 257–267.

44. L. Vekas, D. Bica, and M. V. Avdeev. (2007). Magnetic nanoparticles and concentrated magnetic nanofluids: Synthesis, properties and some applications, *China Particuol.*, **5**, pp. 43–49.

45. X. He, E. Y. H. Park, A. Fowler, M. L. Yarmush, and M. Toner. (2008). Vitrification by ultra-fast cooling at a low concentration of cryoprotectants in a quartz micro-capillary: A study using murine embryonic stem cells, *Cryobiology*, **56**, pp. 223–232.

46. J. Yan and J. Liu. (2008). Nanocryosurgery and its mechanisms for enhancing freezing efficiency of tumor tissues, *Nanomedicine*, **4**, pp. 79–87.

47. M. C. Daniel and D. Astruc. (2004). Gold nanoparticles: Assembly, supramolecular chemistry, quantum-size-related properties, and applications toward biology, catalysis, and nanotechnology. *Chem. Rev.*, **104**, pp. 293–346.

48. K. Sefiane, J. Skilling, and J. MacGillivray. (2008). Contact line motion and dynamic wetting of nanofluid solutions, *Adv. Colloid Interface Sci.*, **138**(2), pp. 101–120.

49. Y. H. Jeong, W. J. Chang, and S. H. Chang. (2007). Wettability of heated surfaces under pool boiling using surfactant solutions and nano-fluids, *Int. J. Heat Mass Transf.*, doi:10.1016/j.ijheatmasstransfer.2007.09.023.

50. D. T. Wasan and A. D. Nikolov. (2003). Spreading of nanofluids on solids, *Nature*, **423**, pp.156–159.

Chapter 7

Use of Ultrasound Thermometry to Study Natural Convection in Opaque Nanofluids

Kaufui V. Wong, Bradley L. Bon, Santina Vu, and Sing Samedi
Department of Mechanical and Aerospace Engineering, University of Miami, Coral Gables, FL 33124, USA
kwong@miami.edu

Buoyancy-induced flows in rectangular enclosures using nanofluids were investigated. The effects of mass fraction concentration of nanoparticles, enclosure aspect ratio, and inclination were observed. The nanofluid under investigation was a water-based alumina nanofluid. Since water exhibits an anomalous density extremum near 4°C, the additional effect of buoyancy force reversal will also be observed. The opacity of nanofluid does not permit the use of particle image velocimetry, laser-induced fluorescence, or any other means of flow visualization or visual temperature measurement of the local fluid temperature. Therefore, to investigate the temperature field, a

This chapter has been reprinted from the *Proceedings of IMECE2007*, 2007 ASME International Mechanical Engineering Congress and Exposition, Kau-Fui V. Wong, Bradley L. Bon, Santina Vu, and Sing Samedi, Study of nanofluid natural convection phenomena in rectangular enclosures, Copyright 2007, with permission from ASME.

Nanotechnology and Energy
Edited by Kaufui V. Wong

ISBN 978-981-4774-24-6 (Hardcover), 978-1-315-16357-4 (eBook)
www.panstanford.com

non-invasive method, namely ultrasound thermometry, will be used to observe the temperature field. The experimental enclosure was validated using water as the initial fluid; measured values of the local fluid temperature were compared with numerical simulations utilizing COMSOL Multiphysics. Nanofluid mass fractions of 10% and 25% were used for comparative purposes of the effects of concentration on the temperature field. Buoyancy force reversal effects were witnessed in both 10% and 25% concentrations. The nanofluid also prolonged the multicellular effects that occur in buoyancy inversion flows. A Rayleigh number inversion was observed for the 25% mass fraction nanofluid. The multicellular regime transitions to boundary layer regime at about Ra = 1E + 07 when the aspect ratio is 2.625 and at about Ra = 2E + 08 when the aspect ratio is 1.000, for different concentrations of nanofluid. For these concentrations of nanofluid and aspect ratio equal to 2.625, instability in the core region occurred at about Ra = 1.2E + 07.

Nomenclature

U	dimensionless X velocity
V	dimensionless Y velocity
U_c	characteristic velocity [m/s]
X	dimensionless X coordinate
Y	dimensionless Y coordinate
Φ	dimensionless temperature
A	aspect ratio
T_m	fluid temperature at density maximum [°C]
T_H	hot wall temperature [°C]
T_C	cold wall temperature [°C]
R	buoyancy force inversion characteristic parameter
G	gravitational acceleration [m/s^2]
β'	nonlinear volume expansion coefficient for water with density maximum [°K^{-2}]
ΔT	temperature difference between hot and cold surfaces [°C]
d	distance between hot and cold surfaces [m]
H	height of enclosure [m]
ρ_{eff}	effective density of nanofluid solution [kg/m^3]
$C_{p,eff}$	effective heat capacity of nanofluid solution [J/kg-°K]

μ_{eff}	effective dynamic viscosity of nanofluid solution [kg/m-s]
k_{eff}	effective thermal conductivity of nanofluid solution [W/m-°K]
F	mass fraction of nanoparticles in solution
$\rho_{f,X}$	density of water near hot/cold surface (x = H,C) [°C]
ρ_{n}	density of alumina (Al_2O_3) [kg/m^3]
Ra	Rayleigh Number
Pr	Prandtl Number

7.1 Introduction

Natural convection arises when local temperature variations in a fluid or gas give rise to density variations. When these density variations become large enough so that buoyancy forces overcome the viscous forces, the lower-density fluid rises. Much research has been done concerning the various phenomena that occur in free convective flows. Many numerical simulations have been conducted to investigate the influence of nanoparticles on convective heat transfer [1–6]. Analytical solutions have been obtained for very simple geometries as well as for limiting extremes of various flow parameters. Yet the majority of free convective flows that occur cannot be described through analytical solutions. Many methods have been developed to approximate the solutions for various geometries, such as the local non-similarity method, integral method, and perturbation methods. However, the success of most of the research is the result of experimentation and numerical simulation.

Theorists have also developed models and correlations to account for the effects of nanoparticles [7–10]. Some experimental work has been conducted with nanofluids under the influence of buoyancy-driven flows [11], yet further experimentation and validation is required to understand convective transport in nanofluids.

This chapter studies the rectangular enclosure, a geometry for which there is no analytical solution. Only for limiting values of key parameters (A, Ra) can the flow field and heat transfer be analytically approximated with the use of perturbation expansions [12]. Natural convection in an enclosure is a very complicated flow that depends on many aspects of the geometry. For a rectangular enclosure, the aspect ratio and Rayleigh number are key to identifying the flow regime present in the enclosure. There are three regimes present when laminar flow is of primary concern. The first is the conduction regime. In this regime, the main mode of energy transport is through thermal conduction. Convective effects are negligible in comparison to the conductive transport. Temperature varies linearly in the enclosure as it does in conduction. This generally occurs for low Rayleigh numbers ($Ra < 1000$) [13]. The second regime is the transition regime; here the viscous effects of the moving fluid near the heated and cooled walls begin to influence the core region. This regime is convectively dominant in the core and generally results in the formation of convective cells, which can vary in number depending on the vigor of the flow and aspect ratio. The third regime for laminar flow is called the boundary layer regime. In this regime, the boundary layers are well formed and can be distinctly identified in the temperature field. The core region is horizontally uniform in temperature and vertically stratified. The large gradients in the temperature field are only present near the walls where the boundary layers are present.

In addition to the already complex flow that occurs within rectangular enclosures, the use of fluid, particularly water, can introduce additional phenomena. Water exhibits a density maximum near 4°C. If the wall temperatures bracket the density maximum temperature, then additional flow effects are produced. A parameter that is generally used to assess the influence of the density extremum is termed R; this parameter is the ratio of the characteristic temperature differences. R is defined as follows [13]:

$$R = \frac{T_\mathrm{m} - T_\mathrm{c}}{T_\mathrm{h} - T_\mathrm{c}} \tag{7.1}$$

where T_m refers to the temperature at which the density maximum occurs, T_h and T_c correspond to the temperatures of the heated and cooled surfaces, respectively. This parameter also introduces additional regimes that indicate the overall influence of the buoyancy force inversion. Counter-rotating cells are present for $0 < R \leq 1$. For the special value of $R = 0.5$, the counter-rotating cells isolate the two isothermal surfaces and in effect reduce the overall heat transfer between the two surfaces. For $R > 0.5$, the cell that forms adjacent to the cold surface is predominantly larger than the one that forms at the heated surface. For $R < 0.5$, the cold cell diminishes in size and the hot cell is the dominant flow structure. When $R < 0$, then $T_c > T_m$, and thus the normal flow direction is recovered. However, when $R > 1$, then $T_m > T_h$ and, therefore, the fluid near the cold wall rises as fluid near the heated wall sinks. The purpose of this experiment is twofold: The first is being able to understand the effects of nanoparticles on free convective flow in enclosures. The second purpose is to investigate whether water-based nanofluids will also exhibit buoyancy force reversal effects.

7.2 Mathematical Model

The continuity, momentum, and energy equations can be non-dimensionalized to give the following:

$$\frac{\partial U}{\partial X}+\frac{\partial V}{\partial Y}=0 \tag{7.2}$$

$$\sqrt{\frac{Ra}{Pr}}\left(U\frac{\partial U}{\partial X}+V\frac{\partial U}{\partial Y}+\frac{\partial \bar{P}_{\mathrm{m}}}{\partial X}-|\Phi-R|^2+R^2\right)=\frac{\partial^2 U}{\partial X^2}+\frac{\partial^2 U}{\partial Y^2} \tag{7.3}$$

$$\sqrt{\frac{Ra}{Pr}}\left(U\frac{\partial V}{\partial X}+V\frac{\partial V}{\partial Y}+\frac{\partial \bar{P}_{\mathrm{m}}}{\partial Y}\right)=\frac{\partial^2 V}{\partial X^2}+\frac{\partial^2 V}{\partial Y^2} \tag{7.4}$$

$$\sqrt{RaPr}\left(U\frac{\partial \Phi}{\partial X}+V\frac{\partial \Phi}{\partial Y}\right)=\frac{\partial^2 \Phi}{\partial X^2}+\frac{\partial^2 \Phi}{\partial Y^2} \tag{7.5}$$

$$U=\frac{u}{U_{\mathrm{c}}},\ V=\frac{v}{U_{\mathrm{c}}},\ U_{\mathrm{c}}=\sqrt{g\beta'\Delta T^2 D}$$

$$X=\frac{x}{D},\ Y=\frac{y}{D},\ \Phi=\frac{T-T_{\mathrm{c}}}{T_{\mathrm{h}}-T_{\mathrm{c}}},\ A=\frac{H}{D} \tag{7.6}$$

$$\bar{P}_{\mathrm{m}}=\frac{P}{\rho U_{\mathrm{c}}^2},\ Ra=\frac{g\beta'\Delta T^2 D^3}{\upsilon\alpha},\ Pr=\frac{\upsilon}{\alpha}$$

The buoyancy terms in Eqs. (7.3) and (7.4) have been modified to account for the buoyancy force inversion effect. The new β' is a nonlinear relation of the volume expansion coefficient to temperature that accounts for the density extremum [14]. Figure 7.1 represents the model used in COMSOL Multiphysics. The no-slip condition was applied to all boundary surfaces. The axis limits used in the model were:

$$0 \le X \le A \tag{7.7}$$

$$0 \le Y \le 1 \tag{7.8}$$

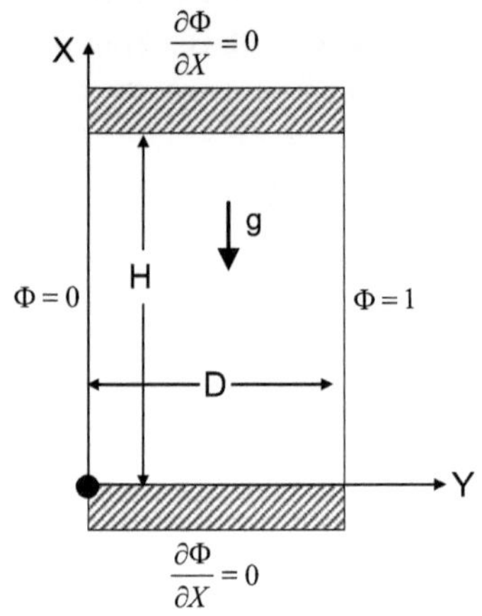

Figure 7.1 Mathematical model of experiment used in numerical simulation in COMSOL Multiphysics.

7.3 Experiment Design

Much of the research that has already been done concerning free convective flows for various geometries has dealt with air, water, or other transparent gases and fluids. However, there are some novel fluids such as nanofluids for which the thermal-fluid behavior is not completely known. Conventional methods of flow field visualization do not lend themselves to the investigation of nanofluids since they are opaque. Most optical methods that have been used or are currently used rely on the fluid being transparent. Therefore, a different method must be used to investigate the flow field with the same non-invasive nature of the current laser-based methods.

Ultrasound thermometry provides a means of obtaining local fluid temperatures without disturbing the flow structure. Ultrasound thermometry has been investigated as a means of obtaining temperature measurements within fluid-filled cavities for fluids that are opaque [15]. The underlying principle behind ultrasound thermometry is that the variation of the speed of sound with temperature provides a means of ascertaining the local temperature when the transit time of a sound pulse for a known distance traversed is measured.

Figure 7.2 is a representation of the return signal from the transducer. The initial peak seen at location (1) is the initial reflection from the interface between the transducer and the contact surface. The second peak at location (2) is the reflection from the interface between the inner surface of the contact wall and the fluid. The third peak at location (3) is the reflection from the fluid interface with the inner surface of the back wall. The final reflection at location (4) is the reflection from the outer surface of the back wall. For the experiment, the main concern is the time difference between the occurrence of the peaks at locations (2) and (3). This time difference represents the time the sound wave takes to traverse two widths of the enclosure. This time will vary with temperature since the speed of sound for the fluid in question will vary with temperature as well.

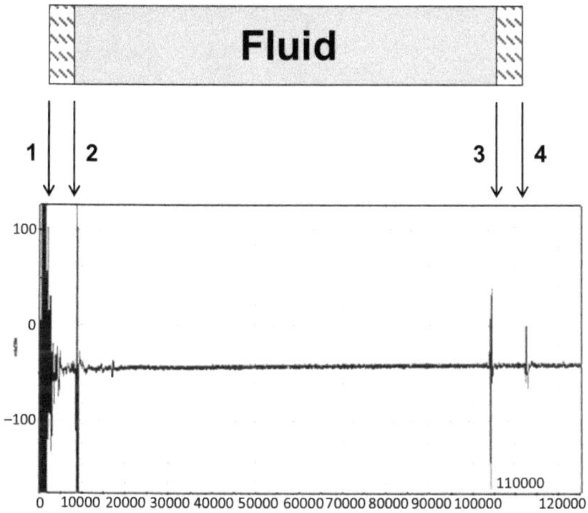

Figure 7.2 Sample signal of received sound pulse during experiment (millivolt versus nanosecond). Each peak corresponds to an interface that is encountered during sound transmission. Ultrasound pulse transmitted normal to page.

The experimental setup can be divided into three main components, the first being the experimental enclosure. Both the heated and cooled surfaces of the enclosure were Lytron 4-pass aluminum cold plates. The test section of the enclosure was made from 3/8″ lexan. For sealing purposes, gaskets were made for the contact interface between the cold plates and the lexan test section. For adequate sealing, an equal amount of compressive force needs to be applied to the entire enclosure apparatus. Therefore, compression plates made from T6-6061 aluminum were used in conjunction with threaded rods to provide a means of compressing the enclosure setup. In addition, since the aspect ratio needed to be varied, the use of compression plates allowed the use of other test sections. In between the back surface of the cold plates and the compression plates was a layer of insulation along with 1/2″ pieces of fiber board. These additional pieces help to significantly reduce the heat loss through the compression plates. Insulation was also placed on the surfaces of the test section that were not needed for experimentation.

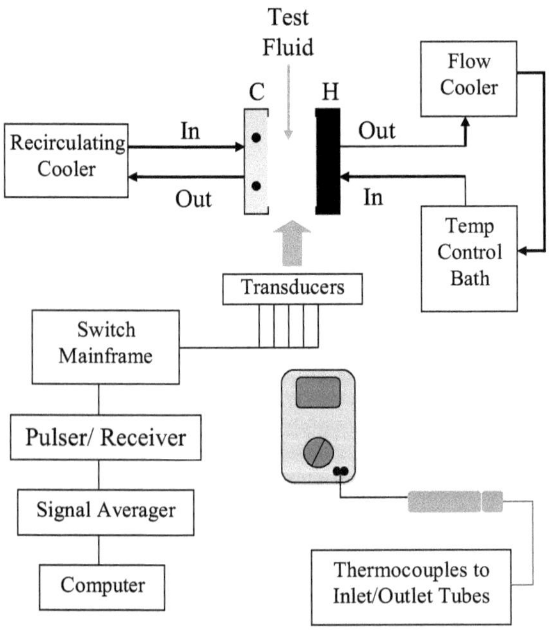

Figure 7.3 Diagram of experimental setup.

The heating and cooling fluid supply is the second major component of the experiment. For this experiment, the cold side was always maintained at 0°C ± 0.5°C. Therefore, for the cold fluid supply, a recirculating chiller was used with a 70/30 mixture of ethylene glycol and water. The heating fluid supply utilized a thermoregulator in conjunction with an OMEGA RCTB-3070 Flow Cooler. The addition of the flow cooler to the heating fluid supply allowed the heated side of the enclosure to be lowered to 0°C as well. For experimental purposes, the hot side of the enclosure was set to 8°C, 24°C, 32°C, 40°C, 48°C, and 56°C. The temperatures of the hot and cold sides of the enclosure were monitored with thermocouples placed at the inlet and outlet of each respective temperature-controlled surface. A cold junction compensator was used to keep the thermocouples calibrated. Voltage output from the thermocouples was read from a multimeter.

The data-acquisition system was modeled after the experiment performed by Fife [15]. It consists of three components. The first is the ECLIPSE Signal Averager, which averages several hundred signals during the course of one data point acquisition. The signal averager triggers the pulser/receiver to send the excitation pulse through the Switch Mainframe. This mainframe then routes the received excitation pulse through an installed high-voltage switch card to the selected transducer. The transducer is operated in pulse–echo mode. This essentially turns the transducer into a transmitter and receiver. The returned signal is then recovered using the software supplied by the signal averager.

The relationship between fluid temperature and transit time for the sound wave was experimentally determined by generating calibration curves for each test fluid. A smaller aspect ratio enclosure was brought to a uniform temperature after which five arbitrary vertical locations were chosen, and the mean transit time for the specified enclosure temperature was recorded. Data were collected for various temperatures in the range of 0–56°C, and then a high-order best fit approximation is used to obtain an expression relating transit time to local fluid temperature.

During experimentation, both the heated and cooled surfaces were brought to their respective temperatures. Once the temperatures of the inlet and outlet for both plates were stable, a period of

4500 s passed before data could be collected. This assured that the flow field had developed, and that the temperature field had reached steady state. After reaching steady state, data acquisition began by placing the transducer at the prespecified locations along the periphery of the exposed surface. For the small and larger enclosure, a total of 70 and 150 data points were collected, respectively.

Figure 7.4 Test sections used during experiment. Small enclosure (left) has a 10 × 7 array of data points. Large enclosure (right) has a 10 × 15 array of data points. Circular depressions are discrete testing points in the experiment domain.

7.4 Results

The numerical model was validated by comparing the generated solution for large aspect ratio and low *Ra* with the known analytical solution. This particular setup assures that the flow is in the conduction regime, which can be analytically solved [13]. The resulting RMS differences for the velocity and temperature profiles between the numerical model and the analytical solution were zero.

The experiment was initially run with water as the test fluid. Results from the numerical simulation for the normalized temperature were compared to those measured using ultrasound thermometry. The RMS difference between the experimental data and the numerical simulation was calculated. Figures 7.5 and 7.6 show the comparison between the two sets of data. The RMS difference value obtained was 7.57%.

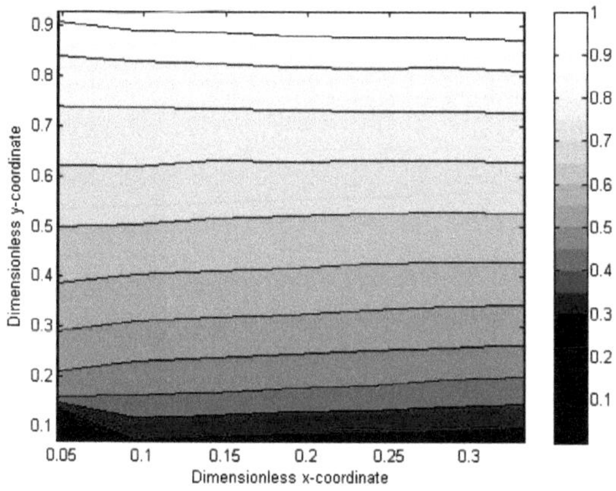

Figure 7.5 Experimental temperature field plot for water with ΔT = 24°C and A = 2.625.

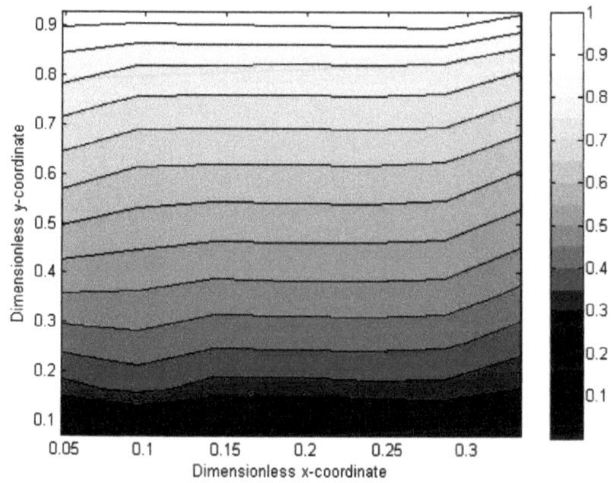

Figure 7.6 Numerical temperature field plot for water with ΔT = 24°C and A = 2.625.

Since the density maximum effect has not been accounted for in nanofluids, the Rayleigh number definition must be redefined to account for mass fraction nanoparticle concentration as well as for large density variations. The typical Rayleigh number approximation for flows with the presence of a density maximum has been previously defined in Eq. (7.6). However, the volume expansion coefficient for nanofluids has not been previously investigated.

Therefore, the following formulation for the Rayleigh number was used to characterize the results obtained from the experiments:

$$Ra = \left(\frac{g\rho_{\text{eff}}^2 c_{\text{p,eff}} d^3}{\mu_{\text{eff}} k_{\text{eff}}} \right) \frac{|\rho_{\text{f,H}} - \rho_{\text{f,C}}|}{\rho_{\text{f,C}} + \frac{\phi}{1-\phi}\rho_{\text{n}}} \tag{7.9}$$

With

$$\rho_{\text{eff}} = (1-\phi)\rho_{\text{f}} + \phi\rho_{\text{n}} \tag{7.10}$$

$$c_{\text{p,eff}} = (1-\phi)c_{\text{p,f}} + \phi c_{\text{p,n}} \tag{7.11}$$

$$Pr = \frac{\mu_{\text{eff}} c_{\text{p,eff}}}{k_{\text{eff}}} \tag{7.12}$$

The values used for the effective thermal conductivities and the kinematic viscosities of the alumina nanofluid are from Wong and Bashkar [16]. The authors measured the specific heats of the nanofluids using standard procedures. Table 7.1 shows the Rayleigh numbers calculated for the various experimental runs. The Rayleigh number used was that formulated above in Eq. (7.9). For nanofluid with mass fraction concentration of 10%, the multicellular flow field is present up to ΔT = 24°C for A = 2.625. At ΔT = 32°C for A = 2.625 (Ra = 1E + 07 for 10% mass fraction), the nanofluid shows transition to the boundary layer regime. At ΔT = 40°C for A = 2.625 (Ra = 1E + 07 for 25% mass fraction), the nanofluid shows transition to the boundary layer regime. Water with the same ΔT shows insignificant cellular formation because R = 0.16. Therefore, the cold cell that forms because of the buoyancy force reversal only occupies approximately 16% of the flow field. In comparison, the cold cell is still a considerable portion of the flow field for the nanofluid. For A = 1.000, nanofluid also exhibited properties of prolonging cellular formation to greater ΔTs. Specifically for the nanofluid with 10% concentration, cellular formation was still present at ΔT = 24°C. For 10% mass fraction at ΔT = 32°C and A = 1.000, transition to the boundary layer regime takes place at about Ra = 2E + 08. In the 25% concentration nanofluid, the presence of multiple cells was still present up to ΔT = 40°C. For 25% mass fraction at ΔT = 48°C and A = 1.000, transition to the boundary layer regime takes place at about Ra = 2E + 08. For the higher ΔTs, the core region of the enclosure exhibits vertical stratification in the temperature field.

At these higher ΔTs, the boundary layer regime has been achieved within the enclosure.

When $A = 2.625$, the nanofluid with 10% concentration showed instability in the core region for $\Delta T \geq 40°C$, and for the nanofluid with 25% concentration, instability was visible at $\Delta T = 56°C$. These both occurred at about $Ra = 1.2E + 07$. The effects were more pronounced when the enclosure was inclined. Therefore, the additional instability induced by the inclination increased the turbulent instability present in the enclosure. The large enclosure showed stable stratification up to the maximum temperature imposed on the hot surface. Another interesting observation is that for $\Delta T = 8°C$, the cold cell for both nanofluid concentrations extends all the way to the hot surface. This behavior is not common for multicellular flows where the cause for cell formation is buoyancy reversal. Only for $R > 0.5$ could the cold cell reach the hot surface but would only occur near the top of the hot wall. But for all the experimental runs, the maximum value of R achievable was $R = 0.5$. Rayleigh numbers are listed in Table 7.1.

Table 7.1 Rayleigh numbers (Ra) for nanofluids of various concentrations for different aspect ratios as well as for corresponding ΔT using Eqs. (7.9–7.12)

	Mass Fraction = 10%		**Mass Fraction = 25%**	
ΔT [°C]	$A = 2.625$	$A = 1.000$	$A = 2.625$	$A = 1.000$
8	6.7E+06	1.21E+08	3.81E+06	6.89E+07
24	9.24E+06	1.67E+08	8.72E+06	1.58E+08
32	1.09E+07	1.97E+08	7.73E+06	1.40E+08
40	1.23E+07	2.22E+08	8.85E+06	1.60E+08
56	1.50E+07	2.71E+08	1.19E+07	2.15E+08

Temperature fields for both 10% and 25% mass fraction concentrations were plotted to allow for comparison of the various flow regimes present. The x and y coordinates used during experimentation were non-dimensionalized with respect to characteristic dimension of the experimental enclosure. For 10% concentration alumina nanofluid, Figs. 7.7 to 7.18 display representative results, whereas Figs. 7.19 to 7.30 display representative results for 25% concentration alumina nanofluid.

Temperature Field Plots

Alumina Nanofluid: 10% Concentration

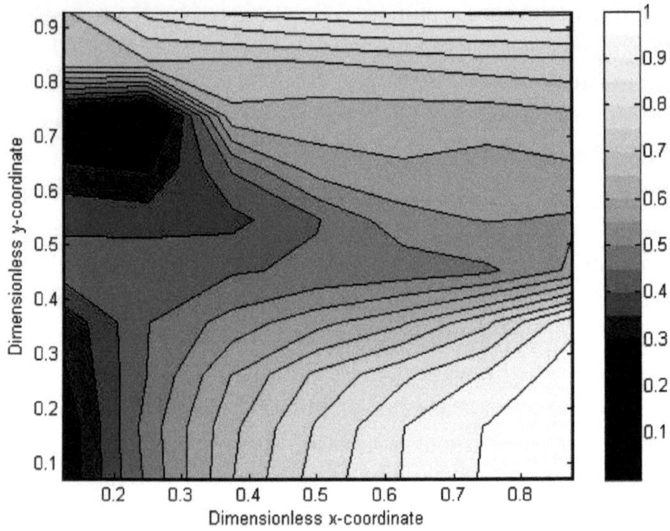

Figure 7.7 $T_h = 8°C$, incline = 0°, $A = 2.625$.

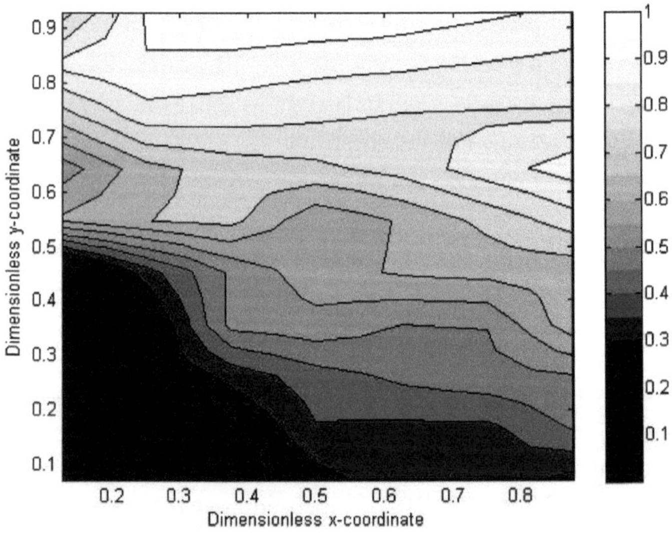

Figure 7.8 $T_h = 8°C$, incline = 30°, $A = 2.625$.

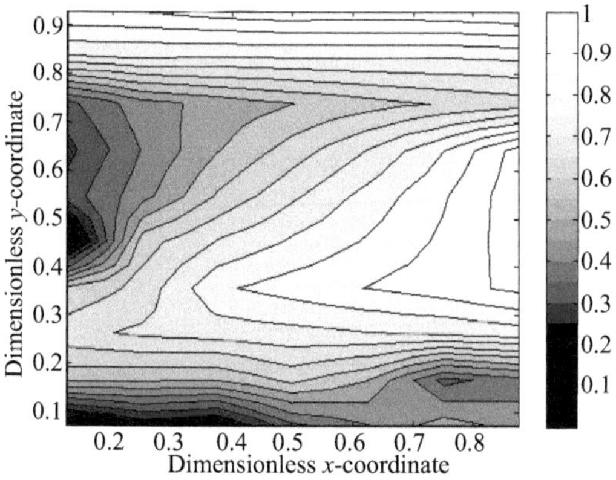

Figure 7.9 T_h = 24°C, incline = 0°, *A* = 2.625.

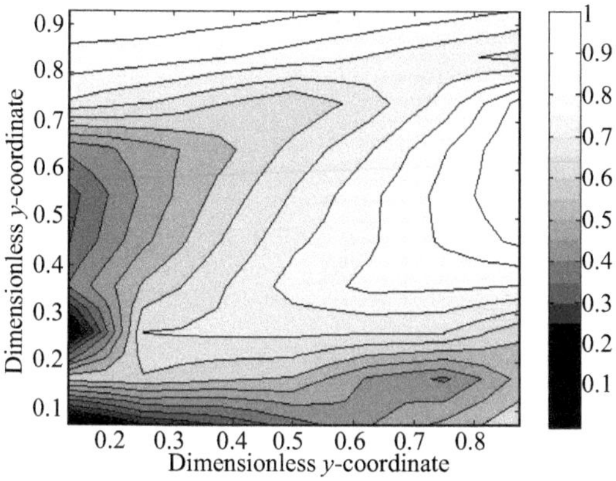

Figure 7.10 T_h = 24°C, incline = 30°, *A* = 2.625.

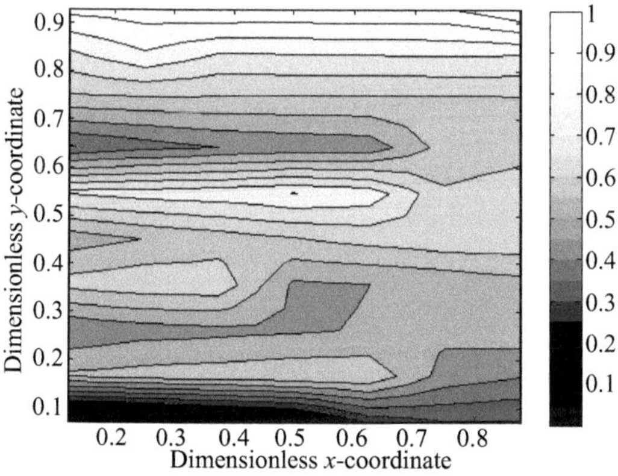

Figure 7.11 T_h = 40°C, incline = 0°, *A* = 2.625.

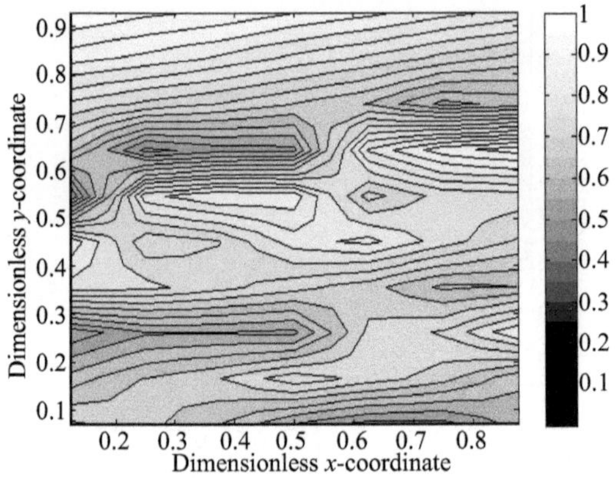

Figure 7.12 T_h = 40°C, incline = 30°, A = 2.625.

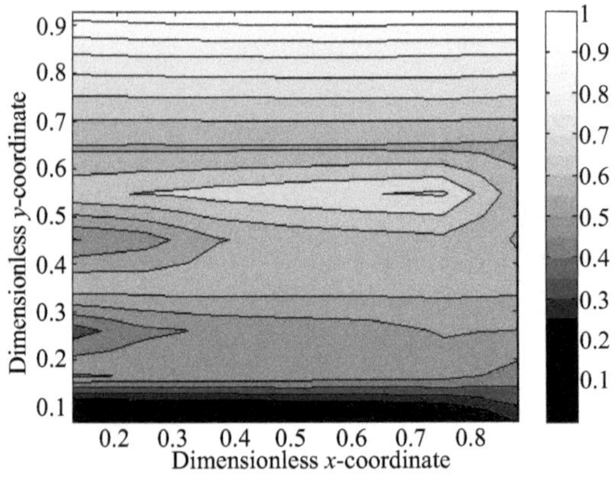

Figure 7.13 T_h = 56°C, incline = 0°, A = 2.625.

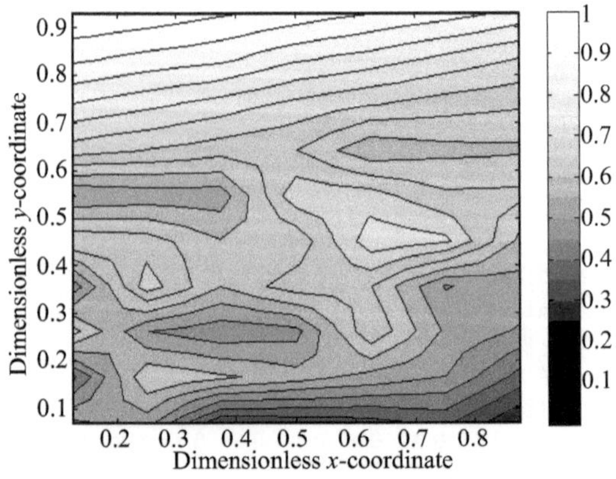

Figure 7.14 T_h = 56°C, incline = 30°, A = 2.625.

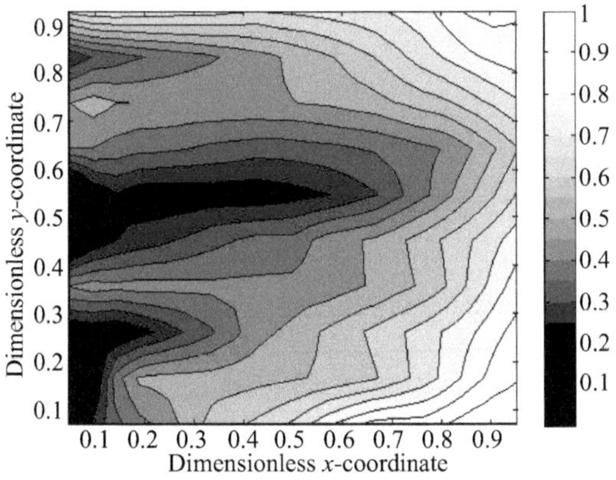

Figure 7.15 T_h = 8°C, incline = 0°, A = 1.000.

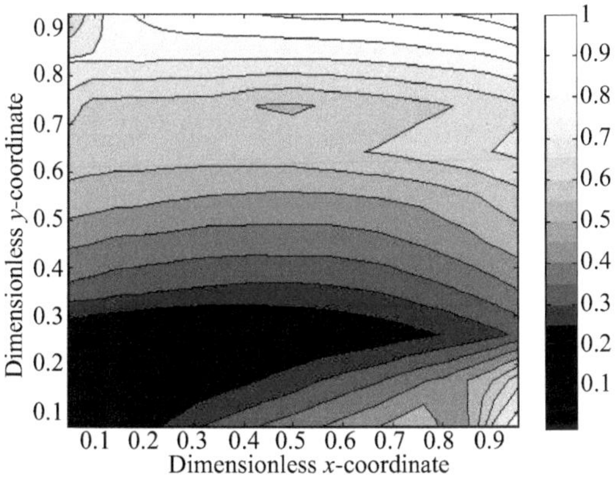

Figure 7.16 T_h = 24°C, incline = 0°, A = 1.000.

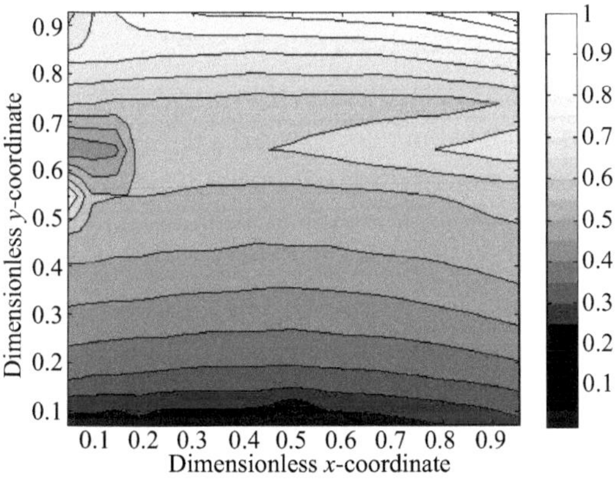

Figure 7.17 T_h = 40°C, incline = 0°, A = 1.000.

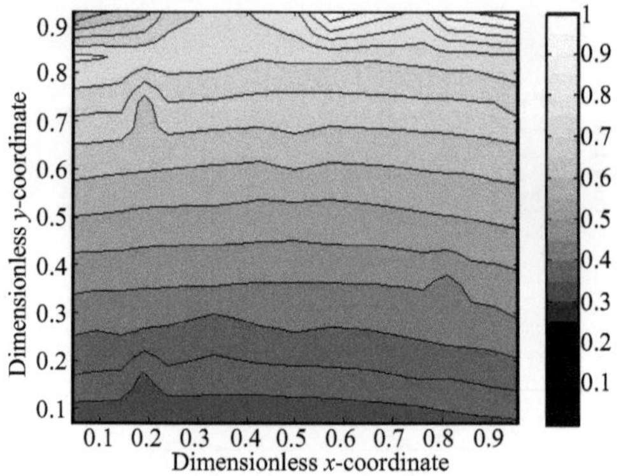

Figure 7.18 T_h = 56°C, incline = 0°, *A* = 1.000.

Alumina Nanofluid: 25% Concentration

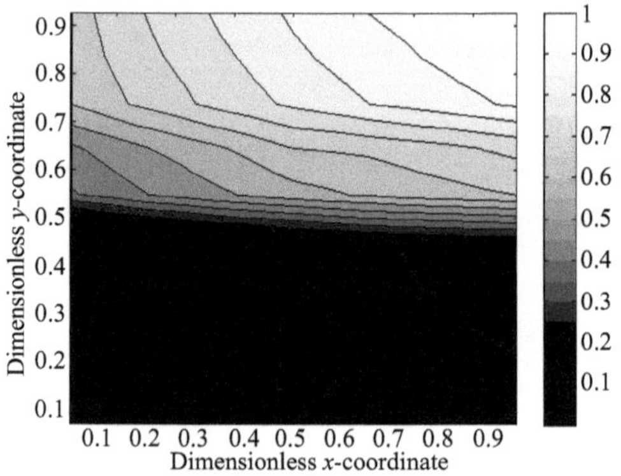

Figure 7.19 T_h = 8°C, incline = 0°, *A* = 2.625.

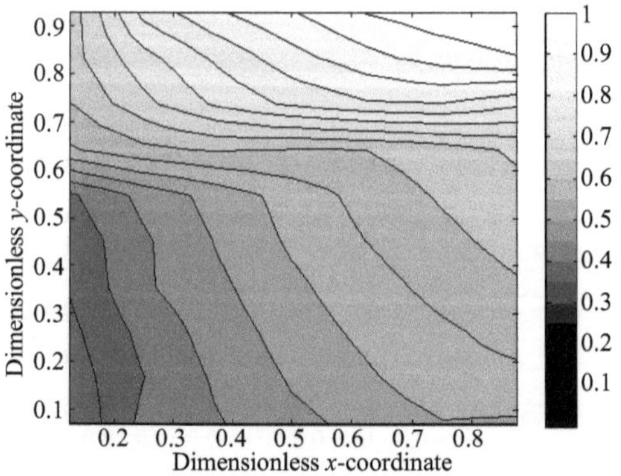

Figure 7.20 T_h = 8°C, incline = 30°, *A* = 2.625.

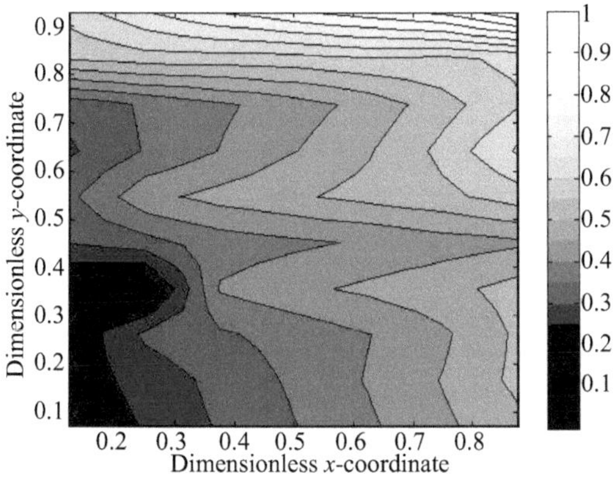

Figure 7.21 T_h = 24°C, incline = 0°, A = 2.625.

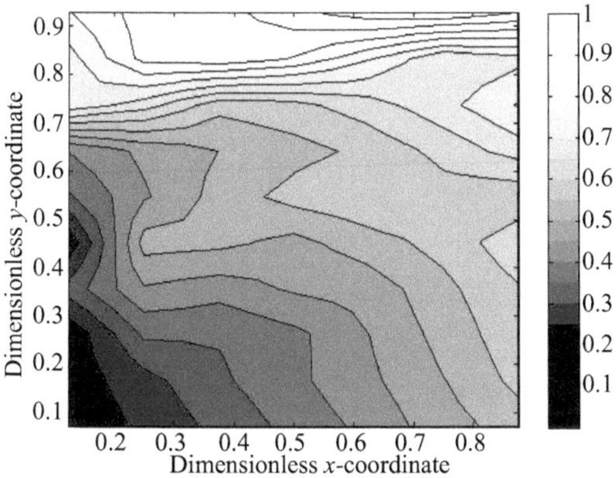

Figure 7.22 T_h = 24°C, incline = 30°, A = 2.625.

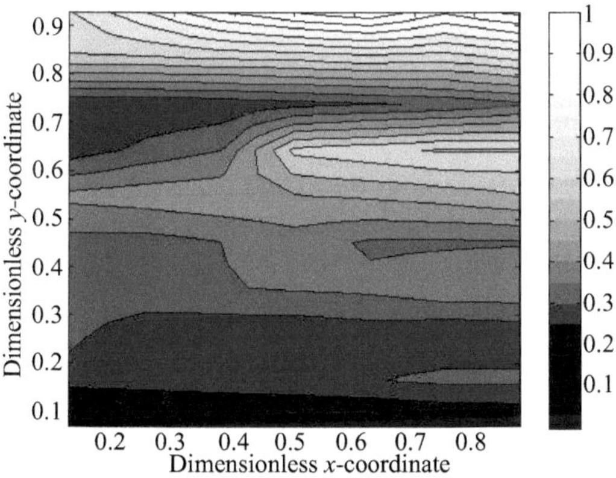

Figure 7.23 T_h = 40°C, incline = 0°, A = 2.625.

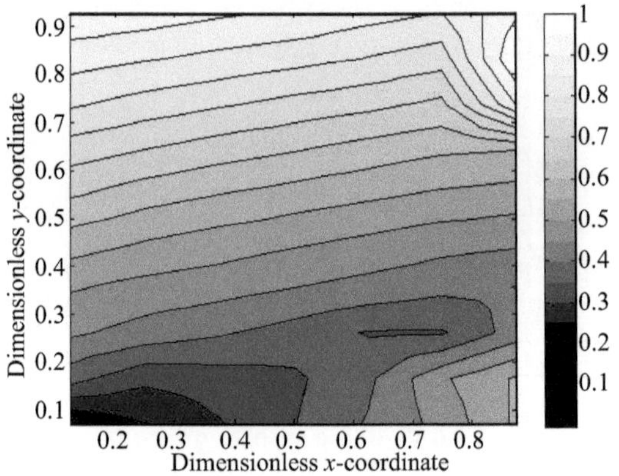

Figure 7.24 T_h = 40°C, incline = 30°, A = 2.625.

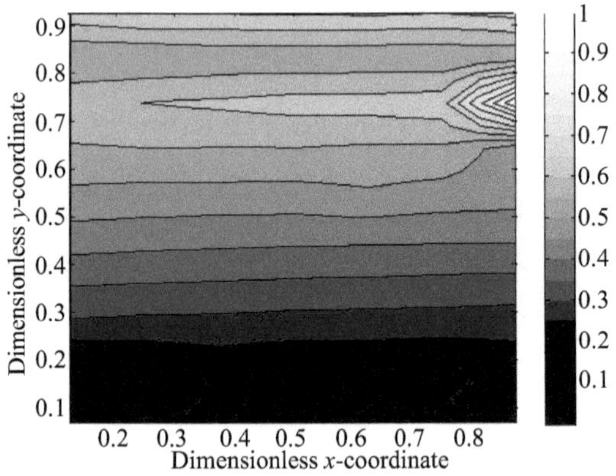

Figure 7.25 T_h = 56°C, incline = 0°, A = 2.625.

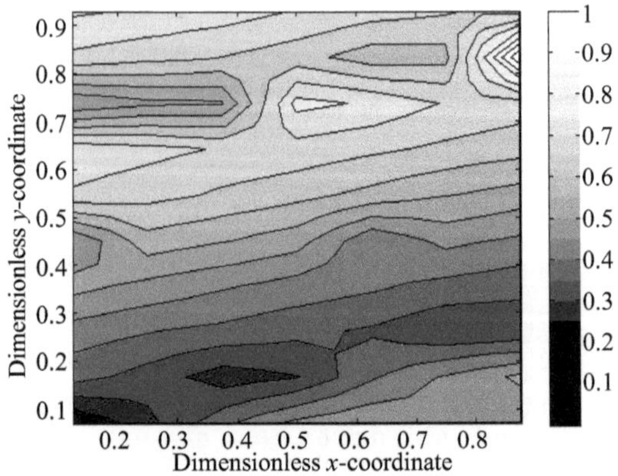

Figure 7.26 T_h = 56°C, incline = 30°, A = 2.625.

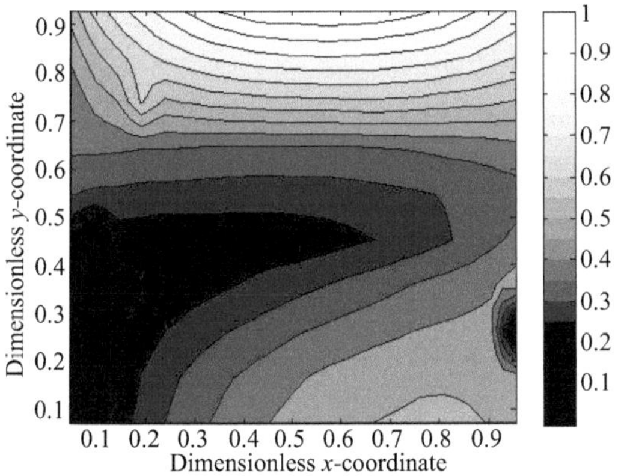

Figure 7.27 T_h = 8°C, incline = 0°, *A* = 1.000.

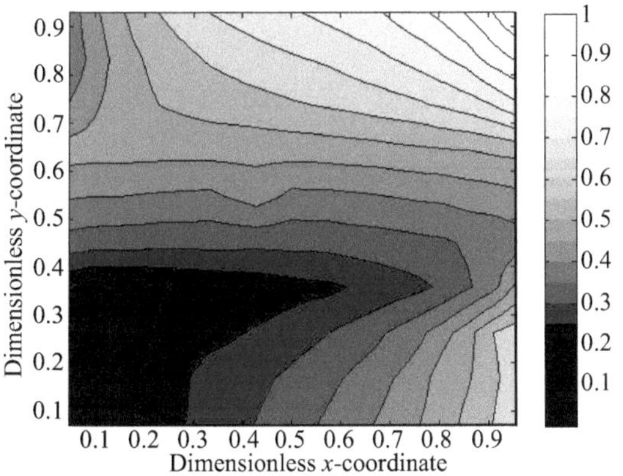

Figure 7.28 T_h = 24°C, incline = 0°, *A* = 1.000.

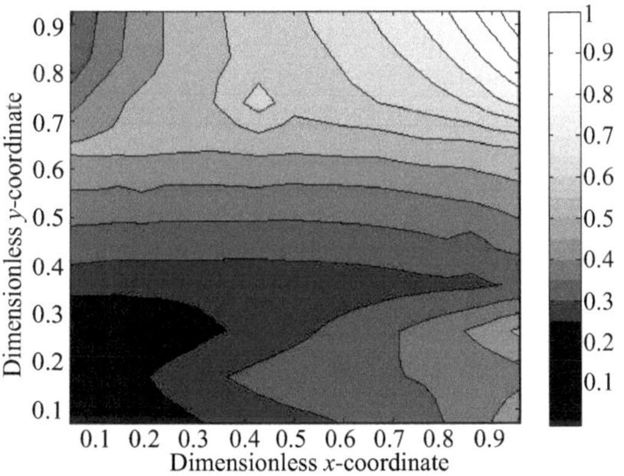

Figure 7.29 T_h = 40°C, incline = 0°, *A* = 1.000.

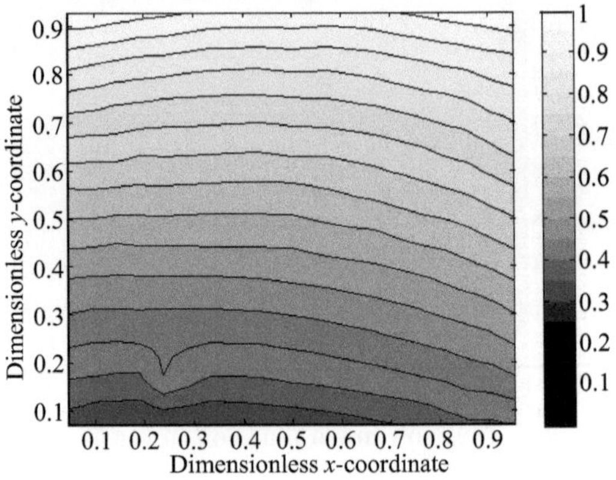

Figure 7.30 T_h = 56°C, incline = 0°, A = 1.000.

7.5 Discussion and Conclusion

Clearly, the addition of nanoparticles drastically affects the free convective behavior of the fluid. Nanofluids of varying concentrations were shown to extend the multicellular flow to larger temperature differences beyond what is expected of water, which is the base fluid. This can be explained by the decrease in the specific heat of the nanofluid with increase in the concentration of alumina nanoparticles. The results do confirm that water-based nanofluids will exhibit buoyancy force reversal effects that are due to the density extremum that water possesses near its freezing temperature. For A = 2.625 and 10% mass fraction alumina nanofluid, the multicellular flow regime exists to *Ra* of about 1E + 07, while the regime exists to the same *Ra* also for 25% mass fraction. For 10% mass fraction at ΔT = 32°C and A = 1.000, transition to the boundary layer regime takes place at about *Ra* = 2E + 08. For 25% mass fraction at ΔT = 48°C and A = 1.000, transition to the boundary layer regime takes place at about *Ra* = 2E + 08. The values of *Ra* are higher than those corresponding to A = 2.625 because the temperature gradients are less than those for A = 2.625.

It is noted that there is an *Ra*-number inversion for 25% concentration around 32°C because there is an inversion of the effective specific heat of the nanofluid corresponding to the inversion of the specific heat of water from about 30°C to 45°C. There is no corresponding *Ra*-number inversion for 10% concentration at around the same temperature range. This would explain the fact that the multicellular regime extends till 40°C before transitioning to the boundary layer regime for the 25% concentration, as compared to only 32°C for the 10% concentration, for the small enclosure with A = 2.625. The multicellular regime extends till 48°C before transitioning to the boundary layer regime. The current work also confirms that temperature measurements of the flow field can be made to assess convective regimes and flow phenomena that induce significant temperature variations. Further investigation of the heat transfer characteristics of nanofluid under the influence of buoyancy-induced flows in enclosures can be done by using the measured experimental data in conjunction with numerical modeling and optimizing an approximation for the heat flux at the walls. This could later be used as a comparative estimate of the average heat transfer coefficient for the enclosure with various test fluids.

References

1. Tong, W., "Aspect ratio effect on natural convection in water near its density maximum temperature," *International Journal of Heat and Fluid Flow*, **20**, pp. 624–633, 1999.
2. Fusegi, T. and Hyun, J. M., "Laminar and transitional natural convection in an enclosure with complex and realistic conditions," *International Journal of Heat and Fluid Flow*, **15**, pp. 258–268, 1994.
3. Khanafer, K., Vafai, K., and Lightstone, M., "Buoyancy-driven heat transfer enhancement in a two-dimensional enclosure utilizing nanofluids," *International Journal of Heat and Mass Transfer*, **46**, pp. 3639–3653, 2003.
4. Bairi, A., Laraqi, N., and Garcia de Maria, J. M., "Numerical and experimental study of natural convection in tilted parallelepipedic cavities for large Rayleigh numbers," *Experimental Thermal and Fluid Science*, **31**, pp. 309–324, 2006.
5. Hsieh, K. J. and Lien, F. S., "Numerical modeling of buoyancy-driven turbulent flows in enclosures," *International Journal of Heat and Fluid Flow*, **25**, pp. 659–670, 2004.
6. Jou, R. Y. and Tzeng, S. C., "Numerical research of nature convective heat transfer enhancement filled with nanofluids in rectangular

enclosures," *International Journal of Heat and Mass Transfer*, **33**, pp. 727–736, 2006.

7. Wen, D. and Ding, Y., "Formulation of nanofluids for natural convective heat transfer applications," *International Journal of Heat and Fluid Flow*, **26**, pp. 855–864, 2005.
8. Xuan, Y. and Roetzel, W., "Conceptions for heat transfer correlation of nanofluids," *International Journal of Heat and Mass Transfer*, **43**, pp. 3701–3707, 2000.
9. Buigiorno, J., "Convective transport in nanofluids," *Journal of Heat Transfer*, **128**, pp. 240–250, 2006.
10. Keblinski, P., Phillpot, S. R., Choi, S. U. S, and Eastman, J. A., "Mechanisms of heat flow in suspensions of nano-sized particles," *International Journal of Heat and Mass Transfer*, **45**, pp. 855–863, 2002.
11. Nnanna, A., Fistrovich, T., Malinski, K., and Choi, S. S., "Thermal transport in buoyancy-driven nanofluids," *ASME International Mechanical Engineering Congress and Exposition, Electronic and Photonic Packaging, Electrical Systems Design and Photonics, and Nanotechnology*, pp. 571–578, 2004.
12. Nansteel, M. W., Medjani, K., and Lin, D. S., "Natural convection of water near its density maximum in a rectangular enclosure: Low Rayleigh number calculations," *Physics of Fluids*, **30**, pp. 312–317, 1987.
13. Gebhart, B., Jaluria, Y., Mahajan, R. L., and Sammakia, B., *Buoyancy-Induced Flows and Transport*, Hemisphere Publishing Company, New York, 1988.
14. Gebhart, B. and Mollendorf, J. C., "A new density relation for pure and saline water," *Deep-Sea Research*, **24**, pp. 831–848, 1977.
15. Fife, S., Andereck, C. D., and Rahal, S., "Ultrasound thermometry in transparent and opaque fluids," *Experiments in Fluids*, **35**, pp. 152–158, 2003.
16. Wong, K. V. and Bhaskar, T., "Transport properties of alumina nanofluids," *Proc. IMECE 2006*, Paper 2006_13282, Nov. 2006, Chicago, IL.

Chapter 8

Nanotechnology Impact on the Automotive Industry

Kaufui V. Wong[a] and Patrick Andrew Paddon[b]
[a]*Department of Mechanical and Aerospace Engineering, University of Miami, Coral Gables, FL 33124, USA*
[b]*Department of Mechanical Engineering, College of Engineering, University of Southern California, Los Angeles, CA 90089, USA*
kwong@miami.edu

Nanotechnology has been implemented widely in the automotive industry. This technology is particularly useful in nanocoatings, improved fabrics and structural materials, nanofluids and lubricants, tires, as well as preliminary application in smart glass/windows and video display systems. As a special subclass of improved substances, alternative energy materials used in automobiles have seen a boost from advances in nanotechnology and continue to be an active research area. A correlation exists between the areas of the automotive industry that have increased the incorporation of nanotechnology and those that have witnessed increased profit margins via improvements and customer demands.

This chapter has been reprinted from the *Recent Patents on Nanotechnology*, Vol. 8(3), Kaufui V. Wong1, Patrick Andrew Paddon, Nanotechnology impact on the automotive industry, 181–199, Copyright 2014, with permission from Bentham Science.

Nanotechnology and Energy
Edited by Kaufui V. Wong

ISBN 978-981-4774-24-6 (Hardcover), 978-1-315-16357-4 (eBook)
www.panstanford.com

8.1 Introduction

Recent developments in the nanotechnology field have led to innovative patents for implementation in the automotive industry. Such technology has been promoted for use in exterior applications such as for scratch- and wear-resistant coatings, smart windows, high-strength steel bodies, and tires. Interior applications include use in fabrics, head-up displays, alternative energies, and engine lubricants and modifications. Each application from this wide array of subsystems utilizes a different beneficial property of nanoparticles or systems.

Enhanced clear coats, for instance, use nanoparticles to alter the curing reaction in order to affect the cross-linking density, which increases both hardness and elasticity, resulting in a reduction in fractures, scratches, and wear from harsh environments. Sunlight and humidity pose as the two largest hazards to automobile coatings, thus nanoparticles' property of high surface area is used to absorb ultraviolet (UV) rays. Coatings can also be used to protect against corrosion of the steel car body, an example of which consists of a three-layer system of zinc, chromium, and nano-silica particles in a self-healing method. Another technique uses nano-silica, nano-titania, nano-clay, and carbon nanotubes (CNTs) to boost electro-coating properties in order to strengthen the barrier against electrolytes. Also using current advanced technology such as polycarbonate (PC), which has many advantageous properties such as high impact strength, toughness, and light weight, and improving upon them with nanocoatings has been undertaken. PC unfortunately also suffers inadequate scratch and chemical resistance as well as susceptibility to UV light. Thus, coatings of various nanoparticle makeup have been applied to produce abrasion resistance, UV protection, antifogging capabilities, and even easy-clean properties [1].

Another new nanotechnology being infused into automobiles is the notion of smart windows. These windows operate based on electrochromism (EC), which are materials whose optical properties may be adjusted by the manipulation of minor electric potential differences. This creates energy-efficient and antiglare windows, mirrors, and displays. The initial issue with this technology was a slow response time of the EC materials, but the use of nano-oxide films has increased the surface area and enhanced the rate of redox reactions [1].

Current methods for recasting high-strength steel parts are inefficient at proper conditions. However, protective nanocoatings have been developed to ensure that production can be economically and structurally done at high temperatures, around 1000°C, which helps avoid the drawbacks of recasting at cold states such as size, accuracy, and spring-back effects [1].

Tires have most likely seen the largest technological advancement in recent years. The two most desired factors in the manufacturing of a tire are low friction or rolling resistance to reduce fuel consumption and high abrasion resistance to ensure a long lifetime. Nano-sized soot particles, called carbon black, have been used to increase a tire's durability as well as fuel efficiency. The impactful characteristics of these particles include rougher surfaces compared to traditional and higher surface energy, which mitigates inner friction for improvements to rolling properties. These particles can also lessen strain vibrations, enhancing traction specifically in wet conditions. Most importantly, these tires have proven to reduce stopping distances by 15–20% and fuel consumption by 5% [1]. The incorporation of nanoscale silicon carbide in tire manufacturing led to an increase in wet skid resistance and decrease in abrasion by up to 50% [2].

On the interior side of innovative applications, new fabrics contain antimicrobial, self-cleaning, and flame-retardant features. Using nanoparticle coatings on standard textiles, nylon, cotton, polyester, and polypropylene yarns produces photocatalytic reactions or biocidical processes that maintain sterile surfaces. Meanwhile, the nanoroughness and low surface energy of the fabric guards against the accumulation of dirt and allows water to roll off easily. Also introducing certain nanostructured materials into textiles reduces their ignitability and increases their time-to-ignition [1].

Display features have seen great improvements from the addition of nanomaterials such as higher efficiencies, chromatic contrasts, low switching times, and color tuning. Much of this is due to the extensive surface area found in nanomaterials and the ease with which organic electrochromic molecules can attach to the inorganic nanoparticles [3]. Readability has also been enhanced, regardless of the viewing angle, as well as increases to background brightness and contrast ratios [4].

Nanotechnology use in alternative energy has also been researched, returning findings in the realm of fuel cells. Nano-porous materials, such as graphene, hold strong potentials for containing and releasing H_2 due to their high surface areas and tunable chemical structures. Surface area benefits of nanoparticles can also be used toward improving battery storage and delivery properties. The energy density of supercapacitors can also be improved with the use of multi-wall and single-wall nanotubes (MWNTs and SWNTs, respectively) [1].

Nanofluids and coatings have been developed to improve upon the most important aspect of the automobile, the engine, in order to increase efficiency. Wear resistance has been improved by coating metal parts with nanostructures to protect against abrasion, and lubricants infused with nanoparticles have led to improved fuel economy, power output, oil burning, and reduced noise, vibration, and pollution of engine [1].

The high strength-to-weight ratio of nanomaterials in general, along with other properties, also allows for the reduction in weight to vehicles. Such adjustments have been made to timing belt covers, fuel line and system components, and cooling systems [1, 5, 6]. These have only been the initial commercialized applications of nanotechnology in the automotive sector.

8.2 Nanocoatings

Nanocoatings have a wide range of applications in vehicles from interior to exterior. One benefit of nanotechnology has been increased resilience to scratches and wear. Nanoparticles' effect on curing reaction has led to increasing the cross-linking density of clear coats, which increases the hardness and elasticity of the coatings and leads to less fractures [1]. Clear coats with silica nanoparticles have improved transparency, gloss, scratch and mar resilience, and a reduction in volatile organic content (VOC), which reduces negative health impacts on organisms [7]. Another method combines metal oxide–based nanoparticles, a polymeric dispersing agent, and a surface-active material to achieve higher abrasion resistive coatings [8]. Further benefits of nanocoatings include weather resistance, chemical resistance, optical properties, antifouling property, antifogging, and antistatic. Such advantages spawn from metal oxide particles of approximate diameter 1–400 nm and a polymer emulsion particle of diameter 20–800 nm [9].

Studies have shown a 20% reduction in the coefficient of friction for coatings with 40% by weight (wt) of nano-silica particles, as well as a wear rate of about 70 times less than normal in terms of wear volume [10]. This improvement has been shown to hold even 48 h after application, while still maintaining or even increasing gloss retention by 10% with 3 wt% of nanofiller [11].

8.2.1 Weather Protection

Polycarbonate (PC) has been a focus for traditional glass replacement due to its high impact strength, toughness, and light weight but has had difficulty in gaining real traction due to its weak resistance to scratch, abrasion, and chemicals. However, a layer of roughly 1 mm thickness nanocoating has yielded abrasion resistance with the use of TiO_2, SiO_2, or Al_2O_3. TiO_2 or ZnO has granted UV protection, since PC also tends to suffer yellowing during exposure, and antifogging was achieved with a sol–gel-based TiO [12]. Weather protection, including UV protection, has been a key area improved upon through the incorporation of nanotechnology in coatings. Aqueous nano-titania sol composed of an acidic nano-titania sol, a dispersant, and an alkalizing agent create a coating applicable for UV or contaminant protection [13]. Metal oxides used in clear coatings create UV blocking effects, while an organic–inorganic coat remains heat and solvent resistant and maintains the transparency properties of glass [14].

Another study found the use of cerium oxide as a UV protection layer and nanoparticles of silicon oxide for an antireflective layer successful for maintaining high transparency, while protecting against the damaging effects of light [15].

Glass plates have been designed with the sole purpose of reducing reflection while maintaining transparency, done so with a coating consisting of silica nanoparticles and a binder and an average coarseness of 2–50 nm, resulting in a decrease of 2% minimum reflection per coated surface at wavelengths between 400 nm and 800 nm [16].

Barrier performance for nonwoven fabrics has been improved through the use of one or more layers of nanodenier continuous filaments and a layer of a tough and resilient substrate. The nanofibers included have an average diameter of less than 1000 nm. The thermoplastic polymers of the filament barrier consist of polypropylene, polyethylene, polyamides, or polyesters. This technology can be applied to car covers for protection against damage from environmental conditions such as humidity and UV energy, which can deteriorate the performance and appeal of painted automobile surfaces. Such fabric could also be used in filtration systems for gas, so as to mitigate the impurities passing into the environment [17]. Nonwoven fabrics such as this, despite superior production efficiency over traditional textiles, suffer with respect to surface abrasion and durability. However, hydro-entangled fabrics show improvements in integrity due to the entanglement of the fibers [18].

Hydrophobic surface coatings have weather-protective qualities, specifically from ice buildup on windshields. Two scientists created a superhydrophobic polymer film with low-density polyethylene (LDPE) and SiO_2 nanoparticles of average agglomerate size 200–300 nm. The 100 μm thick film was created by laminating the LDPE against a layer of the nanoparticles and was tested on the windshield of a car parked overnight in an ice storm. The result was ice accumulation everywhere on the windshield, except the area covered by the film. Table 8.1 shows the hydrophobic results of multiple surfaces produced by a similar method under varying conditions [19].

Table 8.1 Superhydrophobic properties of SiO_2 nanoparticle film

Film #	Lamination Temperature (°C)	Peel Pressure (kPa)	Water Contact Angle (θ_{Static})	Water Slip-Off Angle (θ_{Slip})
2	120	1400	160°	3°
3	125	1400	160°	5°
4	125	69	160°	5°
5	125	69	158°	3°
6	125	69	160°	5°

Source: Ref. [19].

8.2.2 Self-Cleaning/Anti-bacterial

One work studied a hydrophobic coating consisting of carbon nanotubes dispersed within a hydrophobic polymer. This polymer consists of silicones, polyperfluorpolyethers, and a polymer with a monomer repeat of ethylene, propylene, styrene tertrafluoroethylene, vinvylidene fluoride, hexafluoropropylene, and perfluoro (methyl vinyl ether, ethyl vinyl ether, or propyl vinyl ether). The NTs compose anywhere from 10 to 90%wt of the total hydrophobic coating and have aspect ratios of at least 10. A water contact angle of roughly 150° was achieved with a CNT loading of about 50%, allowing for anti-sticking and self-cleaning properties for textiles. It can also be applied for self-cleaning windows and anti-sticking antennas and windows to protect against snow [20].

Cleaning properties have been attained from the use of TiO_2 nanoparticles, specifically in the form of stain-blocking coatings [1]. Water-borne coatings consisting of inorganic nanoparticles such as anionic clays, cationic clays, or layered hydroxy salts, and organic binders utilize positively charged layers to block stains [21].

Silica particles have been modified with the affixation of aldehyde groups to their surfaces, done so by reacting alumina-coated silica particles with an alkyl group with aldehyde functionality. These modified silica particles are then embedded in a polymer composition in order to increase its dirt repellence, scratch-resistance, anti-adhesion, and smooth surface characteristics. These particles can also be embedded in water-based paints for the application to various surfaces such as organic polymers, glass, metal, plaster, ceramics, as well as textiles. The coating's viscosity is also only slightly increased by the silica, thus relatively large amounts can be added [22].

The most prominently used materials for antimicrobial purposes are silver, titanium oxide, gold, copper, zinc oxide, titania nanotubes (TNTs), carbon nanotubes (CNTs), nano-clay, gallium, and liposome-loaded nanoparticles. The anti-bacterial process is enacted via either photocatalytic reactions or biocidal. For example, titania-based agents absorb light, producing photocatalytic reactions that excite charge, which carries and induces oxidation of organic molecules in order to generate hyperoxide radicals that attack the cell membranes of microorganisms and degrade stains. The biocidal process uses silver and gold to damage microorganisms through the interaction between positively charged biocide and negatively charged cell membranes. Silver presents especially strong use due to its high degree of biocompatibility, high resistance to sterilization conditions, and long-term efficacy against various bacteria [1].

Anti-bacteria fabrics have even been created using nanoinfused polymers to generate cell-killing fibrous materials. It has been found that polyphenylene ethynylene (PPE)-based cationic conjugated polyelectrolytes (CPE) and cationic phenylene ethynylene oligomers (OPEs) hold dark- and light-activated biocidal characteristics against bacteria. Coatings on solids have been effectively trapping bacteria and killing it either in the dark or through irradiation with absorbed light. Thus, these demonstrate strong candidacy for attachment to textile fibers through covalent linkages in order to produce antimicrobial properties. This is generally done by dyeing the fabric in a solution of CPE or OPE. The fabrics suitable for such applications include natural fibers such as cotton, silk, or wool, as well as rayon, nylon, and blends thereof previously mentioned. The results of these electrolyte solutions with cotton fibers were 94% cell kill count in the dark and 88% cell kill count for light. The lower kill count for the light case was speculated to be due to polymer inactivation from excess light exposure. DAB CO grafted cotton (DBC) also showed a superior ability at removing bacteria from wet environments, killing 95% of bacteria, thus displaying an efficient source of protection against spills [23].

Coatings consisting of polymeric micelles with hydrophilic neutral corona, complex coacervate core, and radii of roughly 15–30 nm have been used to render surfaces protein resistant, bacteria proliferation resistant, disinfecting, odor suppressive, malodor preventive, and with easy-clean and soil-release properties. The micelles consist of a charged block, a second oppositely charged block, and a hydrophilic and neutral block. Preferably, these micelle compose of first and second polymers, oppositely charged, and with a triblock terpolymer of an anionic, cationic, and neutral hydrophilic block. This composition has a multitude of applicable mediums besides fabrics, including glass, tile, ceramic, metal, plastic, wood, leather, concrete, hair, skin, and oral care, paint coatings, biomedical devices, and crop protection [24].

Textiles, with the aid of yarn fibers, have become permeable to water vapor and impermeable to liquid water. These yarn fibers gain their hydrophobic nature from a coating, additive, finish, or from hydrophobic molecular moieties bound to them. The textile layer yarn consists of either polyester, polyamides, polyvinyl alcohols, lyocell, rayon, viscose, nylon, cotton, linen, flax, hemp, jute, wool, acetates, acrylic, elastane, and silk. Meanwhile, the hydrophobic yarn consists of cotton, polyester, polyamide, acrylic, wool, silk, linen, synthetic, viscose, or elastane fibers. These may be treated with fluorocarbons, hydrocarbons, fluorinated hydrocarbons, silicones, silicon oxides, metals, waxes, paraffins, polysiloxanes, fluorine compounds, hybrid polymer nanolayers, nanocoatings, etc. as water-repellent finishes in order to generate their hydrophobic properties. Wicking of nanofibers from polyester, acrylic, polyamide, or any other hydrophilic fiber with a low surface energy may be a means to achieve water resistance for the inner layer of fabrics [25].

8.2.3 Wear/Erosion Resistance

Wear resistance is especially important within the engine. Thus, ceramic nanostructure coatings have been used to enhance metal parts resistance by reducing the scale of size. Specifically, alumina, alumina–titania, cemented tungsten carbides, and zirconia powder ceramic coatings have exemplified the benefits of reduced grain, particle, or layer sizes on abrasion protection [1]. Carbon nanotube reinforced polymer (CNRP) coatings can also be applied to susceptible surfaces in order to lower stress concentrations, increase reinforcement efficiency, and increase the tensile strength [26].

Erosion protection has been an area for improvement that coatings have specifically targeted. Developments include utilizing an electrolytic plating solution of deposition metal ions of zinc, palladium, silver, nickel, copper, gold, platinum, rhodium, ruthenium, chrome, or derivative alloys mixed with non-metallic nanoparticles of size in between 10 nm and 500 nm at a weight percentage of 1–5%. The solution is electrolytically deposited onto the metal surface, enhancing the water repellency, corrosion resilience, hardness, wear resilience, and lubricity of the surface [27].

Another wear-resistant coating uses a hard backing with a metal alloy matrix of hard particles and nanolayers consisting of metal nitride, metal boride, or metal carbide in a layer effect to improve hardness and wear resistance [28].

Other corrosion-resistive coatings found consist of magnesium hydroxide particles of 200 nm or less average size, and acid-catalyzed sol–gels with silica nanoparticles and polyaniline solutions [29, 30].

Oxidation can damage and corrode metals over long periods of exposure, especially those used in salt and acid environments. A study developed nanocoatings to reduce these damaging effects, which are experienced by components such as hydrogen fuel cell components, fan accessories, inlet–outlet transitions, bearing rods, electrical system components, and exhaust systems. The nanoparticles range in size from 3–20 nm and their impact is seen in Figs. 8.1–8.3, where lower mass gain indicates lower oxidation. As one can see from Fig. 8.1, the surface treatment composed of nanoceria nanoparticles, or cerium oxide, prepared by the study's assignee, Material Interface, produced the largest improvement by reducing the oxidation effect as much as 68% at 442 h.

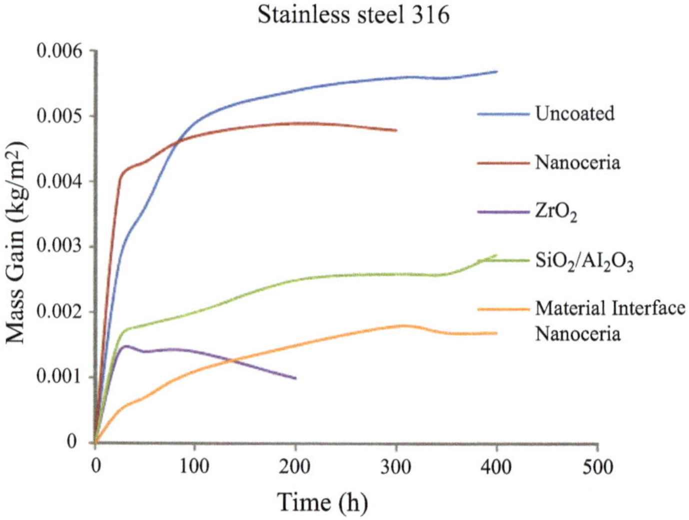

Figure 8.1 Reduction in oxidation of stainless steel 316 by various nanoparticle coatings [31].

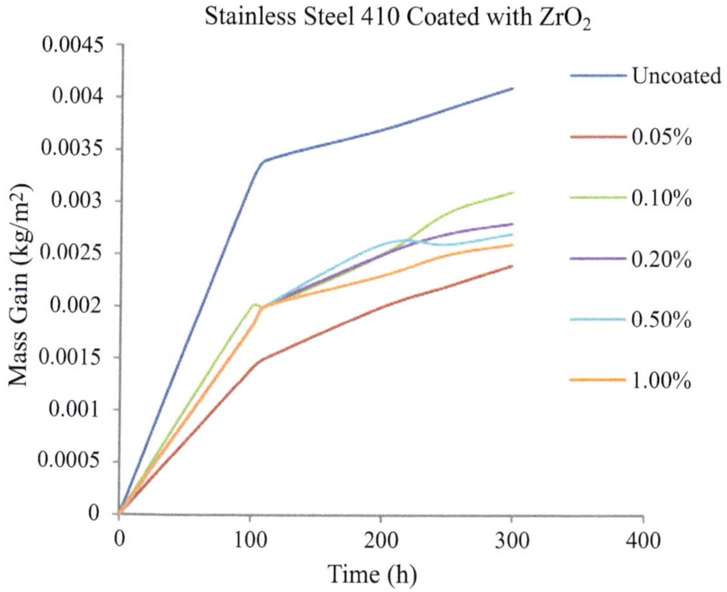

Figure 8.2 Reduction in oxidation of stainless steel 410 by varying concentration of zirconia nanoparticles in protective coating [31].

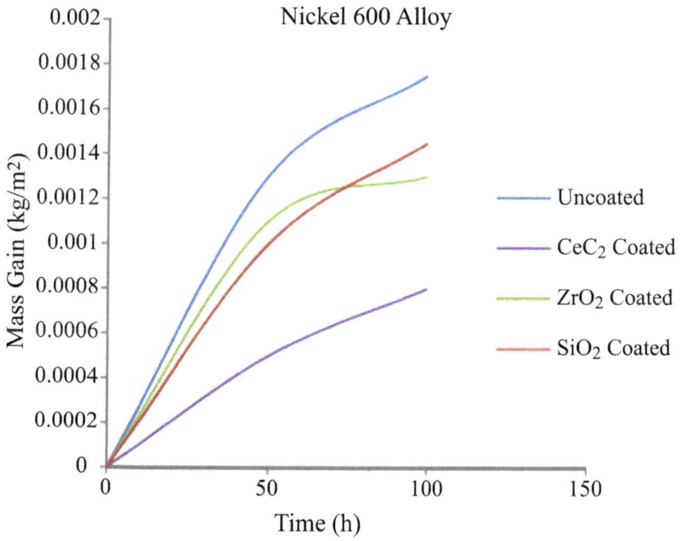

Figure 8.3 Reduction in oxidation of nickel 600 alloy by protective nanocoatings of nanoceria, zirconia, and silicon dioxide [31].

Figure 8.2 shows the impact of varying concentrations of zirconia nanoparticles in a protective coating, with a concentration of 0.05% producing the best protection against oxidation. Meanwhile, Fig. 8.3 compares the oxidation protection of three different nanoparticle coatings on a nickel alloy, with nanoceria providing the largest reduction in mass gain [31].

8.2.4 Fire Retardant

A fire-retardant coating, which also protects from lightning strikes and is thermo-oxidative resistant, consists of pyrograf III carbon nanofiber (CNF). It was developed with CNFs in the form of vapor-grown carbon fibers (VGCF), or a discontinuous graphitic filament, produced from the pyrolyis of hydrocarbons [5].

8.3 Smart Windows

Smart glass was born from developments in electrochromism, which is the ability of certain materials to change colors in response to charges. This presents a valuable technology for use in the automotive industry as the majority of vehicles are exposed to sunlight daily. This exposure leads to energy needs for cooling the interior of the car. This technology was first implemented in rearview mirrors, using an interior electrochromic (EC) reflective element to control the reflectance levels [32]. The technology works by changing optical properties with variable voltages to absorb the incoming light into color centers produced by ions at electrodes, resulting in minimal light being reflected [1].

The next step in development has been the application to windows in order to provide adjustable tinting. Photorefractive polymers have been applied to windows in order to grant it the ability to move charges based on light conditions. Improvements to the earlier designs have incorporated interior and exterior panes separated by the EC cell and protective polymer layer. This polymer layer is an insulator and protects the cell from damage, while improving EC switching time [33].

Dissolving a nanotube-type conducting polymer into an electrochemical reaction cell of a transparent electrode and working electrode has also attained a successful electrochromic device. The resulting large contact area of the polymer and small thickness lead to an increased electrochromism speed [34].

Some smart windows have been designed with photovoltaic (PV) systems, including a PV element, a solar control coating, and a space between the two panes. An electrolytic material fills the space with one surface holding a chromophoric material, so that the system can capture solar radiation to generate electrical energy [35].

Use of indium tin oxide (ITO) as a thin continuous film exhibits alluring properties such as 80–95% optical transparency, high electrical conductivity, and infrared radiation (IR) absorption. IR absorption occurs in the range of 1100–2500 nm, with absorption rates of 12.5, 25, and even 50% for wavelengths greater than 950, 1100, and 1450 nm, respectively. Window films of nano-sized VO_2 exhibit excellent optical transparency as well as efficient IR reflection [36].

Another study also used tin to create a protective film against IR; however, the nanoparticles added were antimony tin oxide, or nano-ATO. The same study also added zinc oxide nanoparticles in combination with the nano-ATO to prepare a multifaceted film that provided UV protection as well. The results of these films are shown in Table 8.2, with two separate methods of preparation, where C Method represents the Czochralski process, which is a method of growth used to obtain single crystals [37].

Table 8.2 UV and IR protection of metal oxide films

Film Type	Retained Transparency [%]	UV Protection [%]	IR Protection [%]
ATO (C method)	83	—	73
ATO (Gravity method)	90	—	70
ATO/ZnO (C method)	80	90	63
ATO/ZnO (Gravity method)	85	95	65

Source: Ref. [37].

The crystalline structured nano-ATO particles ranged in size from roughly 2 to 50 nm and accounted for 10 wt%, while the ZnO nanoparticles ranged in size from 2 to 100 nm and accounted for 5 wt%. As one can see, the tin nanoparticles were effective in increasing the IR protection, while the zinc oxide nanoparticles improved the UV protection. However, the zinc reduced both the IR protection and visible light transmittance [37].

8.4 Improved Material

Improving the physical components of automobiles has also been studied using nanoparticles mixed with traditional materials. Such an instance has been the use of dispersing carbon materials such as graphite, graphite fiber, carbon fiber, CNF, or carbon nanotubes in aluminum. The interface of carbon is hardened using a nanosilicon carbide coating, which results in a lightweight material with high dynamic strength. The application reaches from cars and aluminum wheels to the aerospace and nautical industries as well [38]. Various GM cars include plastic parts doped with nanotubes, which during electrified paintjobs allow for a better surface finish [39].

8.4.1 Durability/Strength

The increased strength of nanoparticle-doped materials has developed a strong interest in the scientific community. Such nanoparticles as CNTs have especially drawn attention specifically for their impressive mechanical, thermal, and electrical properties. Metals such as aluminum or magnesium have been tested with nanoparticles of sizes ranging anywhere from 1 to 200 nm. It is believed that not only CNTs promote increased desired properties, but any nanoparticles of high aspect ratio, including carbides, nitrides, or silicides. One particular study uses CNTs in a concentration range of 0.5 to 10 %wt. This addition of nanomaterial improved the Vickers hardness above that of the original by 40–80% in different concentrations. The CNTs used in the composition were generated by the method of carbon vapor deposition, using acetylene, methane, ethane, ethylene, butene, butane, benzene, or butadiene as the carbon source. This composite material was then formed into a cylinder head, cylinder block, crankcase, and gear wheel. The previously mentioned material can protect against creeping during high temperatures in areas of high bond stress, which typically results in a permanent damage to the strength. Thus, air tightness capabilities of engines can be maintained at higher temperatures and pressures, allowing for an increase in the operational efficiency. The superior heat conductivity of the material can also permit higher heat dissipation to maintain the temperature levels of the engine component within reasonable ranges [40].

Ultra-hard, tough, and corrosion-resistive steel has been fashioned through the inclusion of austenitic nitrogen nanocrystals, an oxide, nitride, or carbide to stop the growth of the crystal grains, and ultrafine ferrite grains. This has created austenitic steels such as chromium, nickel, manganese, or carbon with solid solution nitrogen levels of 0.1–2.0 mass percent [41]. Research has also been conducted on the use of ultrafine nanoferrite crystal grains on the order of 1200 nm or smaller in high-strength steel production, which has shown to vastly improve work hardening properties and impact resistance. The composite structure also contains a second phase of martensite, bainite, or retained austenite, leading to a stress–strain diagram gradient of 3–7% of true strain 5000 MPa or more. These results have not been found by any other conventional techniques for producing high-strength steel sheets. The material's elongation is also improved due to the ferrite particles; however, only when the area ratio of the particles remains between 15% and 90%. Otherwise, deformation properties tend to reduce and cause easy breakage of the material. The implementation of this high-strength and low-weight steel in vehicles reduces the body weight and front overhang, both improving the driving performance [42].

Another improved structural steel alloy exhibits both superior ultrahigh strength from nanometer M2C carbides embedded in a fine lath martensite matrix as well as corrosion resistance from improved chemical partitioning of Cr onto the surface. The steel consists of 0.1–0.5% carbon, 8–17% cobalt, 0–10% nickel, 6–12% chromium, <1% silicon, <0.5% manganese, <0.15% copper, and additives of <3% molybdenum, <0.3% niobium, <0.8% vanadium, <0.2% tintalum, <3% tungsten, <0.2% titanium, <0.2% lanthanum, <0.15%, <0.15% zirconium, <0.005% boron, and impurities of

<0.02% sulfur, 0.012% phosphorus, 0.015% oxygen, and 0.015% nitrogen by weight with the rest consisting of iron [43].

Protective nanocoatings have had an impact on improving methods for recasting high-strength steel parts. Current inefficient conditions have been targeted with nanocoatings to ensure that production can be economically and structurally done at temperatures around 1000°C, which helps avoid the drawbacks of recasting at cold states such as size, accuracy, and spring-back effects [1]. The use of TiO_2 and nano-clays in epoxy systems offer enhanced energy absorption capabilities for materials, specifically for impact loading, as well as improved moisture resistance and superior sound, vibration, and oscillation dampening characteristics [5]. Another epoxy resin composite, but consisting of vapor-grown carbon nanofibers (CGCNF) content of 20 wt%, was used to replace heavier metal structures of poor corrosion resistance. The CGCNF epoxy filler improved the composites strain and resistance to microcracks enough to generate liner-less composite pressure vessels. Opportunities for further application include automotive electronics shielding, exterior panel electrostatic painting, and improving the stiffness of tires [44].

Other nanocomposites that improve structural properties include nano-phased unidirectional laminates of continuous prepare tapes infused with SiC nanoparticles of 30 nm diameter. This improved the flexural strength and stiffness by 32% and 20%, respectively [26]. Surface hardening has been achieved as well through the use of nanocoatings. Carbon materials, such as CNTs, have been coated with nanosilicon carbide in order to infuse the carbon material within aluminum, which greatly enhances the mechanical properties of aluminum relative to its standard state [45].

Engine structural material, currently composed of cast iron due to its high strength, however, has a high density of 7.8 g/cm^3. Thus, nanomaterials such as low-cast aluminum–silicon with a density of 2.79 g/cm^3 have been studied to replace conventional materials in order to reduce car weight. The issue with aluminum alloys is low wear resistance and seizure loads, but decreases in grain size below 100 nm can enhance hardness, flow stress, ductility, and thermal insulation of materials. Coatings have also been looked at as an alternative for strengthening structural materials. Ceramic nanostructures can protect metal parts against abrasion and wear through their reduced size and layer thickness. Such composites include alumina, alumina–titania, cemented tungsten carbides, and zirconia powders [1].

Nanotubes hold the special ability to isolate defections in structures, thus eliminating the risk of chain fractures. This paired with their impressive elasticity could allow for the kinks of post car crash to be unfolded and reveal no remnants of damage. Meanwhile, their conductivity can provide real-time monitoring of material integrity for plastic structures, much like presented in the case of a tire monitoring system.

The physical and mechanical properties of magnesium alloys make them ideal for the automotive industry; however, they suffer weak performance in corrosion and wear properties. Thus, sol-gel-based nanocoatings of Al_2O_3 and CeO_2, of 15 wt%, are applied to the surfaces of the magnesium to enhance the scratch and corrosion resistances, respectively [46]. Protecting aluminum alloys, such as AA2024-T3, is also key for industrial purposes. Using mesoporous silica nanocontainers with the corrosion inhibitor 2-mercaptobenzothiazole in a hybrid sol–gel layer can provide the effective anticorrosive coating needed to allow further application of such metal alloys [47].

Another method discovered for strengthening magnesium alloys infuses the master alloy that is to be added to molten metal with nanoparticles of alumina, Al_2O_3, or silicon carbide, SiC. The metal nanocomposite components created display high strength and creep resistance, proving them useful for structural and high-performance applications. Figure 8.4 shows gains produced over a standard magnesium alloy, AZ31, by adding the above nanoparticles [48].

The results of these tests show that adding SiC nanoparticles to as-cast Mg (AZ31) increases the alloy's ultimate tensile strength by approximately 100% and its yield strength by 50%, as well as achieves a 50% decrease in turning angle during the BLR test, which signifies less creep deformation. The results of adding alumina nanoparticles to the magnesium alloy were lower, but still produced significant numbers with a 26% improvement to yield strength, a 40% improvement to ultimate tensile strength, and a 33% decrease in turning angle. Both master alloys contained 10% nanoparticles of size 50 nm. Other nanoparticles compatible with this technique for strengthening alloys typically include metal oxides, carbides, and metals of size ranging from 1 to 1000 nm. The BLR testing was used as a complementary test for tensile creep as it is more pertinent for Mg alloys in the automotive industry, such as for a cast engine block [48].

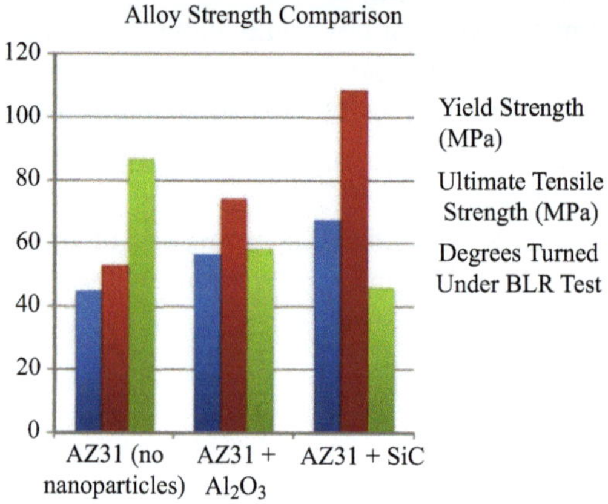

Figure 8.4 Strength of as-cast AZ31 alloy, as-cast AZ31 alloy with Al_2O_3 nanoparticles, and as-cast AZ31 alloy with SiC nanoparticles as measured by yield strength, ultimate tensile strength, and bolt load retention (BLR) test [48].

8.4.2 Thermal Properties

Toyota has been perusing research in the nanotechnology field, and currently employs a nylon-6 matrix of 5 wt% nano-clay heat-resistant timing belt cover. The improvements made to the material were increases in tensile strength by 40%, tensile modulus by 68%, flexural strength by 60%, flexural modulus by 126%, and a raise in the heat distortion temperature from 65 to 152°C [6].

Honda, on the other hand, has used clay-pp nanocomposites for structural seat back components, while Ube created fuel lines and other fuel system composites from clay–nylon 12 nanocomposites. These materials showed superior flammability qualities to that of pure polymers [5].

Components susceptible to and highly exposed to high temperatures, such as engine covers, pose as important areas for nanomaterial incorporation. One study created a polyamide nanocomposite resin of 0.2–5.0 wt% hydrotalcite-based heat stabilizer, 2–10 wt%

nano-clay, 0.3–5.0 wt% phosphite-based thermochromism protector, 1.5–9.0 wt% aluminum powder, and 0.5–3.0 wt% pearl pigment. The inorganic heat stabilizer, hydrotalcite-based, is represented by the formula $Mg_{(1-x)}Al_x(OH)_2A_{n-}(x/n)\cdot mH_2O$, where $0 < x < 0.5$, $0 \leq m \leq 4$, and A_{n-} is an anion with $n-$ valence [49].

8.4.3 Light Weight

GM has designed their "step assists" in the Safari and Astro vans from a polymer nanocomposite for its lightweight, scratch-resistant, and rustproof properties [50]. While the Impala has body side moldings of an olefin, clay nanocomposite that provides weight savings and enhances the surface quality [6].

8.5 Tire Enhancement

Tires, which primarily consist of rubber or elastomeric particles, have possibly seen the largest quantifiable benefit from the use of nanotechnologies. Such improvements as rolling resistance, thermal properties, permeation barrier properties, run-flats, tear and wear resistance, and sensor systems have all been witnessed in recent years. The many benefits of nanomaterials in tires also include their superhydrophobic nature, leading to improved wet surface performance, as well as reduction in negative impacts of temperature on traction since traditional tires are designed for best traction at high temperatures [51].

8.5.1 Tread Enhancement

Tire contact friction has been enhanced through the use of steel nanofibers, leading to a fivefold gain in the coefficient of static friction and also enhanced wear and lifetime characteristics. Tires can also damage roads over time, and with the use of nanofibers in tire design, a "snap back" phenomenon occurs at the cusp of the contact patch, generating less wear on the road. Nanofiber-incorporated tires degrade at a much slower rate than traditional tires when skidding or sliding occurs. This is due to the fact that nanofibers cannot reach dynamic friction. Skidding may even be found desirable since larger stopping forces are exhibited by nanofibers at higher velocities [51].

Another case of nano-enhanced tread uses a polyurethane polymer with either TiO_2 or carbon black nanoparticles ranging from 10 to 40 nm in diameter at a wt% of 1–3, leading to superior flexibility and durability, low rolling resistance, and reduced oil consumption. The 22–25% reduction in rolling resistance alone leads to a 3–8% reduction in fuel consumption, making these tires environmental friendly [52].

8.5.2 Gas Barrier Properties

Improvements to water and gas barrier properties have also been found through the use of a metal-reinforcing layer covered in a composite elastomeric layer, which could include a diene of styrene elastomer and aluminous particulate, creating no need for the expensive butyl liner barrier used in current tire designs. The weight percentage of alumina particulates included ranges between 0.5% and 50%. These composite materials provide reductions to the oxygen transfer rate (OTR) and water vapor transfer rate (WVTR) [53].

8.5.3 Mechanical Properties

The inclusion of functional graphene sheets (FGS) in rubbers has shown improvements in the gas barrier properties, mechanical properties, thermal stability, and electrical conductivity. Increases in tensile strength have been marked at sixfold in polydimethylsiloxane (PDMS) and natural rubber and 6- and 12-fold for Young's modulus, respectively, all with <5 wt% FGS composition. FGS has also increased the toughness of natural rubber by 29% to 5426 MPa, but at the cost of total elongation. Compared to other clay nanoplatlets, FGS has superior aspects: Its carbon structure retains better interfacial compatibility with elastomeric matrices, and its greater flexibility reduces fatigue and debonding from the matrices [54, 55]. Table 8.3 shows the enhancements to tire characteristics with differing compositions in comparison to standard tire composites of examples 2 and 3. All examples contain 100wt. reactive rubber, except example 16, which consists of 80wt. reactive and 20wt.

general rubber, 80wt. carbon black, 1.7wt. sulfur, 1.5wt. N-t-butylbenzothiazole sulfonamide (NS), 3wt. ZnO, both P#2 Oil and Silica with 5wt., and both Stearic Acid and a coupling agent with 1wt. The variable factors are shown in the second column and their effect is denoted in the subsequent columns, as compared to the base index of 100 for example 2. Air permeation resistance improves as the number decreases, while both the wear resistance and tensile strength increase as their respective numbers increase [55].

Table 8.3 Comparing properties of differing tire compositions

	Composition Variables	Air Permeation Resistance	Wear Strength	Tensile Strength
Ex. 2	—	100	100	100
Ex. 3	5wt. cloisite 15A	100	92	84
Ex. 10	5wt. reactive layered silicate	75	115	104
Ex. 11	7wt. reactive layered silicate	74	110	115
Ex. 12	10wt. reactive layered silicate	52	105	129
Ex. 13	15wt. reactive layered silicate	65	105	134
Ex. 14	20wt. reactive layered silicate	45	110	139
Ex. 15	30wt. reactive layered silicate	10	110	144
Ex. 16	10wt. reactive layered silicate	41	102	146

Source: Ref. [55].

These impressive improvements to wear, tear, and air permeation resistance of rubber are due to the polymer in a raw rubber material with a reactive layered silicate. The reactive functional group can consist of either aryl halide, hydroxl group, benzyl halide, amine and phenol group, while the material between the particles of layered silicate is either aryl halide, alkyl halide, hydroxyl group, benzyl halide, amine, tertiary phosphorous, carboxyl group, phenol group, maleic acid, or maleic anhydride. These functional groups reacting with the polymer between the silicate increase the dispersion of raw rubber material, which increases the tensile strength, leading to greater dispersibility of layered silicate, which is key to the improvements in rubber properties [56].

$Al(OH)_3$ inorganic nanoparticles of sizes from 5 to 10 nm have also been used in enophile-functionalized polymers to improve the bonding of rubber, tensile strength, tear strength, and 10% better wet traction [57]. Another study found the improvement in tensile and tear strength to be quantified at 30% or more over non-spherical copolymer-enhanced rubber. These polymeric nanoparticles consisted of two multiblock polymers of polyalkenylbenzene and one monomer from conjugated dienes, alkenylbenzenes, alkylenes, ethylene oxide, methyl methacrylate, nitriles, or acrylates. This results in a nanoparticle with an average diameter of less than 100 nm. These compounds are under consideration to be used in treads as well as side walls due to their higher dynamic modulus, lower temperature dependence, and lower hysteresis values at high temperatures. These characteristics improve cornering, handling, dry, wet, and snow traction, rolling resistance, dispersion, and aging for tire composites. Tin has been especially used as a functionalized nanoparticle in composites due to its enhancement of thermal and mechanical aging [58]. Similar results were found with a flower-like nanoparticle composition of a polyaklenylbenzene core and a polyconjugated diene surface layer, having nanoparticle of mean diameter less than 100 nm [59].

A nano organic–inorganic composite of styrene–butadiene–acrylate copolymer and silica showed superior tensile and tear properties to that of conventional products. Increased compatibility between the two was achieved with a silane compound and functional group as a coupling agent. Manipulating the ratio between the two in order to maximize the compatibility has had the ability to reduce rolling resistance while increasing wet traction capabilities at the same time [60].

The properties of tires have also been enhanced through the inclusion of a nanocomposite made up of part heterocyclic catioinic, part hydrophobic cationic mediator, which is used to exfoliate a clay of mica or synthetic mica. A loading under 15 wt% showed an improvement of 800% to tensile strength, as well as improved gas permeability and cure properties, making the composite a strong candidate for use in the inner liners, sidewalls, tread, and curing bladder of tires. Other components that use butyl rubber also see benefits, such as airbags and damping mounts for engines. Curing capability and time were some of the improvements made to the method of production [61, 62].

The use of carbon nanotubes in tire compositions has been explored, ranging from use in natural rubber, styrene–butadiene rubber (SBR), and polybutadiene rubber (BR). The carbon nanotubes, or graphitic nanofibers (GNF), replace the typical reinforcement particles of carbon black or silicone in order to improve the tensile stress, electrical properties, wear resistance, and elasticity. Increases of 35% and 37% for tensile stress and stress at 300% strain, respectively, were obtained with a composition of natural rubber, 5 %wt of CNT, 45 %wt of carbon black, 2 %wt of elemental sulfur, 4 %wt zinc oxide, and an accelerator of 0.4 %wt. This result was compared against a case of using carbon black only as an additive, and the CNTs used were multi-wall with an average diameter of 60 nm. The next experiment used similar ratios for the previous mentioned materials, but with 25 %wt silicone in place of carbon black and 5 %wt of CNT. This resulted in a 32% and 41% increase for tensile stress and stress at 300% strain, respectively, compared to using silicone with particle size of 32 nm solely. Thus, it is found that the inclusion of small amounts of CNTs in place of carbon black or silicone can produce superior tensile strength and resistance to deformation or destruction [63].

Nano-sized soot particles, carbon black, have been found to drastically improve the durability and fuel efficiency of tires. The gains have been attributed to coarser surfaces relative to conventional tires, thus promoting a higher surface energy and interactions with the rubber matrix resulting in lower inner friction and improved rolling properties. The smaller particles inherently reduce strain vibration, allowing for superior traction in wet road conditions. CNT-doped styrene–butadiene rubbers have shown increases in tensile strength, tear strength, and hardness of 600%, 250%, and 70%, respectively. Certain nanoparticles also provide specific results, where lamellar nano-organoclays such as montmorillonite produce higher stiffness and improved thermoplastic stability and silane-treated silica can prominently reduce rolling resistance. Other particles used for tire enhancement include nano-alumina, CNFs, and graphene, which can all lead to reduced stopping distances of 15–20% and increased fuel efficiency by 5% [1].

The inclusion of nanosilicon carbide in elastomers has produced tires with higher wet skid resistance and staggering reductions to abrasion by as high as 50% [2]. Further improvements to wet skid performance have been attained using polysulfidic organosilane triethoxysilylpropyl disfulfide (TESPD) with a special solution of polymerized statistical styrene–butadiene copolymers as a hydrocarbon polymer matrix. This silica technology also improves the fuel economy [64]. Studies with nanosprings in elastomers have also shown to enhance mechanical strength and significantly decrease hysteresis loss, reducing fuel consumption [65].

8.5.4 Thermal Properties

Pristine nanographene platelets (NGPs) have also been used in rubber compositions to enhance the thermal properties of tires due to having a thermal conductivity as high as 5300 W/m-K for a single-layer graphene. Thus, the thermal conductivity of neat rubber could witness a fivefold increase by adding only 0.01 %wt of single-layer graphene. The preferred NGP amount of 1–15 %wt raised the thermal conductivity by 4000 times, to a range of 0.5–500 W/m-K. For comparison, carbon black holds a thermal conductivity of roughly 1 W/m-K. The flexibility of the NGPs also allows for deformation under low strain and hardening at high deformation. These NGPs typically have lengths between 300 nm and 3000 nm, and thicknesses as small as 0.335 nm. Butyl rubbers can also benefit from the incorporation of NGPs, which can significantly enhance the rubbers' permeation resistance. The NGPs even showed higher efficacy in improving thermal conductivity of rubber compared to the previously studied graphite oxide nanoparticles. The thermal conductivity values from the NPG-enhanced rubbers were the greatest yet recorded (2011) for rubber or elastomeric materials [66].

Nanozeolite particles have drawn special interest due to their negative thermal expansion coefficients. This phenomenon reduces the overall thermal expansion of rubber compounds, reducing physical property degradation with increasing temperatures. The compound generated consists of a diene-based elastomer and 1–10

%wt of a nanozeolite with sizes less than 1000 nm. The nanozeolite exhibits a negative coefficient of thermal expansion over part of the temperature range, 0–200°C, and silica to alumina molar ratio of 300:1. This rubber compound can be integrated into multiple facets of tires, such as tread, sidewall, apex, chafer, sidewall insert, wire coat, or inner liner. This particular study utilized the composition for a run-flat apex, meant to support the sidewall of a run-flat in the event of a deflation [67].

8.5.5 Property Monitoring System

Nanosensors have been designed for embedment into elastomeric materials with the intention of measuring the physical properties and relaying the information via radio frequency. The sensors, ranging from nano to micro, are intended to be small enough so as not to obstruct the elastomer, while collecting and signaling data in the following fields: threshold temperatures, stress, strain, shear, vibration, pressure, tread wear, sidewall flex, and chemistry. The tires used for incorporation may be solid, pneumatic, or airless. Signaling would derive from changes in the sensor, such as activation, deactivation, a transmission frequency change, or a change in the time between transmissions. More in-depth designs convert the energy generated by the flexing sidewalls into electricity with piezoelectric technology to power the sensor. The data collected may also be stored as informational history and used for future reference to design modified tires [68].

8.6 Fabrics

Nanoapplications have also made their way into the interior of automobiles, namely for hygienic and cleanliness purposes for fabrics. The fabric's fibers are tightly knitted with the coatings, as many of the applications are coatings especially formulated for adhesion with fabrics. Thus, many of the studies discussed will, in fact, be nanocoatings designed for application to fabrics during production.

Self-cleaning fabrics utilize the nano-roughness of surfaces to protect against dirt, since the minimal distance between summits does not allow for adhesion of larger dirt particles. This creates a nonstick surface, which also facilitates low surface energy allowing water to easily roll off. Superhydrophobic surfaces include CNTs, silica, and fluoro-nanoparticles coated to nylon, cotton, or polyester fabrics [1].

Flame-retardant characteristics have also seen improvements from the implementation of nanoparticle fillers, including layered silicates and CNTs that render fabrics less ignitable and even self-extinguishable in post-flame conditions. Polyhedral oligomeric silsesquioxane (POSS) nanoparticles increase the ignition time of polypropylene yarns, cotton, and knitted polyester. The elevated cost has kept this technology from being administered to most daily fabrics [1].

Polymerized amidine with a CO_2-responsive monomer, created by atom transfer radical polymerization (ATRP), grafted onto a surface proves efficient at creating an easy-to-clean fabric. Carbon dioxide bubbling allows for easy water-based cleaning, and hydrophobic restoration can again be attained by treatment with air or nitrogen (N_2) [69].

8.7 Head-up Display

Video display systems in cars are becoming more prominent with increasing technological advancements reducing the costs for their incorporation. The next step has been enhancing the properties of these displays for better user experience. Improvements to be addressed include view angle, definition, power consumption, and temperature stability. The field emission display (FED) witnesses lower power consumption, higher brightness, and superior definition with the implementation of CNTs as electron emitters [70].

Monitoring and alerting sensor systems have been developed for the purpose of protecting against drowsy drivers. Nanoparticles are used, due to their minimal light scattering effect that maintains glass transparency, for a transparent windshield head-up display. The driver's condition is determined through a monitoring sensor that measures duration of eye closure against a threshold time or gaze location that may signify head drooping. Another sensor tracks the frequency of steering adjustment. Once the driver is deemed in a drowsy condition, luminescent graphics will appear on the windshield display in order to capture the attention of the driver. An ultraviolet light excites the substrate with light-emitting particles, sized 0.5–500 nm, to generate the display. The fluorescent material categories chosen include inorganic nano-sized phosphors, organic molecules and dyes, semiconductor-based nanoparticles, and organometallic molecules [71].

The same technology has been put forward to the development of a head-up navigation display system. This system would display navigational instructions, an indication to the appropriate destination building, a directional indicator portraying the appropriate heading, the distance to the destination, all while permitting vision through the windshield [72].

In addition, the same technology is used to develop a full rear window head-up display to assist with parking. The graphic includes a camera system, a pair of lines projected outward along the surface from the automobile in order to approximate the current heading, and a distance measurement, while still maintaining the windows transparency [73]. Another application of this technology is that of a traffic infrastructure indicator, which monitors street signs, including speed limit, school zone, and cautionary signs, and maintains their content on the display for a set period. Other indicators examined are traffic lights, railroad crossings, traffic message, and low visibility conditions where the screen displays a low visibility aid [74].

Protective shield coatings can be applied to electronic device windows, such as navigation displays. These electromagnetic interference (EMI) coatings typically consist of nanoparticles of EMI conductive or absorptive materials, such as steel, nickel, tin, silver, Monel, and gold. The sizes typically have a maximum of 100 nm, with a preference of below 20 nm, and come in various shapes. These nanoparticles are combined with an appropriate polymer and solvent, typically materials such as acrylics, epoxies, silicones, and polyurethanes. Plating the coating with an additional layer of nickel, silver, or copper can improve performance, but at the cost of transparency [75].

Recently, advanced Plexiglas film withholds polyvinyl butyral resin and phenanthrenequinone molecules as a photosensitive dopant. The design uses this photosensitive dopant as an interlayer holographic medium between two sheets of glass, allowing a transparent window with the ability to produce head-up holographic displays with the diffractive elements, or possibly capture light for solar cells [76]. The size of nanoparticles used in a polymer for display depends on the type of material. Titania, for example, has a high refractive index, so the particle diameters should not surpass 100 nm. While zinc oxide, clays, and magnesium silicate particles have a lower refractive index and can be of sizes up to 200 nm without overly scattering light [77]. One study used TiO_2 nanoparticles of 300–400 nm in size with poly(styrenecodivinylbenzene) methyl methacrylate [P(StcoDVB)MMA] as a shell and carbon black nanoparticles of 30–50 nm size with a shell of poly(methacrylic acid) [PMAA]. The negatively charged TiO_2 and cationically charged carbon black were dispersed in tetrachloroethylene uniformly to create an electrophoretic display. Tests showed these particles to move quickly and reversely in response to an alternating electric field [78].

The filler, on the other hand, serves the purpose of the medium for which the electrophoretic particles must traverse through in order to alter optical states. It also improves color intensity control over the display element and generates resistance in the movement of the particles, which enhances the stability, reduces particle settling, and allows threshold creation [79]. The optical states can be controlled through temperature measurement systems, which in turn control the driving pulse that determines the optical state of the glass. Another design measures the ambient light level in order to drive the optical state pulse. A final method makes the driving pulse dependent on the age of the electro-optic medium [80].

8.8 Alternative Energy

Nanotechnology has also made an impact on the alternative fuel regime, improving preexisting systems and granting the possibility of proposed theories. Such theories include the absorption of hydrogen into alloys for storage and transportation purposes.

8.8.1 Fuel Cells

In general, there are six types of fuel cells: alkaline, polymer-electrolyte membrane (PEFCs), direct methanol (DMFC), phosphoric acid, molten carbonate, and solid oxide. In regards to the storage of solid-phase H_2, carbon nanomaterials and metal organic framework (MOFs) have high enough surface areas and tunable chemical structures to grant high intake. Graphene's ultimate physisorption value allows the hydrogen uptake to be proportional to the specific surface area. However, recent findings show the greatest H_2 uptake at only 6.5%, not enough to produce an economical fuel cell quite yet. Current fuel cells use platinum (Pt)-based catalysts for both the oxidation of fuel and reduction of oxygen, where activity increases with decreasing particle size. The high cost of Pt catalysts has led to the search for cheaper alternatives, including Fe or Co with metal nanoparticles embedded in nitrogen-doped graphenic carbon. Nonmetal catalysts studied include N-doped carbon nanomaterials of SWNTs, graphene, mesoporous graphitic array, and carbon quantum dots [1].

Fuel cells have enjoyed improvements from the recent incorporation of nanostructured components, specifically electrodes. They observe higher precious metal utilization rate at the electrodes, increases in power density, while still achieving lower costs. Electrocatalysts are formed from oriented platinum arrays and platinum alloy nanotubes. This is especially useful for proton exchange membrane fuel cells (PEMFCs) as the electrocatalyst durability holds high importance. Support-less platinum nanotubes (PtNTs) and platinum alloy nanotubes (PtPdNTs) both show impressive durability and high catalytic activity, resulting in the eradication of degradation in most typical platinum catalysts. The NTs hold dimensions of 5–100 nm diameter, 0.1–100 μm length, and 1–10 nm wall thickness. They hold the ensuing advantages relative to traditional catalysts: no support requirements at high potentials, start–stop, or fuel starvation, vastly reduced mass transfer, greater protection against dissolution or agglomeration, greater mass activity, increased electrode utilization, greater electrochemical surface area, and improvements in the electrode charge and mass transportation. Platinum NTs may be generated with the help of silver, or selenium nanowires. The improved activity of PtNT catalysts is possibly due to the exposure of specific crystal components, and further reductions in wall thickness can produce increased mass activity. PTPdNTs show superior mass activity, 1.4 and 2.1 times greater than Pt/C and PtB, respectively, and a specific activity 5.8 and 2.7 times greater, respectively. Meanwhile, the loss of Pt electrochemical active surface area (ECSA) of PtNTs amounts to approximately 20% after 1000 cycles, in comparison to a loss of about 51% and 90% for PtB and Pt/C, respectively [81].

Nano-sized, non-stoichiometric transition metal oxides have also displayed superior electrocatalytical properties to that of traditional platinum catalysts in polymer-electrolyte membrane (PEM) fuel cells. The formula, Me_XO_Y, represents the metal oxide with Me consisting of molybdenum (Mo), vanadium (V), titanium (Ti), niobium (Nb), and/or tungsten (W), while the values X and Y ensure the oxides are Magneli-phases. The catalyst's particles range from diameters of 2–5 nm preferably and are supported by large-surface-area carbon particles to assure electrical conductivity across the entire electroactive layer. The catalyst surface area is treated with a perfluorosulfonate ionomer in order to increase proton transfer to the electrolyte membrane, while the carbon structure is treated with a hydrophobic polymer to protect against water clogging the pores of the gas diffusion layer [81].

The operation rate of platinum can be further improved by replacing conventional carbon powders in PEMFCs with dual-walled carbon nanotubes (DWNTs), which eliminates the issue of carbon particle isolation in the electrode layer. The assembly composes of an anode, cathode, and PEM and is often referred to as a membrane electrode assembly (MEA). With the use of hydrogen (H_2) as fuel, it becomes oxidized in the anode, while oxygen is reduced in the cathode, generating the transfer of protons and electrons from the anode to the cathode through the PEM. Water is produced on the cathode; hence why the hydrophobic layer of carbon particles is so pertinent to ensure reactant gases can reach the catalyst. The Pt Ru/DWNT anode catalyst produced a 63% improvement in the direct methanol fuel cells (DMFCs) best performance compared to carbon black. Specific and mass activity of platinum also saw further improvement from the use of DWNTs, presumably due to the high electrical conductivity and surface area, as well as small diameter. Orienting the CNT film may also pose benefits in the nature of improved electrical conductivity due to zero energy loss when electrons transfer along the tubes opposed to across, increased gas permeability, and enhanced mass transport from superhydrophobicity [82].

Hydrocarbon nanostructures, including silicon nanowires (SiNW), silicon nanodots (SiND), NTs, and/or nano-onions, can be used in a hydrogen storage device that also consists of a sonicator with the purpose of releasing the hydrogen gas. These nanostructures are embedded in a solvent of alkyl halide, aromatic hydrocarbon, chloroform, methylene chloride, methyl iodide, benzene, toluene, xylenes, and mixtures. Another design uses a non-reactive solvent such as castor oil, mineral oil, silicone oil, polyalhaolefin, low melt wax, ethylene glycol, water, and combinations. The benefit of this storage system for automobile applications is that it is lightweight, compact, operates at ambient temperature and atmospheric pressure, requires low energy input, superb stability, and most importantly improves safety and reduces cost for the materials [83].

8.8.2 Batteries/Supercapacitors

Batteries have seen improvements in response to the incorporation of nanomaterials such as nanowires due to their short ion diffusion distance, high conductivity, and impressive stress tolerance. One example uses a crystalline nanowire of diameter less than 70 nm with the formula $Na_XA_YMnB_ZO_NC_A$, where A denotes H, Li, Be, Mg, K, Ca, Rb, Sr, Cs, Ba, Fr, and Ra; B stands for V, Cr, Mo, Tc, Fe, and Ru; and C signifies N, P, S, F, or Cl. This storage device holds a capacity of a minimum of 80 mAh/g, a minimum energy density of 800 kWh/kg, and roughly less than 30% degradation after 100–2000 charge/discharge cycles. These wires displayed an impressive reversible capacity of 128 mAh/g, superior cycle ability of 82% capacity preservation for 500 cycles, and an alluring rate capability for a sodium ion battery. Further performance enhancements can be achieved through carbon coating, doping, and adjustments to particle size [84].

Battery performance and cost reduction are achieved from improved efficiency by enhancing catalysts in such ways as reducing electrical resistance. This can be done by adding carbon fiber, CNTs, silica, alumina, zirconia, titania, and polysilsesquioxane, all with the intention of strengthening and improving the adhesion of the electrical resistance layer [85]. Nanoparticle improvements in battery characteristics include energy storage, delivery rates, and stability. The use of nano-sized $FePO_4$ particles in place of standard low-electricity conductive materials increased the storage capacity and sped up energy delivery due to the higher surface area, permitting increased intercalation of Li ions. Sulfur has also been used to enhance the capacity of Li-ion batteries by an order of magnitude through its use as a cathode. The capacity fades in standard materials due to the loss of active materials from polysulfide ions dissolution in contact with electrolytes. However, nanostructuring the sulfur inside mesoporous carbon capacitances, such as CNF, graphene oxide, or pyrolzed PAN/graphene, immobilized the sulfur to mitigate these effects [1].

Electrodes of metal and CNT thin films also create high-powered supercapactitors. The porous structure increases the surface area and allows solvated ions to easily contact the electrode, thus increasing the storage capacity of the supercapacitor. Reducing the equivalent series resistance of a supercapacitor has been noted as the source to achieve high power density. CNTs have produced power densities of 8 kW/kg and 20 kW/kg for designs utilizing MWNTs and SWNTs, respectively. The layer of CNTs can be coated without the need for a binder, leaving the contact resistance between the active material and substrate much lower than typical designs [86].

Nanostructures in capacitors allow for enhanced charge collection due to their increased surface areas, while also reducing degradation from the multitude of stresses associated with high voltage and current situations. Improvements to be made for electrical storage devices include current density, voltage rating, power transfer properties, and frequency response [87]. MWNTs, SWNTs, metal oxide nanoparticles, and conducting polymers have all revealed positive effects on the energy density of supercapacitors; however, their poor stability and high cost have left them inefficient for storage devices. Graphene, on the other hand, has revolutionized the field with high energy density graphene-based electric double-layer capacitors (EDLCs) attaining improved capacity [1].

The massive push for nanotubes and composites in the battery industry is largely due to the material's vastly superior electrical conductivity. Three-dimensional (3D) composites especially hold superior qualities to that of standard two-dimensional materials, which typically have a conductivity equal to 0.075 e-6 S/cm, whereas the 3D nanocomposite has a conductivity of roughly 0.408 S/cm, which is an increase of over one million fold [88].

The separators of batteries are also crucial to their efficiency by allowing maximum usable power within the battery and ensuring minimal energy loss on recharging. Both can be attained by lowering the air resistance property of the separators. Table 8.4 shows the material characteristics of various separators composed of micro- and nanofibers, and Table 8.5 shows their electrical properties. The nanofibers of the microporous polymeric battery separator are composed of at least one polymer of polyacrylonitrile, cellulose, polypropylene, polyethylene, polybutylene, polyamide, polyvinyl alcohol, polyethylene terephthalate, polybutylene terephthalate, polysulfone, polyvinyl fluoride, polyvinylidene fluoride, polyvinylidene fluoride-hexafluoropropylene, polymethyl pentene, polyphenylene sulfide, polyacetyl, polyurethane, aromatic polyamide, semi-aromatic polyamide, polypropylene terephthalate, polymethyl methacrylate, polystyrene, and blends, mixtures and copolymers thereof. CE2 stands for Celgard 2325 separator, which is used as a comparative example of prior technology.

Table 8.4 Mechanical properties of separators composed of micro- and nanofibers

Separator #	MD Tensile (kg/cm^2)	CD Tensile (kg/cm^2)	Gurley Air Resistance (s/100 cc)	Mean Flow Pore Size (mm)
12	1128	1128	218	0.61
13	778	778	153	0.71
18	354	354	351	0.78
19	296	296	212	0.89
CE2	1700	150	620	0.028

Source: Ref. [89].

The machine direction (MD) and cross direction (CD) indicate the strength of the materials, which provide the advantage of allowing for a faster manufacturing process, tighter winding, and increased durability during use. The miniscule pore size mean indicates the capability to permit a large quantity of recharge cycles for a battery. Increasing the number of nanofibers, which have an average diameter of less than 700 nm, relative to microfibers reduces the overall pore size mean, thus enabling accurate targeting of specific measurements. As one can see from Table 8.5, the micro- and nanofiber separators have a reduced air resistance, which reduces the overall effect of heat on the battery in addition to the previously mentioned benefits [89].

Table 8.5 Electrical properties of separators composed of micro- and nanofibers

Separator #	Layers	Impedance (Ohm)	Ionic Resistivity (Ohm-cm)
18	1	3	733
19	1	2	466
12	1	6	1153
13	1	9	1497
CE2	1	2	1583

Source: Ref. [89].

8.8.3 Innovative Designs

An entirely new propulsion system has been designed with the help of nanotechnology, using nano-perforated plates of 70 nm as a medium across which to provide a temperature gradient while at atmospheric pressure. The force generated provides a linear thrust powerful enough to propel a vehicle. This would result in an electric propulsion system with zero moving parts, only mechanisms for maintaining the temperature gradient and for steering. These radiometric plates, whether aligned vertically or horizontally, could generate translational motion and lift, respectively. For example, accelerating a car of 1500 kg from 0 to 100 km/h in 10 s would take a plate of 1 m^2 surface area and a temperature gradient of 12 K. The radiometric membrane required for such an invention would have to conduct heat efficiently longitudinally, have poor electric conductivity, have a high Young's modulus for mechanical stresses, and be chemically and thermally stable. The suitable materials include silicon carbide, diamond-like carbon, aluminum nitride, silicon nitride, and tungsten. Nanocomposites and nanodot-based coolers can also be used as low-dimension thermoelectric membranes. This radiometric propulsion system uses Einstein's theory of radiometric forces combined with present developments in nanotechnology [90].

Nanogenerators are another innovative design with the intention of offering alternative, renewable systems for generating electricity to run automobile electronic systems. Until 2013, such a system has only been tested at a smaller scale in bicycles. Nanogenerators were integrated into the inner surface of tires in order to harvest energy from the compression of tires during rotation. An acceleration of 30 m/s^2 and travel distance of 12 mm for the generator created an output voltage near 1.5 V and a current of 25 nA [91].

8.9 Nanofluids/Lubricants

Nanoparticles have been responsible for a prominent improvement in the fluids industry, including lubricating oils and coolants. Providing lubricating oils of lower viscosity grade, while still maintaining wear protection, can enhance fuel economy. The efficiency in production of petroleum can be enhanced with the use of Zeolites for directed catalysis, as well as for their ability to select particular desired molecular products [92]. Reductions in energy consumption and improvements to environmental impacts can be made with the use of stabilizing Pt-group metal particles as catalysts. These particles enable increases in reaction speed, yield, and durability, while reducing loading levels. They also enable the planned design and control of catalytic behaviors, contrary to the current approach of trial and error [93]. The characteristics that influence wear and friction effects are particle shape, size, and concentration levels.

8.9.1 Lubricants

Gold particles of diameter 20 nm have returned the best lubricating results, while dialkyldithiophosphate-modified copper enhances anti-wear properties through the creation of an anti-friction film on metal surfaces. Diamond and inorganic fullerene (IF) particles also show anti-wear improvements through the generation of ball-bearing effects, created by the nanospherical particles moving between surfaces. They also conduct surface polishing and surface hardening. Meanwhile, CuO nanoparticles employ both the ball-bearing effects at high temperatures and high viscosity effects at low temperatures, both improving the anti-wear behavior of oils [1].

Nanopowders of metal, alloy, oxide, or nonmetal with diameters of 0.1–100 nm at concentrations between 0.1 and 10 %vol have resulted in the following effects. The same production method can yield a superior pressure load and wear-resistant nano-lubricating oil, as well as a cooling nanofluid with improved efficiency. The low viscosity of this fluid allows a decrease in the operating rate or volume of a circulation pump, resulting in lower energy consumption. Frictional heat is also reduced, leading to improved functionality of oil, in turn leading to reduced damage of gears. Specific materials and average particle sizes used for the nanopowder include a fullerene at 1 nm, silver at 7 nm, copper at 25 nm, graphite at 55 nm, and carbon black at 60 nm. Data showed that the graphite oil maintained the greatest load resistance, while fullerene and silver showed the least. In contrast, the silver and fullerene oils displayed superior wear resistance, while copper and graphite showed poor results. The load resistance appeared to increase with increasing particle concentration, while wear resistance demonstrated the reverse relationship. Particle size also maintains a relationship with resistances, as larger diameters display lower load resistances. The highest recorded reduction in power consumption was 7% due to the oil's high viscosity properties, which is anticipated to increase on larger scales [94].

The lubrication of zirconium nanopowder doped oils in engine operations can be achieved when the pistons move and allow the ceramic particles to bond to the metal surfaces. The heat generated during engine operation would be high enough to cure the ceramic powder, producing hard and smooth boundaries on the cylinder walls, piston rings, and tops, thus improving fuel efficiency, power output, oil life, and reducing engine noise, vibration, and pollution levels [1].

Other lubricants include nanoparticles of rod, sphere, or plate shape with sizes of less than 50 nm and aspect ratios for the rods ranging from 2 to 20. The infusion of approximately 0.1–5.0 %wt cerium oxide nanoparticle solutions into natural oil lubricants allows for the eradication of such pollutants as sulfur and phosphorus, which are typically in compounds used as anti-wear products. At the same time, similar or even enhanced friction performance is achieved with zero negative effects on the corrosive characteristics of oil, thus allowing for the application of the said lubricant to both spark-ignited and compression-ignited internal combustion engines (ICE). This product can improve the frictional coefficient, reduce the wear for low viscosity compositions, and reduce emissions of CO and hydrocarbons from engines using current lubricant compositions [95].

The efficiency of a diesel engine has been increased by the use of an improved lubricating oil that utilizes cerium dioxide nanoparticles, as well as by adding reverse-micellar additives to the fuel source. The cerium dioxide nanoparticles improve the fuel efficiency on two fronts: first within the fuel itself and second to the engine lubricant to reduce friction. A diesel engine operating with cerium dioxide nanoparticles of average size 6 nm in the fuel and oil can witness improved mileage, lower engine wear, and lower pollution levels of soot, sulfur, hydrocarbons, and NO_X. The proposed particles produce a 30% increase in lubricity [96].

Another experiment tests the anti-wear properties differentiating between nanopowders of silicon, silica, silicon nitride, graphite, diamond, copper, nickel, zinc, aluminum, or alumina. Mixed with oil, these films improved the sliding of gears, reduced the noise and vibration of bearings, reduced wear damage, improved the sliding of rings in ICE, and improved the seal. The improvement in the seal resulted in reductions to pressure losses, harmful gas and particulate emissions, and heat dispersion. The particle sizes are no greater than 15 nm and can be dispersed within oil at roughly 30% of the total volume [97].

NGPs have also been used to improve lubricant properties, dispersed at a proportion of 0.001–75 %wt. The average thickness of such platelets is less than 10 nm, with a preference of less than 1 nm. The improved properties include friction, wear, viscosity, electrical and thermal conductance, and thermal stability. The entire composition consists of the NGPs, a neat fluid, and a surfactant, wherein the neat fluid can be chosen from petroleum distillates, synthetic petroleum oils, greases, gels, oil-soluble polymer compositions, or vegetable oils. The surfactant serves the purpose of properly dispersing and maintaining the NGPs throughout the fluid, and may be chosen from a sulfonate surfactant, a sulfosuccinate, a sulfosuccinamate, dioctyl sulfosuccinate, bistridecyl sulfosuccinate, or dimethylbutyl sulfosuccinate with a weight of roughly 0.1–30%. The NGP may be chosen from natural graphite, synthetic graphite, highly oriented pyrolytic graphite, carbon or graphite fiber, carbon or graphitic nanofiber, or meso-carbon micro bead. NGPs with lengths of 1 μm or greater can be used to cover working parts with wide surface areas, while lengths of less than 500 nm suffice for re-circulating systems. NGPs simultaneously improve both the thermal conductivity and fluid viscosity of lubricants. Due to this phenomenon, typical thickeners such as silica gel and fatty acid soaps can be replaced by these nanoplatelets. NGPs have proven more effective at improving thermal conductance than both CNTs and graphite nanoparticles. At equal weight fractions, NGP-enhanced fluid displayed a thermal conductivity of 7.5 times greater than that of its graphite nanoparticle counterpart. NGP contents of 20 and 45 %wt displayed record level thermal conductivities for lubricants of 12.1 and 33.4 W/m-K, respectively. NGP grease levels, at 20.2 W/m-K, proved 14 times greater than opposing compositions of graphite nanoparticles at 30 %wt compositions. NGPs of mean thickness less than 1 nm, primarily single-layer graphene, provide vastly superior frictional reductions [98].

Nanodiamonds are networks of carbon atoms bonded in tetrahedral arrangements forming a nanoparticle. The benefit they present is the ability to divide metal surfaces, thus protecting against the welding together of nanoparticles during sliding. This reduces the overall friction of the system, which poses advantages for use in lubricants where the particle size is less than 100 nm but has a preferred size between 4 nm and 6 nm. The carbon nanoparticle content in lubricants is believed to be most desirous at about 40–105 ppm, but adequate friction reduction is seen anywhere between 10 ppm and 500 ppm. This composition proves fitting for engine or gear oil, transmission fluid, differential fluid, power steering fluid, and hydraulic fluids. It is also believed that water enhances the friction-reducing properties of nanodiamonds; however, it is typically thought to have negative impacts in oil applications [99].

Self-lubricating coatings have been developed with the help of nanotechnology for application to pistons and cylinders of an ICE and the mating surfaces of transmission gear assemblies, thus reducing friction and wear. Such coatings can include nanoparticles and a base material of chromium, nickel, cobalt, zinc, metal nitrides, metal carbides, or carbon-based materials. The nanoparticles come from solid industry lubricant materials such as zinc dialkyldithiophosphate (ZDDP), molybdenum disulfide (MoS_2), tungsten disulfide (WS_2), nibium dielininde ($NbSe_2$), gold, silver, lead, tin, or salts. For this particular application, the average size of nanoparticles is roughly 200 nm [100].

8.9.2 Coolants

More compact cooling systems for car radiators can be achieved with higher efficiency coolants loaded with nanoparticles, thus further reducing car weight and improving fuel efficiency. Using an ethylene glycol-based fluid with nanoparticles, such as CuO and Al_2O_3, in place of the conventional mix of water and ethylene glycol creates advantages such as lower-pressure operation and a higher boiling point, reducing the required frontal area of a radiator by 10%. This reduces aerodynamic drag and increases fuel savings by 5%. Reduction in friction to pumps and compressors can provide an additional 6% in fuel savings. The enhancement in thermal conductivity comes from the increases in volume fraction of the spherical particles, in surface area to volume fraction, and sphericity. A loading of 4 vol% CuO, with a 30 nm diameter, increased the thermal conductivity by 20% in ethylene glycol and a similar result was achieved for Al_2O_3 in water. MgO nanoparticles create the largest increases in thermal conductivity and lowest viscosity effects, while 1 vol% of CNTs improves conductivity up to 175% [1].

The nanoparticles studied in this experiment consisted of SWNT, DWNT, MWNT, nonmagnetically sensitive particles of carbon, graphite or graphene, and magnetically sensitive particles such as Fe, Co, Fe_2O_3, and Fe_3O_4. The anionic surfactant for the nanofluids is sodium dodecylbenzene sulfonate (NaDDBS), while the cationic consists of cetyl trimethylammonium bromide (CTAB). Exposing the magnetically sensitive fluids to magnetic fields resulted in almost perfect alignment, significantly improving thermal, mechanical, and electrical properties. For example, the thermal conductivity of the nanofluids saw an increase of 50–90% after exposure to a magnetic field [101].

Silicon carbide (SiC) exhibits one of the highest bulk thermal conductivities in ceramics at roughly 120 W/m-K. The beneficial characteristics of an SiC and water nanofluid include stability, superior thermal conductivity, while maintaining a relatively minimal viscosity increase. This relationship among viscosity, temperature, and percentage of SiC can be seen in Fig. 8.5. SiC particle water nanofluid exhibits a 28% increase in thermal conductivity above water alone for a concentration of 7.4 vol%. The advantages of SiC nanofluids can be applied to both heating and cooling applications. SiC/water nanofluids witness decreases in improvement with further particle concentration, but SiC/water-ethylene glycol nanofluids maintain almost constant heat transfer coefficients with increased loading. Potential applications include heat exchangers and radiators for engines, where the enhanced efficiency in cooling leads to higher productivity and reduced energy consumption [102].

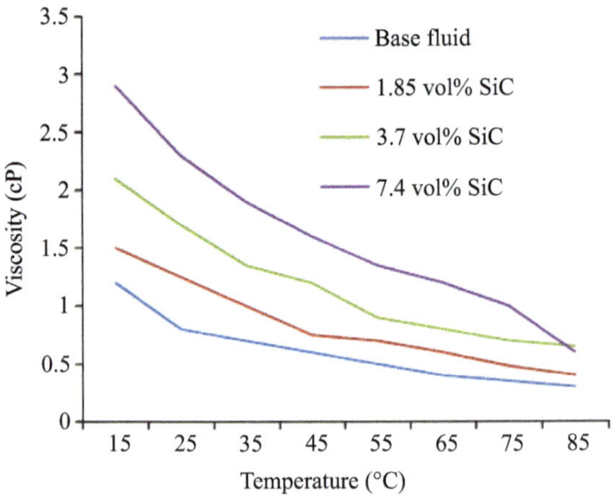

Figure 8.5 Relationship between viscosity and temperature for multiple water-based fluids enhanced with silicon carbide nanoparticles [102].

Heat transfer efficiency has been improved using greater than 15 wt% of titanium, vanadium, chromium, cobalt, nickel, iridium, zirconium, niobium, molybdenum, rhodium, palladium, tantalum, tungsten, platinum, silver, and gold. The remainder of the enhancement powder comprises a transition metal oxide such as titanium oxide (TiO_2), copper oxide, iron oxide, and molybdenum oxide (MoO_2). This additive resulted in reductions in fuel consumption in excess of 10%, as well as dropped the idle rpm for any vehicle [103].

8.9.3 Multipurpose

Various nanofluids can also be used in new high efficiency systems for automobile fuel consumption. Such a system utilizes the concept of preheating the fuel before combustion in order to increase fuel efficiency, which can be administered to natural gas, diesel, liquefied petroleum gas, liquefied natural gas, compressed natural gas, cold compressed natural gas, etc. The system can be mounted to preexisting fuel systems without ruining the structure. Higher molecular weight gases saw a greater percentage of savings. Increasing the initial fuel temperature from 30°C to 200°C generated 40% fuel savings and considerably increased the rate of energy production. The savings ranged anywhere from 1 to 90%, depending on the type of fuel, type of combustion chamber, feasibility of attaching a fuel burning system, available heat supply, as well as other factors [104].

A multipurpose nanofluid was generated with nanodiamonds, granting it characteristics suitable for use as a coating, lubricating oil additive, polymer plastic, a sealing agent, paint additive, textile, or any heat-dissipating products. The particle sizes range from 1 to 100 nm, based on application, and the nanodiamond aggregation was dispersed in an aqueous metal hydroxide solution of K, Ca, Na, Mg, Al, Zn, Fe, Ni, Sn, or Pb. The fine grain and high stability of the fluid grant it reliability for sealing pores for an oxide film, thus enhancing corrosion resistance, anti-contamination, stability of dye color, and weather or UV protection. Its application as a polymer additive can help increase the resin's elasticity, high and low temperature resistance, chemical resistance, and reduce the coefficient of friction. Its characteristics as a heat-dissipating additive grant it application for coating car radiators, or LEDs. The frictional moment at contact surfaces can be reduced by roughly 20–40% with the use of nanodiamonds in lubricating oils, resulting in lower temperatures and reduced wear of mechanical parts [105].

Alterations to fuel compositions using metal nanoparticles have also been explored for increased combustion efficiency. The particles make up anywhere between 0.01 ppm and 500 ppm in a liquid fuel, have spherical shape, and a size ranging from 1 to 990 nm. The metal or metal oxide particles consist of any of the following: magnesium, calcium, strontium, barium, cerium, titanium, zirconium, iron, ruthenium, osmium, cobalt, rhodium, iridium, nickel, palladium, platinum, copper, silver, gold, tin, zinc, aluminum, as well as mixed metal particles, mixed metal oxide particles, and mixed metal-metal oxide particles of the previously mentioned metals. The fuel compositions tested were gasoline, reformulated gas, oxygenated gas, diesel, biodiesel, jet and marine fuel, alcohol, bioalcohol, and kerosene. These nanoparticle fuel compositions can enhance combustion, reduce harmful emissions, and improve catalytic chemical oxidation of fuel in order to increase engine power. The nanoparticles also promote various other benefits on the internal combustion engine, such as decrease in the sensitivity between research octane number (RON) and motor octane number (MON) levels for ethanol, which leads to better fuel performance. The particles were also coated onto metal components to generate lubricity and mitigate carbon deposition, both reducing engine maintenance. The increased surface area of the nanoparticles also spawns a more extensive combustion of the fuel. The impact on emissions is also beneficial, as the fuel is designed to decrease the amount of hydrocarbon, carbon monoxide, nitrogen oxide, and molecular oxygen in automobile exhaust, while also increasing carbon dioxide emissions to a desirable degree. For example, a decrease of roughly 53% in NO_X emissions was attained in diesel fuel doped with zinc oxide particles of size 1–20 nm [106].

8.10 Discussion and Conclusion

This current review of the contribution of nanotechnology to the automotive industry has uncovered many recent patents. The industry has committed to investments and thus innovations in nanocoatings, better fabrics and improved materials, nanofluids and lubricants, and tires. There seems to be less activity in the categories of smart windows and video display systems, most likely due to the expense of smart glass and "intelligent" electronic systems. This would undoubtedly increase the higher risk-to-reward ratio, thus rendering these two categories less attractive investments. It is clear that financing and investments made by the industry have fueled the research and development of nanotechnology in automotives.

As a special subclass of improved substances, materials for alternative energy used in automobiles have been stimulated by advances in nanotechnology and continue to be an active research area. Alternative energy materials, improved or augmented by nanotechnology, are not an exclusive domain of the automotive industry. Hence, they will be affected by nanoscience research and nanotechnology developments of other industrial sectors, such as energy and utility organizations, as well as the government.

8.11 Current and Future Developments

The key focus of nanotechnology research should be developing the science into a commercially and economically viable position. Only then will its current and potential benefits be realized for numerous industries, such as the automotive industry. The wide range of applications for nanotechnology in automobiles creates a unique situation for this industry, where small leaps in several different fields can accumulate to a major leap for a single product.

An area that has major potential for important applications to modern automobiles is that of electronics and optics. The developments in heads-up display systems, or windshield displays, counteract the distractions created by innovations such as the mobile phone and current built-in display systems by allowing drivers to keep their eyes on the road while navigating their electronic system. A monitoring system is the second half of this technology, which can ensure the driver's alertness while operating the automobile. The implementation of smart glass can greatly reduce the number of automobile accidents in the future. However, with the current pace of autonomous vehicle development, these may become unnecessary features.

The second major field in nanotechnology development is that of materials, for both strength and weight reduction purposes. Increasing efficiency in an automobile can be as simple as reducing weight, which when done by incorporating nanocomponents in place of standard materials, can also mean increasing strength. This not only means a more fuel-efficient vehicle, regardless of human or computer operated, but a safer one as well.

This leads into the final, and of late the most popular, important field in the automotive industry—fuel efficiency. Recent developments have proven that nanomaterials hold characteristics far superior to that of current commercially used materials, not only in regards to battery technology but fuel cells as well. This has made the use of alternative energy sources for personal transportation a viable enterprise. Nanocoatings, lubricants, and materials have also offered advancements to the efficiency of the internal combustion engine, which can have a more immediate impact commercially.

Future developments coupled with current technology will provide exponential growth to the automotive industry. However, before any major impact can be seen, research and development must go toward increasing the efficiency of commercial nanomaterial production and implementation.

References

1. Mohseni, M., Ramezanzadeh, B., Yari, H., and Gudarzi, M. M., "The role of nanotechnology in automotive industries," in *New Advances in Vehicular Technology and Automotive Engineering*, Carmo, J. P. and Ribeiro, J. E., eds., InTech, 2012, DOI: 10.5772/49939.
2. Wang, M.-J., Reznek, S. R., Kutsovsky, Y., and Mahmud, K., "Elastomeric compounds with improved wet skid resistance and methods to improve wet skid resistance," US Patent 6469089, 2002, http://www.patentstorm.us/patents/6469089.html, accessed on 30 January 2013.
3. Vidotti, M. and Cordoba de Torresi, S. I., "Nanochromics: Old materials, new structures and architectures for high performance devices," *J. Braz. Chem. Soc.*, **19**(7), pp. 1248–1257, 2008.
4. Vlachopoulos, N., Nissfolk, J., Moller, M., Briancon, A., Corr, D., Grave, C., Leyland, N., Mesmer, R., Pichot, F., Ryan, M., Boschloo, G., and Hagfeldt, A., "Electrochemical aspects of display technology based on nanostructured titanium dioxide with attached viologen chromophores," *J. Int. Soc. Electrochem.*, **53**(11), pp. 4065–4071, 2008.
5. Hussain, F., Hojjati, M., Okamoto, M., and Gorga, R. E., "Review articles: Polymer-matrix nanocomposites, processing, manufacturing, and application: An overview," *J. Composite Mater.*, **40**(17), pp. 1511–1575, 2006.

6. Stewart, R., "Nanocomposites: Microscopic reinforcements boost polymer performance," *Plastics Engineering*, **60**(5), 2004.
7. Lin, J., "Coated substrate having enhanced scratch and mar resistance," E.I. Du Pont De Nemours and Company, EP 2027224 A2, 2009.
8. Cayton, H. R., Brundkemeyer, M., Lenz, P., Muray, P., Sawitowski, T., and Schulte, K., "Enhanced scratch resistance of articles containing a combination of nano-crystalline metal oxide particles, polymeric dispersing agents, and surface active materials," Nanophase Technologies Corporation, EP 1773511 A2, 2007.
9. Niguma, T., Nakabayashi, A., and Takanohashi, H., "Aqueous organic-inorganic hybrid composition," Asahi Kasei Chemicals Corporation, Tokyo, JP, Patent No.: US 8,013,055, 2011.
10. Zhang, H., Tang, L., Zhang, Z., Gu, L., Xu, Y., and Eger, C., "Wear-resistant and transparent acrylate-based coating with highly filled nanosilica particles," *Tribol. Int.*, **43**(1–2), pp. 83–91, 2010.
11. Ahmadi, B., Kassiriha, M., Khodabakhshi, K., and Mafi, E. R., "Effect of nano layered silicates on automotive polyurethane refinish clear coat," *Prog. Organic Coatings*, **60**(2), pp. 99–104, 2007.
12. Edwards, J. and Lowry, K., "Stable nano titania sols and a process for their production," Tioxide Europe Limited, EP 2477947 A1, 2012.
13. Choe, S.-G., Hwang, H., Lee, K.-G., and Shin, C.-H., "Transparent color coating composition containing nanosize dispersed pigments, coated substrate and method for preparing same," Nepes Co., Ltd., EP 2383316 A2, 2011.
14. Arfsten, J. N., Delcourt, L. A., Langermans, H. A., Thies, J. C., and Vrijaldenhoven, P. W. A., "Coating system," DSMIP Assets B.V., EP 1984765 A1, 2008.
15. Thies, C. J., "Anti-reflective coated glas plate," DSMIP Assets B.V., EP 1984762 A1, 2008.
16. Carter, N. M., Mayhorn, J., and Zucker, J., "Nonwoven industrial fabrics with improved barrier properties," Polymer Group, Inc., EP 1549790 A4, 2007.
17. Moody, R. L. and Rick, A., "Three-dimensional nonwoven fabric with improved loft and resiliency," Polymer Group, Inc., EP 1684972 B1, 2011.
18. Gardner, S. J., Qi, Y., Hum N.-X., Zhang, Q., and Song, G., "Superhydrophobic nano-fabrics and coatings," Xerox Corporation, EP 2210921 A1, 2010.

19. Lyons, A. M. and Xu, Q., "Polymer having superhydrophobic surface," Research Foundation of the City University of New York, US20130251948 A1, 2013.
20. Klijn, A. T., Twene, D., and Mestach, D. E. P., "Stain blocking water borne coating composition," Nuplex Resins B.V., EP 1713870 A1, 2006.
21. Mitina, V. K. and Peter, W., "Modified silica particles and dirt repellent polymer compositions comprising them," Basf Se, EP 2408864 A1, 2012.
22. Whitten, D., Ista, L., Corbitt, T., Ogawa, M., Schanze, K., Lopez, G., and Nagarajan, R., "Materials incorporating antimicrobial polymers," Stc. Unm, EP 2486184 A2, 2012.
23. Stuart, A. M. C., De, K. A., Fokkink, R. G., and Van, D. B. S., "Complex coacervate core micelles as surface modification or surface treatment," Rhodia Chimie, EP 1667744 A1, 2006.
24. Carr, C. and Emden, O. V., "Breathable fabric," Lightex Limited, EP 1778035 A2, 2007.
25. Siddiqui, N. A., Sham, M.-L., Tang, B. Z., Munir, A., and Kim, J.-K., "Tensile strength of glass fibres with carbon nanotube-epoxy nanocomposite coating," *Compos. Part A Appl. Sci. Manuf.*, **40**(10), pp. 1606–1614, 2009.
26. Abys, A. J., Fan, C., Kudrak Jr., E. J., Li, J., and Xu, C., "Electrolytic deposition of metal-based composite coatings comprising nano-particles," Enthone, Inc., EP 2242873 A1, 2010.
27. Bruce, R. W., Gray, D. M., Krishnamurthy, A., Srinidhi, S., and Srinivasan, D., "Wear resistant coatings," General Electric Company, EP 1927670 A1, 2008.
28. Abrami, S., Tang, G., and Tipon, P., "Coating compositions including magnesium hydroxide and related coated substrates," Prc-Desoto International, Inc., US 20120315466 A1, 2012.
29. Akid, R. and Wang, H., "Anticorrosion sol-gel coating for metal substrate," Sheffield Hallam University, EP 2526219 A1, 2012.
30. Drummond, J. P. and Lynam, N. R., "Vehicular rearview mirror system," Donnelly Corporation, US 8362885 B2, 2013.
31. Kerber, S. J., "Nanoparticle surface treatment," Material Interface, Inc., US 8568538 B2, 2013.
32. Skryabin, L. I. and Tulloch, E. G., "Electrophotochromic smart windows," Dyesol Ltd., EP 1012661 B1, 2005.
33. Choi, S.-J., Hwang, Y., Kim, S.-H., Kwon, W.-J., Lee, S.-B., and Park, S.-A., "Electropolymerization method for preparing nanotube type

conduction polymer using porous template and method for preparing electrochromic device," Lg Chem, Ltd., EP 1495360 B1, 2008.

34. Weidner, E. S., "Photovoltaic window assembly with solar control properties," Pilkington Group Ltd., EP 2553723 A2, 2013.

35. Yang, W., Li, S., and Gutowski, W.S., "Multi-functional nano-materials for advanced automotive applications," *Chem. Listy*, **101**, pp. s1–s72, 2007.

36. Lee, Y. H., So, K. P., Kim, E. S., and Park, Y. W., "Method of hardening interface of carbon material using nano silicon carbide coating," Research & Business Foundation of Sungkyunkwan University and Dayou Smart Aluminum Co., Ltd., EP 2492241 A1, 2012.

37. Shen, Z., Zhigang, S., Soh, W. K., Qiang, S. W., Jiyao, Z., Wang, A., Zhong, J., Yun, S. L. J., Li, Z., Sher, H. S., Cost, X., Chen, J., and Jianfend, C., "Nanometric tin-containing metal oxide particle and dispersion, and preparation method and application thereof," Xiamen Nanotech Co Ltd, Xiamen Tektronix Nano Technology Co., Nanomaterials Technology Pte Ltd., Singapore Nano Materials Technology Co., Ltd., WO 2013143456 A1, 2013.

38. Mamalis, A. G., Vogtlander, L. O. G., and Markopoulos, A., "Nanotechnology and nanostructured materials: Trends in carbon nanotubes," *Precision Eng.*, **28**, pp. 16–30, 2004.

39. Adams, H., Zoz, H., and Dvorak, M., "An engine or engine part and a method of manufacturing the same," Bayer International Sa, EP 2396442 B1, 2012.

40. Katsumura, M., Miura, H., Miyao, N., Ogawa, H., Oda, K., and Mizutani, M., "Nano-crystal austenitic steel bulk material having ultra-hardness and toughness and excellent corrosion resistance, and method for production therof," Nano Technology Institute, Inc., EP 1555332 A1, 2005.

41. Okitsu, Y., "High-strength steel sheet, automotive strengthening member comprising the same, and process for producing automotive strengthening member," Honda Motor Co., Ltd., EP 2224029 B1, 2012.

42. Jou, H.-J., Kuehmann, C. J., Olson, G. B., and Tufts, B. E., "Nano-precipitation strengthened ultra-high strength corrosion resistant structural steels," Questek Innovations LLC, EP 1644540 A2, 2006.

43. Cardoso, P. J. M., "A study of the electrical properties of carbon nanofiber polymer composites," *Universidade do Minho*, 2012.

44. Lee, Y. H., So, K. P., Kim, E. S., and Park, Y. W., "Method of hardening an interface of carbon material using nano silicon carbide coating," Lee Young Hee, US 20120210823 A1, 2012.

45. Hun, H. U., "Polyamide nanocomposite resin composition for automobile engine cover and interior or exterior material with long-term heat resistance," Hyundai Motor Co Ltd., KR 10-2004-0023904 A, 2004.

46. Phani, A. R., Gammel, F. J., and Hack, T., "Structural, mechanical, and corrosion resistance properties of Al_2O_3-CeO_2 nanocomposites in silica matrix on Mg alloys by a sol-gel dip coating technique," *Surf. Coat. Technol.*, **201**(6), pp. 3299–3306, 2006.

47. Borisova, D., Mohwald, H., and Shchukin, D. G., "Influence of embedded nanocontainers on the efficiency of active anticorrosive coatings for aluminum alloys Part 1: Influence of nanocontainer concentration," *ACS Appl. Mater. Interfaces*, **4**(6), pp. 2931–2939, 2012.

48. Feng, Z., Qu, J., Santella, M. L., Pan, T. Y., Roche, A. D., and Yu, S. T., "Method of making nanoparticle reinforced metal matrix components," Feng, Z., Qu, J., Santella, M. L., Pan, T. Y., Roche, A. D., Yu, S. T., US 20120315399 A1, 2012.

49. Oriakhi, C. O., "Commercialization of nanotechnologies," Massachusetts Institute of Technology, 2004, http://dspace.mit.edu/handle/1721.1/17803, accessed on 30 January 2013.

50. Jones, S. and Fearing, R., "Apparatus for friction enhancement of curved surfaces," The Regents of the University of California, US 20070289696 A1, 2007.

51. Li, J., Ma, T.-J., Shen, J.-R., Ye, Y.-P., Yi, Y.-H., and Zhang, H., "Method for manufacturing a composite tire of polyurethane tread and radial carcase," Hangzhou Hanma Tyre Technology Ltd., EP 1529788 B1, 2007.

52. Yener, O. D., Guiselin, O., and Bauer, R., "Applications of shaped nano alumina hydrate as barrier property enhancer in polymers," Saint-Gobain Ceramics & Plastics, Inc., EP 2209842 A1, 2010.

53. Mercuri, R. A., "Production of exfoliated graphite," Mercuri Robert Angelo, US 20130022530 A1, 2013.

54. Adamson, H. D., Aksay, I. A., O'Neil, C. D., Ozbas, B., Prud'Homme, R. K., and Register, R. A., "Functional graphene-polymer nanocomposites for gas barrier applications," The Trustees of Princeton University, EP 2067146 A2, 2009.

55. Song, H.-S., "Rubber composition for tires," Kumho Tire Co., Inc., EP 1972466 A1, 2008.

56. Bohm, G. G. and Wang, X., "Nano-sized inorganic metal particles, preparation therof, and application therof in improving rubber properties," Bridgestone Corporation, US 7560510 B2, 2009.

57. Foltz, V. J. and Wang, X., "Multi-layer nano-particle preparation and applications," Bridgestone Americas Holding, Inc., US 7238751 B2, 2007.

58. Hall, J. E., Ozawa, Y., and Wang, X., "Polymeric nano-particles of flower-like structure and applications," Bridgestone Corporation, US 7205370 B2, 2007.

59. Choi, H. K., Kwag, G., Kim, H., Lee, S., and Park, H., "Organic-inorganic nano composites and preparation method therof," Korea Kumho Petrochemical Co., Ltd., US 8242204 B2, 2012.

60. Foltz, V. J., Fudemoto, H., and Wang, X., "Nano-composite and compositions manufactured therof," Bridgestone Corporation, US 7572855 B2, 2009.

61. Foltz, V. J., Fudemoto, H., and Wang, X., "Nano-composite and method therof," Bridgestone Corporation, US 7601772 B2, 2009.

62. Kim, Y.-N., "Rubber composition comprising carbon nanotubes as reinforcing agent and preparation thereof," Kh Chemicals Co., Ltd., EP 1465945 A4, 2006.

63. Heinrich, G., Huinink, H., and Kaliske, M., "Materials science and engineering in tyre development for optimising tyre performance," *Chem. Listy*, **101**, pp. s1–s72, 2007.

64. Liu, J., Lu, Y.-L., Tian, M., Li, F., Shen, J., Gao, Y., and Zhang, L. "The interesting influence of nanosprings on the viscoelasticity of elastomeric polymer materials: Simulation and experiment," *Adv. Funct. Mater.*, **23**(9), pp. 1156–1163, 2012.

65. Zhamu, A. and Jang, B. Z., "Pristine nano graphene-modified tires," Nanotek Instruments, Inc., US 7999027, 2011.

66. Agostini, G., Corvasce, F. G., and Lechtenboehmer, A., "A rubber composition including nano-zeolite and a tire with a component having such a composition," The Goodyear Tire & Rubber Company, EP 1772484 B1, 2009.

67. Merat, F., Wilson, P., and Rensel, J., "Elastomeric article with wireless micro and nano sensor system," Bridgestone Americas Tire Operations, LLC, EP 2275286 A1, 2011.

68. Li, M., "Synthesis and swithcability study of amidine-containing vinyl monomers and their polymers," *Open Access Dissertations and Theses*, Paper 6685, 2012.

69. Chung, D.-S., Lee, H.-W., and Park, S.-H., "Method of fabricating field emission display employing carbon nanotubes," Samsung SDI Co., Ltd., EP 1283541 A2, 2003.

70. Szczerba, J. F., Cui, D., and Seder, T. A., "Driver drowsy alert on full-windshield head-up display," GM Global Technology Operations LLC, US 8344894 B2, 2013.
71. Seder, T. A., Szczerba, J. F., and Cui, D., "Point of interest location marking on full windshield head-up display," GM Global Technology Operations LLC, US 8358224 B2, 2013.
72. Szczerba, J. F., Seder, T. A., and Cui, D., "Rear parking assist of full rear-window head-up display," GM Global Technology Operations LLC, US 8350724 B2, 2013.
73. Seder, T. A., Szczerba, J. F., and Cui, D., "Traffic infrastructure indicator on head-up display," GM Global Technology Operations LLC, US 8395529 B2, 2013.
74. Finley, M., Mcbride, R., and Severance, C. L., "Nano inks for imparting EMI shielding to windows," Parker-Hannifin Corporation, EP 2196078 A1, 2010.
75. Yan, Y. and Chen, Z., "Platinum and platinum based alloy nanotubes as electrocatalysts for fuel cells," The Regents of the University of California, EP 2008322 A2, 2008.
76. Matusevich, V., Tolstik, E., Kowarschik, R., Egorova, E., Matusevich, Y., and Krul, L., "New holographic polymeric composition based on plexiglas, polyvinyl butyral, and phenanthrenquinone," *Opt. Commun.*, **295**, pp. 79–83, 2013.
77. Jacobs, J. M., Pullen, A. E., Whitesides, T. H., Drzaic, P. S., Morrison, I. D., Wang, J., and Gray, C. L., "Electrophoretic displays using nanoparticles," E Ink Corporation, US 6721083 B2, 2004.
78. Yao, X.-N., Wang, J.-P., Zhang, X.-X., and Wang, X.-C., "Preparation and characterization of high contrast electrophoretic display suspension containing charged black and white nanoparticles," *SPIE Proceedings, Second International Conference on Smart Materials and Nanotechnology in Engineering*, 749352, 2009.
79. Danner, G. M., Amundson, K. R., Herb, C., and Zhang, L., "Electrophoretic display element with filler particles," E Ink Corporation, US 8115729, 2012.
80. Abramson, J. J., Amundson, K. R., Cronin, J. E., Danner, G. M., Duthaler, G. M., Gates, H. G., Honeyman, C. H., Knaian, A. N., Morrison, I. D., O'Neil, S. J., Paolini Jr., R. J., Pullen, A. E., Wang, J., Zalesky, J. L., and Zehner, R. W., "Electro-optic displays, and methods for driving same," E Ink Corporation, US7787169 B2, 2010.
81. Schoonman, J., Ten, H. S., and Van, E. K. R., "Nanosized catalysts for the anode of a PEM fuel cell," Stichting PowerPlus, EP 1808920 A1, 2007.

82. Chen, Z., Yan, Y., Wang, X., Li, W., Waje, M., Goddard, W., and Deng, W.-Q., "Carbon based electrocatalysts for fuel cells," The Regents of the University of California, US 8247136 B2, 2012.

83. Teo, B. K., "Hydrogen storage using hydrocarbon nanostructures and sonication," The Board of Trustees of the University of Illinois, US 2012010046 A1, 2012.

84. Cao, Y., Choi, D., Liu, J., Nie, Z., Wang, W., Xiao, L., and Yang, Z., "Nanomaterials for sodium-ion batteries," Battelle Memorial Institute, US 20120295147 A1, 2012.

85. Adachi, S., Izuhara, D., Kidai, M., Nakamura, M., Shimoyama, N., and Uete, T., "Film electrode composite element and production method therefor, and fuel cell," Toray Industries, Inc., EP 1858096 B1, 2011.

86. Du, C. and Pan, N., "High power density supercapacitors with carbon nanotube electrodes," The Regents of the University of California, EP 1825619 A2, 2007.

87. Duff, Jr. W. B., "Electrical charge storage device having enhanced power characteristics," Duff, William B., Jr., EP 1616360, 2006.

88. Ajayan, P., Ashkari, D., Cao, A., Nejhad, M. N. G., and Veedu, V. P., "Three-dimensionally reinforced multifunctional nanocomposites," Rensselaer Polytechnic Institute, US8148276 B2, 2012.

89. Morin, B. G. and Schaeffer, J. L., "Single-layer lithium ion battery separator," James L. Schaeffer, Brian G. Morin, US 20130078525 A1, 2013.

90. Kremen, S. H., Scandurra, M. Dr., "Radiometric propulsion system," Kremen, Stanley H., Scandurra, Marco Dr., EP1728093 A2, 2006.

91. Kumar, B. and Kim, S.-W., "Energy harvesting based on semiconducting piezoelectric ZnO nanostructures," *Nano Energy*, **1**, pp. 342–355, 2012.

92. Ratner, M. and Ratner, D., *Nanotechnology: A Gentle Introduction to the Next Big Idea*, Prentice Hall, p. 208, 2002.

93. Lloyd, S. M., Lave, L. B., and Matthews, H. S., "Life cycle benefits of using nanotechnology to stabilize platinum-group metal particles in automotive catalysts," *Environ. Sci. Technol.*, **39**, pp. 1384–1392, 2005.

94. Choi, C., Oh, J. M., and Jung, M. H., "Preparation method of lubricating oil and lubricating oil produced thereby," Korea Electric Power Corporation, US 8349774 B2, 2013.

95. McLaughlin, M. J. and Mathur, N., "Nanoparticle additives and lubricant formulations containing the nanoparticle additives," Afton Chemical Corporation, US 8333945 B2, 2012.

96. Reed, J. K., "Method of conditioning an internal combustion engine," Cerion Technology, Inc., EP 2074201 A1, 2009.
97. Coccollone, F., "Antifriction, antiwear compound," Nanotek S.R.L., EP 2250243 A1, 2010.
98. Jang, B. Z. and Zhamu, A., "Nano graphene-modified lubricant," Nanotek Instruments, Inc., US 8222190 B2, 2012.
99. Factor, A., Markovitz, M., and Verdino, G., "Lubricant with nanodiamonds and method of making the same," Nanotech Lubricants, LLC, US 20100029518 A1, 2010.
100. Adhvaryu, A., Davies, L. V., and McGilvray, A. N., "Self-lubricating coatings," Caterpillar Inc., US 7998572 B2, 2011.
101. Hong, H. and Peterson, G. P. B., "Alignment of carbon nanotubes comprising magnetically sensitive metal oxides in nanofluids," Hong Haiping, Peterson G P Bud, US 20120235080 A1, 2012.
102. France, D. M., Heifetz, A., Routbort, J., Singh, D., Smith, D. S., Timofeeva, E., and Yu, W., "Heat transfer fluids containing nanoparticles," U. Chicago Argonne, LLC, US 20110001081 A1, 2011.
103. Huange, Y.-H. and Chen, H.-W., "Heat transfer enhancing agent," Yen-Hao Huang, Hung-Wei Chen, US20130119302 A1, 2013.
104. Canacik, A. and Donmez, A., "Fuel preheating system," Yenbu Makine Sanayi Ve Ticaret A.S., US 8360770 B2, 2013.
105. Park, T. H., Lee, K. T., and Lee, J. S., "Nano-diamond dispersion solution and method for preparing same," Neoenbiz, US 20120304545 A1, 2012.
106. Sanders, J. K., Tock, R. W., and Yang, D. J., "Nano-sized metal and metal oxide particles for more complete fuel combustion," Sanders, J. K., Tock, R. W., and Yang, D. J., EP 2164932 A1, 2010.

Chapter 9

Carbon Nanotubes Used for Renewable Energy Applications and Environmental Protection/Remediation

Kaufui V. Wong and Benoit Bachelier
Department of Mechanical and Aerospace Engineering, University of Miami, Coral Gables, FL 33124, USA
kwong@miami.edu

Carbon nanotubes (CNTs) are surprisingly ubiquitous in their use for renewable energy applications as well as for environmental protection and remediation. This is a result of their large surface areas and their unique mechanical, electrical, and chemical properties, and, in no small part, due to its relatively easy manufacturability. CNTs are activated by certain chemicals. They are used in energy storage and solar cells. They may be used to react with solids, liquids, and gases. Hence, they are employed for waste water treatment as well as liquid and gaseous cleanup. They may be used to remove metals and life pathogens.

This chapter has been reprinted from the *Proceedings of IMECE2013*, 2013 ASME International Mechanical Engineering Congress and Exposition, Kaufui V. Wong and Benoit Bachelier, Carbon nanotubes used for renewable energy applications and environmental protection/remediation: A review, Copyright 2013, with permission from ASME.

Nanotechnology and Energy
Edited by Kaufui V. Wong

ISBN 978-981-4774-24-6 (Hardcover), 978-1-315-16357-4 (eBook)
www.panstanford.com

9.1 Introduction

Carbon materials are a class of adsorbents widely used in engineering. CNTs have high surface areas and good electrical, chemical, mechanical, and conducting properties. Their hollow and layered nanosize allows them to be good absorbers. They also are hydrophobic and prone to aggregation due to the strong van der Waals interactions. CNTs can be used to eliminate organic pollutants from environment. Their affinity to adsorb molecules is mainly due to hydrophobic, pi–pi H-bonding, and electrostatic interactions. They interact strongly with the benzene ring of aromatic compounds. Absorption of aromatic compounds on CNTs is mainly due to the pi–pi bond interaction. The pi bonds are described in Ref. [1], and the hydrogen bond is described in Ref. [2].

The surface property of CNTs can affect the adsorption of aromatic-containing compounds. The presence of oxygen and hydrogen within the surface groups crucially affects the adsorption. Due to the van der Waals force of interactions, CNTs can provide large external surface areas that can immobilize large biological contaminants, including bacteria and viruses. CNTs have good adsorption properties due to the numerous adsorption areas [3]:

- The internal cavity
- The intra-aggregate spacing
- The inter-wall spacing
- The external surface

Moreover, the surface areas can be increased by purification of pristine CNTs. CNT purification can be performed by treating them with acidic compounds (e.g., HNO_3) [4], basic compounds (e.g., KOH, NH_3) [5, 6], gaseous compounds (e.g., air, CO_2, and ozone) [3], or simply by thermal treatment [7]. Those treatments produce different surface properties influencing the uptake of contaminants.

CNTs are hydrophobic. In some cases, it is an important property as it allows nanotubes to float very well. This property is used for "carbon nanotube sponges" [8]. They are oleophilic; they absorb oil spilled in water. But for other uses, it is possible to change this property by functionalization. Functionalization results in physical or chemical modification; certain molecules or functional groups are attached to the nanotubes' sidewalls without significantly changing their desirable properties. This makes nanotubes more easily dispersible in liquids.

However, complete water-soluble CNTs could lead to difficulties when using them as adsorbent media. This is why for capturing bacteria, CNTs have to be semi-dispersible and partially hydrophobic because they exhibit greater affinity toward bacteria than both hydrophobic and weakly hydrophilic CNTs [9].

9.2 Activation of CNTs

The surface area has an important role for adsorptions on CNTs [10]. Even if CNTs are known to have high surface areas, it is possible to increase this surface area, thus increasing the molecule adsorption. One of the purposes of chemical treatment is to create micropores on the walls of the multi-walled CNTs (MWCNTs), thus increasing the specific surface area of CNTs.

9.2.1 CNTs Activated by NH_3 (NH_3-MWCNTs)

Investigations about environmental pollution, carried out across the world, confirm the presence of chlorophenols (CPs) in many ecosystems: surface and ground waters, bottom sediments, atmospheric air, and soils [11]. Treated MWCNTs can be used to remove CPs from water.

Mesopore spaces play a crucial role in the adsorption process among the adsorption sites available in CNTs; the inter-wall spacing is not large enough for CPs. Therefore, only the inner cavity, external surface, and intra-aggregate spacing are available for adsorption [12]. The NH_3 treatment increases the mesopore volume. For pristine MWCNTs, the mesopore volume is 0.401 cm^3/g, whereas for NH_3-MWCNTs, it is 0.419 cm^3/g [4]. Thus, NH_3-MWCNTs possess a larger absorption capacity for CPs. The surface chemistry also plays an important role in the absorption of CPs. NH_3 increases the pi–pi bond interactions between the aromatic ring of CPs and the graphitic structure of MWCNTs [13]. Therefore, the adsorption of CPs is increased. Finally the hydrophobic interaction is increased by NH_3.

9.2.2 CNTs Activated by KOH

Another common activation agent is KOH. These CNTs, activated by KOH, could be used for N_2 absorption. In its inert form, nitrogen is harmless and abundant, but in reactive form, it has an important impact on the environment and human health. This nitrogen-reactive form is due to food and energy production, and it accumulates in soil, water, atmosphere, and coastal oceanic waters [14]. Before activation by KOH, it is important to wash CNTs. It allows removing impurities as Mg, O, and Al [15]. The specific surface area of CNTs is increased because CNTs are open-ended and the walls of CNTs are etched, forming many micropores and mesopores. With activation by KOH, the MWCNTs turn much shorter, and their inner hollow core becomes much larger.

Yong et al. [16] carried out experiments of KOH activation on raw CNTs, purified CNTs, and activated CNTs, and they measured the *Brunauer, Emmett, and Teller (*BET) specific surface area of each one [16]. They found that the specific surface area of the MWCNTs was improved by a factor of 12.0 after the KOH activation. This leads to important increases in N_2 adsorption.

Table 9.1 Parameters of pore structure of MWCNTs in different states

Sample	BET (m^2/g)	V_{mi} (cm^3/g)	V_{me} (cm^3/g)
Raw MWCNTs	65.7	0.0088	0.1182
Purified MWCNTs	270.4	0.0557	0.5546
Activated MWCNTs	785.2	0.1654	1.0438

Source: Ref. [16].
V_{mi}: micropore volume; V_{me}: mesopore volume

9.3 Renewable Energy Sources

9.3.1 Hydrogen Storage

In the current state of water and energy nexus problems, there is an accelerating demand for green and efficient energy. One consequence is that the scientific community is more and more focused on hydrogen research. Production of electricity via hydrogen is very efficient because the process is not limited by the Rankine cycle [17].

CNTs are considered good adsorbents for hydrogen because of their high surface area and percentage of pore volume. The hydrogen adsorption is proportional to both parameters, and it is also known that gas adsorption depends on pressure and temperature for any adsorbent material [18]. Hydrogen is a gas, so the adsorption operates in the space unoccupied by CNTs, as the inner walls and the intra-aggregate spacing [17]. During the last decades, lots of studies were carried out to improve hydrogen storage using CNTs [19–29]. Some results are listed in Table 9.2.

Dillon et al. [23] were the first to publish experimental data about hydrogen adsorption on CNTs. They found that SWCNTs could have high energy storage at temperatures and pressures near room conditions. Their results show that SWCNT with diameter between 1.6 nm and 2 nm could reach the target H_2 adsorption density of 6.5wt%. Soon after this work, other studies reported optimistic results for hydrogen storage in CNTs. In their study, Chen et al. [17] found that a high adsorption capacity (4–6 wt%) can be reached only at very low temperatures.

Gayathri et al. [22] found that the presence of structural defects in CNTs increases the adsorption binding energy by 50%, which has an important effect on hydrogen storage.

Table 9.2 Results of hydrogen storage on SWCNT with thermodynamic conditions

Adsorbent type	Hydrogen storage (wt%)	Temperature (K)	Pressure (MPa)	References
SWCNT	11	80	10	[24]
SWCNT	2	80	10	[25]
SWCNT	6.5	300	16	[26]
SWCNT	5–10	300	0.04	[27]
SWCNT	10	300	0.04	[23]
SWCNT	4	300	12	[28]

Source: Ref. [18].

These studies show a large range in the results of hydrogen storage. As stated by the review of Ong et al. [29], there is controversy over claims that CNTs display great performance as hydrogen storage since the high storage capacity could not be reproduced [31]. Recently, after all these optimistic results listed earlier, new studies showed very low hydrogen storage capacities of CNTs at room temperature [30–33].

Tibbetts et al. [30] studied the adsorption of hydrogen on various CNTs at pressures up to 11 MPa and temperatures from 280 to 1500°C. Through their experiments, they found that none of the CNTs they measured showed good hydrogen adsorption. The largest sorption observed was about 0.1wt%.

Hydrogen storage with the use of CNTs has not met up to expectations of the United States Department of Energy (USDOE) [29].

It would seem that hydrogen, as a gas, could be stored in less space than when adsorbed by a solid such as CNTs. If space is an issue, hydrogen can be stored as a liquid under pressure. If safety is the issue with hydrogen storage, another process would be necessary to release the adsorbed hydrogen for use after storage. It does not seem logical that CNTs would be very useful for the storage of hydrogen.

9.3.2 Solar Cells

As renewable and alternative source for electricity production, solar cells have a great potential because of their inexhaustible source. The photons adsorbed from the sun are converted into electricity in the photovoltaic devices. The photon's energy is usually twice the energy required by an electron to be eliminated. Thus, the remaining energy is lost as heat. Low overall power conversion efficiency is one of the main drawbacks of organic solar cells.

Adsorption of light by organic solar cells leads to the generation of exciton (tightly bound electron–hole pairs). The limit of the power conversion efficiency of organic solar cells is mainly due to the high exciton binding energy, the short exciton diffusion length, the large absorber's bandgap between the lowest unoccupied molecular orbital (LUMO) level and the highest occupied molecular orbital

(HOMO) level of the organic absorbers, the low open-circuit voltage, and the low carrier mobility [34].

Researchers are working on solar cells to improve their efficiency and reduce their price.

The main strategy to implement CNT into solar cells consists of mixing CNTs with a solution of an electron-donating conjugated polymer. This mixture is then spin coated onto a thin transparent conductive electrode (60–120 nm) [35]. These electrodes are usually glass, plastic, covered with indium tin oxide and a layer of poly (3,4ethylenedioxythiophene):poly(styrenesulfonate) [36]. Moreover, to improve the CNTs' solubility into solvents for incorporation in a matrix, the physical and electrical properties of CNTs can be modified by functionalization, using for example nucleic acids.

It was shown that polymers mixed with CNTs are an efficient and rapid way for exciton dissociation, due to their high electron affinity. The excitons are dissociated into free electrons with high yield. It allows for obtaining higher energy conversion efficiency [37].

Kymakis et al. [36] were the first to study CNTs in organic solar cell as electron acceptor. The conjugation of CNTs and polymer poly(3-octylthiophene) (P3OT) in solar cell was studied by Scharber et al. [37]. They found that the interaction between CNTs and polymer allows the charge separation of the photogenerated excitons in the polymer, thus resulting in an increase in electron mobility to the negative electrode through the CNTs. Moreover, it increases by 10 the conductivity of the composite, thus indicating percolation paths within the material [38]. The photovoltaic performance of P3OT devices was highly increased with SWCNTs doping, revealing an increase in photocurrent and in the open-circuit voltage. It was demonstrated [39] that the introduction of just a small amount of CNTs in the polymer matrix highly improves the photovoltaic properties.

Kymakis et al. [39] and Geng et al. [40] studied the interaction between SWCNTs and poly(3-hexylthiophene) (P3HT). SWCNTs have good homogeneous dispersions in P3HT matrices due to their high solubility with P3HT, thus leading to a large interface between SWNTs and polymers. This large interface guarantees efficient utilization of absorbed light for generation of excitons. In their experiments, SWCNTs are used to increase the crystallinity of P3HT, which is important to increase the performance of the photovoltaic devices; they are also used as electron acceptors and electron transport tunnel, hence efficiently increasing the exciton dissociation and charge transport [39, 40].

Since organic solar cells are flexible and have low production costs as compared to silicon solar cells, the organic ones have attracted much research interests as well as commercial interests. However, there is enough experience at this time that it is known that organic solar cells do not have a long life since they deteriorate in the sun. Research into long-lasting, yet inexpensive organic solar cells is an active area.

9.4 Wastewater Treatment

9.4.1 Removal of Heavy Metals

Heavy metals are often found in wastewater electronics assembly plants, battery recycling plants, printed circuit board factories, painting, dying, photography, battery manufacturing, etc. Heavy metals are not biodegradable; they accumulate in the environment and cause numerous diseases and disorders in living organisms [41].

9.4.1.1 Lead removal

Lead and lead compounds are generally toxic pollutants and are non-biodegradable. Different water purification technologies already exist for lead removal, such as sand filtration and ion exchange, but their applications are limited by low efficiencies. Since the last decade, researchers are very interested in the use of CNTs for lead removal due to their exceptional properties listed earlier.

The absorption capabilities of CNTs increase after they are oxidized with nitric acid due to the introduction of functional groups, such as hydroxyl (–OH), carboxyl (–COOH), and carbonyl (>C=O), on the surface of CNTs [42]. These functional groups improve the adsorption capability of lead. It was found that adsorption of lead by CNTs is highly dependent on the pH value of the solution. Decrease in the pH of the solution leads to neutralization of surface charge, which leads to a decrease in the adsorption of cations [43].

Figure 9.1 shows that the removal of lead from water by using MCNTs is higher than the raw CNTs due to the introduction of functional group on the surface of CNTs by oxidation.

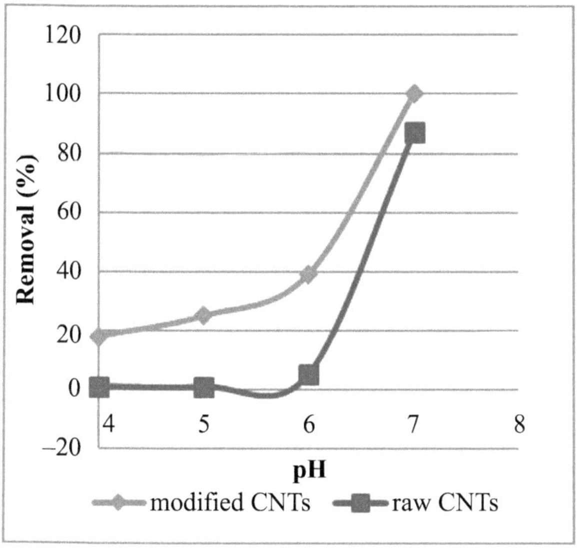

Figure 9.1 The effect of pH on percentage removal of lead at 150 RPM [43].

In addition, it was found that the adsorption capability depends greatly on the morphology of CNTs [44]. In this study, the adsorption properties in water solution of four CNTs with different morphologies (Xylene-Fe-800-C-Hor and Propylene-Ni-750-C-Hor, Benzene-Fe-1150-C-Vert and Methane-Ni-650-C-Vert.) were examined.

The introduction of defects in the carbon network can improve properties of CNTs [45, 46]. But it is important to control those defects because a large amount of defects on CNT surfaces can easily break up the CNTs. Adsorption studies show that the more perfect the structures of CNTs are, the lower adsorption capabilities they have.

Another important parameter is the contact time. Figure 9.2 shows that lead adsorption increases rapidly in the first 10 min and then reaches an equilibrium after 30 min for modified CNTs and 60 min for raw CNTs.

It was shown that CNTs are good adsorbents for lead removal and it is possible to increase adsorption by modifying their properties.

Table 9.3 Properties of CNTs with different morphologies

	Average diameter	**Specific surface area**	**Porews volume**	**Adsorption capacities**	**Characteristics**
Xylene-Fe-800-C-Hor	60 nm	47 m^2/g	0.18 cm^3/g	14.8 mg/g	Straight
Benzene-Fe-1150-C-Vert	20–30 nm	62 m^2/g	0.26 cm^3/g	11.2 mg/g	Straight with little defects
Methane-Ni- 650-C-Vert	30 nm	145 m^2/g	0.54 cm^3/g	82.6 mg/g	Weak with many defects on the walls; layered structures are not perfect
Propylene-Ni-750-C-Hor	30 nm	154 m^2/g	0.58 cm^3/g	59.8 mg/g	Curved, entangled with each other

Source: Ref. [44].

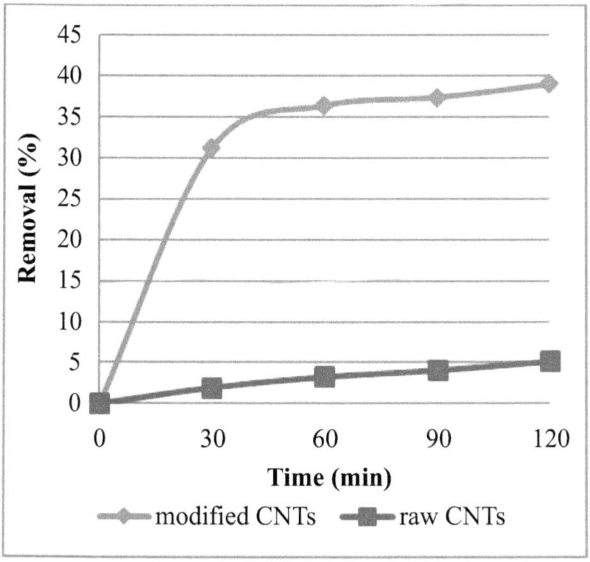

Figure 9.2 Effect of time on percentage removal of lead [43].

9.4.1.2 Chromium removal

Chromium is a naturally occurring element found in soil and volcanic dust and gases. It is found in the environment in two major states: chromium (III) and chromium (VI) [47]. Chromium (III) is essential for human beings and less toxic than chromium (VI). Chromium (VI) is found in various industrial wastewaters and is extremely toxic [48]. Several studies have shown that chromium (VI) compounds can increase the risk of lung cancer [49].

Some studies were carried out to show the capabilities of CNTs to remove chromium from water [50–55]. Raw MWCNTs and different oxidized MWCNTs were studied, and all results show that adsorption depends on pH, concentration, contact time, and temperature.

First, as in lead removal, it was shown through these studies that the removal percentage of Cr (VI) on the oxidized MWCNTs is higher than that of Cr (VI) on raw MWCNTs. In addition, pH appeared to be one of the most important parameters for chromium adsorption process.

Figure 9.3 indicates that the adsorption of Cr (VI) decreases with increasing pH. This is due to the different forms of Cr (VI) at low pH and high pH. The effect of pH on Cr (VI) removal by CeO_2-activated CNTs (CeO_2/ACNTs) is shown in Fig. 9.4. It indicates that Cr (VI) is highly adsorbed for pH 3 to 7.4. For pH < 3 and pH > 7.4, the adsorption decreased strongly [55].

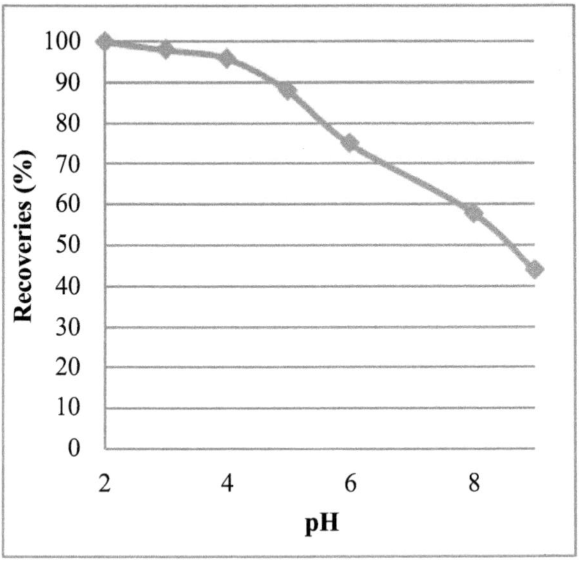

Figure 9.3 Influence of pH on the recoveries of Cr (VI) on MWCNTs (n = 3) [54].

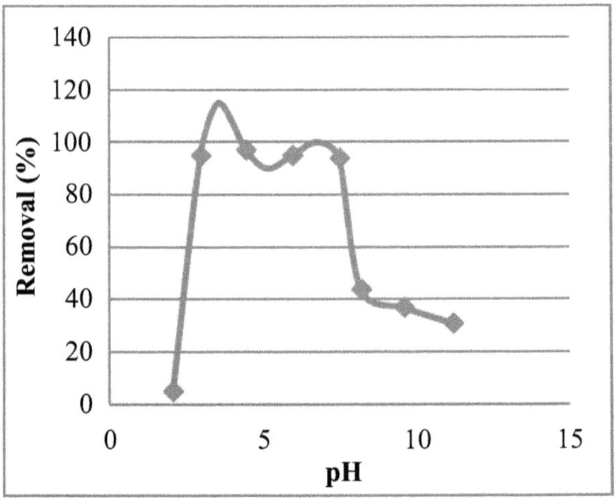

Figure 9.4 Adsorption efficiency for Cr (VI) of CeO_2/ACNTs (%) as a function of pH values (C_i = 10.0 mg/L and 25°C) [55].

The dosage of MWCNTs also has an impact on adsorption results [50]. For different values of pH, the adsorption of Cr (VI) increases with an increase in MWCNT concentration. Because of increasing MWCNT dosage, the number of available sites for binding Cr (VI) increases. As it was shown for the dosage, by increasing the temperature and the contact time, the adsorption is increased [51].

Different thermodynamic parameters were studied to understand the adsorption process of the adsorption of Cr (VI) on MWCNTs: free energy change ΔG, enthalpy change ΔH, and entropy change ΔS. These parameters can be calculated as follows:

$$\ln k_{\mathrm{d}} = \frac{\Delta S^{\circ}}{R} - \frac{\Delta H^{\circ}}{RT} \tag{9.1}$$

$$\Delta G^{\circ} = \Delta H^{\circ} - T\Delta S^{\circ} \tag{9.2}$$

These parameters were calculated for different concentrations of chromium at different temperatures.

Table 9.4 Thermodynamic parameters of Cr (VI) adsorption on MWCNTs at different Cr (VI) initial concentrations and temperature

C_0 (mol/L)	T (K)	ΔH°	ΔS°	ΔG°
	293.15	48.96	0.219	−15.25
6.73×10^{-5}	306.15	48.96	0.219	−18.1
	316.15	48.96	0.219	−20.29
	293.15	51.26	0.223	−14.23
$9.62.10^{-5}$	306.15	51.26	0.223	−17.13
	316.15	51.26	0.223	−19.37

Source: Ref. [49].

The positive values of ΔH° indicate that the adsorption process is an endothermic reaction. The positive values of ΔS° indicate the increased randomness at the solid–solution interface. Finally, the negative values of ΔG° confirm that the adsorption process is spontaneous, which becomes more negative with an increase in temperature. These results show that higher adsorption occurred at higher temperature because ions are easily desolated, which helps in adsorption. Moreover, the ΔG° values are more negative at low concentrations; this indicates that adsorption is higher at low initial concentrations [50].

9.4.1.3 Nickel removal

Nickel salts are commonly used in silver refineries, electroplating, zinc base casting, storage battery industries, printing, and in the production of some alloys, which discharge a significant amount of nickel in various forms in the environment. At higher concentrations, Ni^{2+} has important impact for human health (lungs and bone cancer, chest pain, tightness of the chest, dry cough, and shortness of breath) [56]. Hence, it is essential to remove Ni^{2+} from industrial wastewaters before discharge into natural water sources.

An Ni^{2+} removal experiment was carried out with MWCNTs with outer diameter (dp) < 10 nm and SWCNTs with dp < 2 nm [56, 57]. The CNTs were oxidized by NaClO solutions and employed as sorbents to study characteristics of nickel (II) adsorption from aqueous solution. In other studies, triphenylphosphine-linked MWCNTs [58] and MWCNTs oxidized with nitric acid [59] were employed for Ni^{2+} removal.

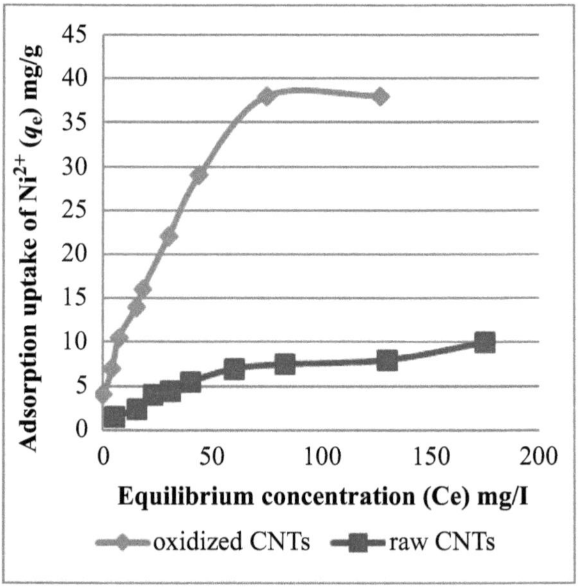

Figure 9.5 Effect of equilibrium concentration on the adsorption uptake of nickel ions for oxidized CNTs [59].

It is known that oxidation treatment improves adsorption due to the presence of different functional groups formed on the CNT surface, as mentioned earlier, during the oxidation, which improves their hydrophilic and cation exchange properties. Thus, an increase in equilibrium concentration increases the adsorption of Ni^{2+} on CNTs.

Figure 9.6 shows that Ni^{2+} adsorption capacities are greater for SWCNT oxidized by NaClO (NaClO/SWCNTs) than those reacted by NaClO/MWCNTs. Moreover, it was shown that NaClO/SWCNTs and NaClO/MWCNTs can be reused through many cycles of water treatment and regeneration even if the Ni^{2+} recovery of CNTs slightly decreased (Fig. 9.7). The adsorption capacities of granulated activated CNTs decreased after each cycle [56].

As it was shown above with lead and chromium removal, other parameters have as important impact on adsorption as pH and time contact. Ni^{2+} removal by oxidized CNTs increased with pH and time contact [56–58].

These studies show that oxidized CNTs have better Ni^{2+} adsorption than raw CNTs, and also SWCNTs and MWCNTs keep good adsorption properties after many cycles of water treatment and regeneration. Finally, it appears that oxidized SWCNTs have better adsorption properties than oxidized MWCNTs.

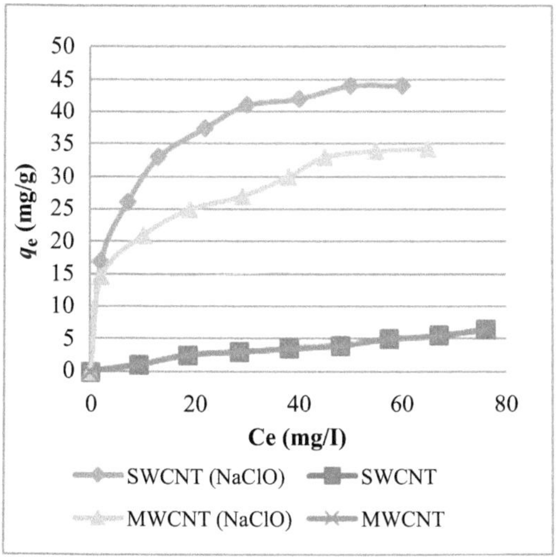

Figure 9.6 Sorption isotherms of Ni^{2+} by SWCNTS and MWCNTS [57].

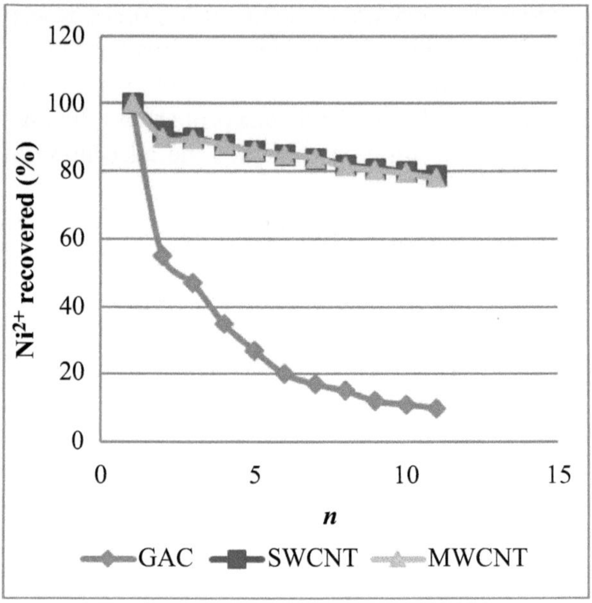

Figure 9.7 Ni^{2+} recoveries of sorbents under various *n* [56].

9.4.2 Adsorption of Phenolic Compounds

Phenol, also known as carbolic acid, is an environmental pollutant. Exposure to phenol can have important effects on the human body. Phenol is widely used in the production of herbicides, wood preservatives, fertilizers, explosives, rubber, paper, and pharmaceuticals [60]. Substitution of the OH groups on CNTs enhances the sorption of phenolic compounds to CNTs; it is due to the enhancement of the pi–pi interactions between the phenolic compounds and the CNTs. Moreover, the hydrophobic interaction has an important influence on the sorption of phenolic compounds; while hydrophobicity decreases, adsorption increases [61].

9.4.3 Removal of Waterborne Pathogens

Waterborne pathogens are very important issues in developing countries because they lead to millions of deaths every year [62]. Carbon nanotube filters could be used as a solution against this, by removing bacteria from water, such as the *Escherichia coli* (*E. coli*), which is a common pollutant of drinking water and could lead to many diseases [63]. Both SWCNTs and MWCNTs are effective for the removal of bacteria, even if the MWCNT filter is 1.5 to 3 orders of magnitude higher than SWCNT filter for bacterial removal over a range of solution chemistries [64]. Study on electrochemical MWCNT microfilter was carried out [62] for bacteriophage MS2 and *E. coli* inactivation and removal. The polluted water was injected through the filter where electrolysis was carried out. It was found that at applied potentials of 2 and 3 V, this filter completely removed bacteria and viruses in the water.

Another experience regarding the SWNT filter shows that the *E. coli* cells are retained inside the SWNT layer by attaching to the fibers where they are inactivated. *E. coli* cells underwent severe damage when they are in contact with SWCNT aggregates or deposit layer, which result in their death [65].

It was found [66] that higher ionic strengths (NaCl) eliminate the electrostatic interactions between viruses and SWNTs, which lead to the increase in removal of viruses. The solution pH also affects viral removal. The interactions between viruses and SWNTs are attractive below the virus isoelectric point found at pH 3.9 and become repulsive over this value [66].

One of the advantages of nanotube filters for bacteria removal is that they can be cleaned, regenerated by thermal or chemical treatment if ceramic is used as their base supports. If reused many times, they keep their full efficiency.

9.5 Air Treatment

9.5.1 Capture of Carbon Dioxide

MWCNTs can have large adsorption capacities of pollutants due to their pore structures and functional groups on their surface achieved by chemical modification or thermal treatment. It was shown that CNTs modified by chemical agents, such as ethylenediamine (EDA), polyethylenimine (PEI), and 3-aminopropyl-triethoxysilane (APTS), have better CO_2 absorption properties than raw CNTs [67].

The modification methods include a thermal treatment using an oven. Then the CNTs were dispersed into flasks containing the chemical agents. These solutions are polymers with many amine groups, which can react with CO_2 to form carbamate in the absence of water; that is the reason for the amount of CO_2 absorbed on CNTs to increase. Carbamates are organic compounds derived from carbamic acid (NH_2COOH) [68]. The absorption depends on temperature; when temperature increases, the absorption decreases. When moisture increases, the absorption increases.

Figure 9.8 shows the CO_2 adsorption capacity of adsorbents, q_e, of 15% CO_2 adsorption via CNTs modified by various amines at different temperatures. For each adsorbent, q_e decreases with temperature, indicating that the CO_2 adsorption process is exothermic. The CNT (APTS) appears to be the better CO_2 adsorbent at all tested temperatures. Moreover, it was shown [70] that CNTs modified by amines are twice more efficient than activated carbon.

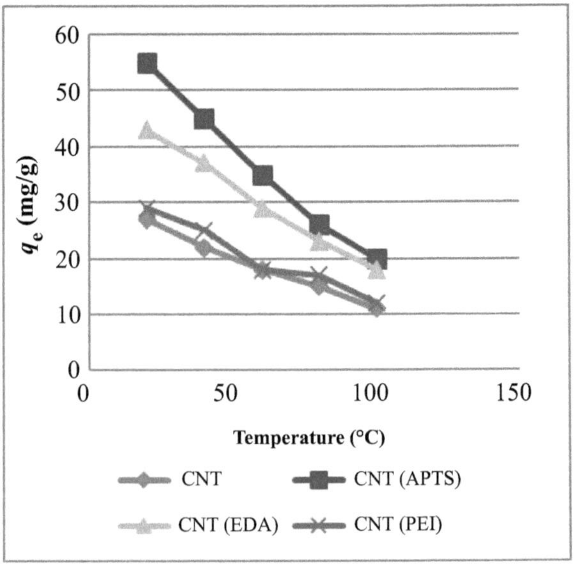

Figure 9.8 Effects of temperature on 15% CO_2 adsorption via CNTs and various amine-modified CNTs [69].

It was shown [71] that the Dubinin–Astakhov equation can perfectly represent the CO_2 adsorption isotherm of SWCNTs at 273 K or above. The Dubinin–Astakhov equation is as follows:

$$\frac{\frac{P}{P_0}}{\left(1-\frac{P}{P_0}\right)V}=\frac{1}{V_0\text{c}}+\left(\frac{\text{c}-1}{V_0\text{c}}\right).\frac{P}{P_0} \tag{9.3}$$

$$\frac{\left(\frac{P}{P_0}\right)}{V}=\frac{1}{V_0\text{b}}+\frac{1}{V_0}.\frac{P}{P_0} \tag{9.4}$$

$$\log(V)=\log(V_0)-\left[\frac{\text{R}T}{\beta E_0}\right]^n\left[\log\left(\frac{P_0}{P}\right)\right]^n \tag{9.5}$$

where P is equilibrium pressure, P_0 is saturation vapor pressure of the gas at the analysis temperature (T), V is the volume adsorbed at equilibrium pressure, V_0 is the monolayer capacity, n is the Astakhov exponent, R is the gas constant, E_0 is the characteristic energy, β is the affinity coefficient of the analysis gas, and b and c are constants.

In this study, it was found that the Astakhov exponent is less than 2 for CO_2 adsorption, indicating that the porosities on SWCNTs are heterogeneous and also that CO_2 molecules are absorbed on both external and internal surfaces of SWCNTs.

9.5.2 Bioaerosol Removal

Exposures to bioaerosols lead to a wide range of health effects with major public health impact as infectious diseases, allergies, and cancer. Common bioaerosol exposure effects are respiratory symptoms and lung function impairment [72]. These diseases can be avoided by controlling indoor air quality. Therefore, there are many different air-cleaning technologies for removing bioaerosols, such as electric filtration, electrostatic precipitation, and ultraviolet germicides [73].

CNTs can also be used for removal of bioaerosols. Experiments were carried out to study the capacities of a non-thermal plasma system that inactivates bioaerosols [73]. This study used corona discharge plasma. For the experiments, the bioaerosols were mixed with diluted clean air into a test column. They compared results between two different coronas, one with CNT electrodes and another using stainless steel electrodes. Results shown in Fig. 9.9 indicate that the use of CNTs significantly increases bioaerosol inactivation.

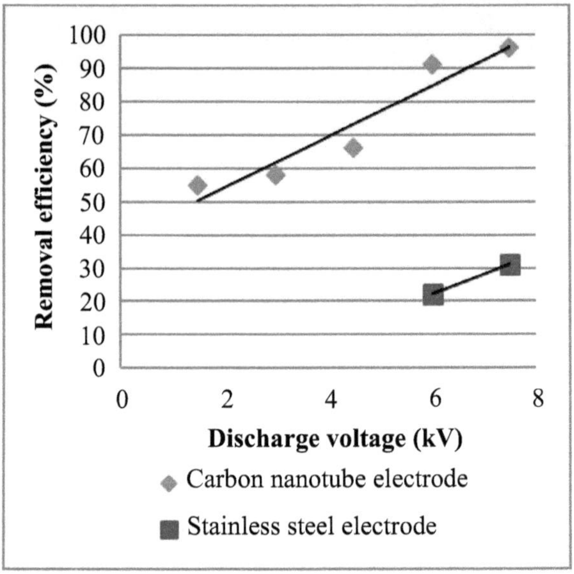

Figure 9.9 Comparison between CNT electrode and stainless steel electrode for removal efficiencies of *E. coli* bioaerosols by using corona discharge plasma [73].

Furthermore, it is known that plasma technologies need a high operating voltage, which leads to generation of high ozone concentration, but the use of carbon nanotube electrodes in the corona discharge system decreases the threshold voltage of plasma, thus decreasing the ozone concentration. Moreover, results show that bioaerosol inactivation increases with the operating voltage and reach 90% at 6 kV.

In another study, CNT filters created by depositing both SWCNTs and MWCNTs on polypropylene and polyamide membrane supports were used to remove bioaerosols in indoor environment [74]. The bioaerosol removal efficiency increases with an increase in the loadings of CNTs. However, a low CNT loading could lead to a high removal efficiency of biological aerosols, thus minimizing the cost of CNT filters [74, 75].

Nevertheless, because of their small size, CNTs can be easily airborne and represent important issues for human beings because their adsorption could be toxic for lungs and cause significant respiratory health effects [76]. Thus, it is important to study the attachment kinetics of CNTs on various membrane supports, to characterize the risk of utilization of CNTs [74].

9.6 Discussion and Conclusion

Research has been done to investigate if hydrogen, as a gas, could be stored in less space than when adsorbed by a solid such as CNTs. If space is an issue, hydrogen can be stored as a liquid under pressure. If safety is the issue with hydrogen storage, another process would be necessary to release the adsorbed hydrogen for use after storage. It does not seem logical that CNTs would be very useful for the storage of hydrogen. CNTs used for solar collectors are used principally for their improved thermal and electrical conductivities. Organic solar cells do not have a long life since they deteriorate in the sun. Research into long-lasting, yet inexpensive organic solar cells is an active area.

In the environmental remediation and protection arena, CNTs are used for their heightened adsorption characteristics, as compared to the macro forms of carbon. In addition, the nano forms are more reactive. They are used in wastewater treatment to remove undesirable solids such as heavy metals and phenolic compounds. For drinking water treatment, they are used to remove waterborne pathogens. For air quality control, CNTs are used for adsorption of carbon dioxide and bioaerosol removal.

References

1. Pi bond. Wikipedia. Retrieved July 11, 2012. http://en.wikipedia.org/wiki/Pi_bond.
2. Hydrogen bond. Wikipedia. Retrieved July 11, 2012. http://en.wikipedia.org/wiki/Hydrogen_bond.
3. Kang, S., Herzberg, M., Rodrigues, D. F., and Elimelech, M., "Antibacterial effects of carbon nanotubes: Size does matter," *Langmuir*, 2008; **24**: 6409–6413.

4. Liao, Q., Sun, J., and Gao, L., "Adsorption of chlorophenols by multi walled carbon nanotubes treated with HNO_3 and NH_3," *Carbon*, 2008; **46**: 544–561.
5. Chen, Y., Liu, C., Li, F., and Cheng, H. M., "Pore structures of multi walled carbon nanotubes activated by air, CO_2 and KOH," *J. Porous Mater.*, 2006; **13**: 141–146.
6. Lee, S. M., Lee, S. C., Jung, J. H., and Kim, H. J., "Pore characterization of multi walled carbon nanotubes modified by KOH," *Chem. Phys. Lett.*, 2005; **416**: 251–255.
7. Lu, C. and Su, F., "Adsorption of natural organic matter by carbon nanotubes," *Sep. Purif. Technol.*, 2007; **58**: 113–121.
8. Gui, X., Wei, J., Wang, K., Cao, A., Zhu, H., Jia, Y., Shu, Q., and Wu, D., "Carbon nanotube sponges," *Adv. Mater.*, 2009; **22**(5): 617–621.
9. Akasaka, T. and Watari, F., "Capture of bacteria by flexible carbon nanotubes," *Acta Biomater*, 2009; **5**: 607–612.
10. Apul, O. G., Shao, T., Zhang, S., and Karanfil, T., "Impact of carbon nanotube morphology on phenanthrene adsorption," 2012; **31**(1): 73–78.
11. Sinkkonen, S. and Paasivirta, J., "Polychlorinated organic compounds in the Artic cod liver: Trends and profiles," *Chemosphere*, 2000; **40**: 619–629.
12. Yang, K. and Xing, B., "Desorption of polycyclic aromatic hydrocarbons from carbon nanomaterials in water," *Environ. Pollut.*, 2007; **145**(2) :529–537.
13. Coughlin, R. W. and Ezra, F. S., "Role of surface acidity in the adsorption of organic pollutants on the surface of carbon," *Environ. Sci. Technol.*, 1968; **2**(4): 291–297.
14. UNEP and the Woods Hole Reseach Center., "Reactive nitrogen in the environment," 2007.
15. Donnaperna, L., Duclaux, L., Gadiou, R., Soneda, Y., and Yoshizawa, N., "KOH activated carbon multiwall nanotubes," *Carbon Sci. Technol.*, 2009; **3**: 120–124.
16. Yong, C., Chang, L., and Hui-Ming, C., "Characterization of multi-walled carbon nanotubes activated by KOH," Shenyang National Laboratory for Materials Science, 2004.
17. Cheng, H., Yang, Q., and Liu, C., "Hydrogen storage in carbon nanotubes," *Carbon*, 2001; **39**: 1447–1454.

18. Darkrim, F. L., Malbrunot, P., and Tartaglia, G. P., "Review of hydrogen storage by adsorption in carbon nanotubes," *Int. J. Hydrogen Energy*, 2002; **27**: 193–202.

19. Banerjee, S. and Puri, I., "Enhancement in hydrogen storage in carbon nanotubes under modified conditions," *Nanotechnology*, 2008; **19**: 155702 (6pp).

20. Iyakutti, K., Kawazoe, Y., Rajarajeswari, M., and Surya, V. J., "Aluminum hydride coated single-walled carbon nanotube as a hydrogen storage medium," *Int. J. Hydrogen Energy*, 2009; **34**: 370–375.

21. Chen, C. and Huang, C., "Enhancement of hydrogen spillover onto carbon nanotubes with defect feature," *Microporous Mesoporous Mater.*, 2008; **109**: 549–559.

22. Gayathri, V. and Geetha, R., "Hydrogen adsorption in defected carbon nanotubes," *Adsorption*, 2007; **13**: 53–59.

23. Dillon, A. C., Jones, K. M., Bekkedahl, T. A., Kiang, C. H., Bethune, D. S., and Heben, M. J., "Storage of hydrogen in single walled carbon nanotubes," *Nature*, 1997; **386**: 377.

24. Darkrim, F., "High adsorptive property of opened carbon nanotubes at 77 K," *J. Phys. Chem. B*, 2000; **104**: 6773–6776.

25. Wang, Q. and Johnson, J. K., "Optimization of carbon nanotube arrays for hydrogen adsorption," *J. Phys. Chem. B*, 1999; **103**: 4809–4813.

26. Yin, Y., Mays, T., and McEnaney, B., "Molecular simulations of hydrogen storage in carbon nanotube arrays," *Langmuir*, 2000; **16**: 10521–10527.

27. Dillon, A. C., Bekkedahl, T. A., Jones, K. M., and Heben, M. J. *Fullerenes*, 1999; **3**: 716–727.

28. Liu, C., Fan, Y. Y., Liu, M., Cong, H. T., Cheng, H. M., and Dresselhaus, M. S., "Hydrogen storage in single-walled carbon nanotubes at room," *Science*, 1999; **286**: 1127.

29. Ong, Y., Ahmad, A., Zein, S., and Tan, S. "A review on carbon nanotubes in an environmental protection and green engineering perspective," *Brazilian J. Chem. Eng.*, 2010; **27**(2): 227–242.

30. Tibbetts, G., Meisner, G., and Olk, C. "Hydrogen storage capacity of carbon nanotubes, filaments, and vapor-grown fibers," *Carbon*, 2001; **39**: 2291–2301.

31. Hirscher, M., Becher, M., Haluska, M., Quintel, A., Skakalova, V., Choi, Y., Dettlaff-Weglikowska, U., Roth, S., Stepanek, I., Bernier, P., Leonhardt, A., and Fink, J., "Hydrogen storage in carbon nanostructures," *J. Alloys Compounds*, 2002; **330–332**: 654–658.

32. Kajiura, H., Tsutsui, S., Kadono, K., Kakuta, M., and Ata, M., "Hydrogen storage capacity of commercially available carbon materials at room temperature," *Appl Phys Lett*, 2003; **82**: 1105–1107.

33. Liu, C., Chen, Y., Wu, C., Xu, S., and Cheng, H., "Hydrogen storage in carbon nanotubes revisited," *Carbon*, 2010; **48**: 452–455.

34. Li, G. and Liu, L., "Carbon nanotubes for organic solar cells," *IEEE Nanotechnol. Mag.*, 2011; **5**(3): 18–24.

35. Sgobba, V. and Guldi, D., "Carbon nanotubes as integrative materials for organic photovoltaic devices," *J. Mater. Chem.*, 2008; **18**(2): 141–248.

36. Cataldo, S., Salice, P., Menna, E., and Pignataro, B., "Carbon nanotubes and organic solar cells," *Energy Environ. Sci.*, 2012; **5**: 5919.

37. Scharber, M. C., Muhlbacher, D., Koppe, M., Denk, P., Waldauf, C., Heeger, A. J., and Brabec, C. J., "Design rules for donors in bulk-heterojunction solar cells-towards 10% energy-conversion efficiency," *Adv. Mater.*, 2006; **18**(6): 789–794.

38. Kymakis, E. and Amaratunga, G. A. J., "Single-wall carbon nanotube/conjugated polymer photovoltaic devices," *Appl. Phys. Lett.*, 2002; **80**(1): 112.

39. Kymakis, E. and Amaratunga, G. A. J., "Carbon nanotubes as electron acceptors in polymeric photovoltaic," *Rev. Adv. Mater. Sci.*, 2005; **10**: 300–305.

40. Geng, J. and Zeng, T., "Influence of single-walled carbon nanotubes induced crystallinity enhancement and morphology change on polymer photovoltaic devices," *J. Am. Chem. Soc.*, 2006; **128**: 16827–16833.

41. Zhao, X., Song, L., He, J., Wu, T., and Qin, Y., "Adsorption characteristics of Ni^{2+} ion onto the diethylenetriaminepentaacetic acid-melamine/polyvinylidene fluoride blended resin," *Int. J. Energy Environ.*, 2010; **1**(1): 121–132.

42. Shimoda, H., Oh, S. J., Geng, H. Z., Walker, R. J., Zhang, X. B., McNeil, L.E., and Zhou, O., "Self-assembly of carbon nanotubes," *Adv. Mater.*, 2002; **14**(12): 859–933.

43. Atieh, M. A., Bakather, O. Y., Al-Tawbini, B., Bukhari, A. A., Abuilaiwi, F. A., and Fettouhi, M. B., "Effect of carboxylic functional group functionalized on carbon nanotubes surface on the removal of lead from water," *Bioinorg. Chem. Appl.*, 2010; Article ID 603978.

44. Li, Y., Zhu, Y., Zhao, Y., Wu, D., and Luan, Z., "Different morphologies of carbon nanotubes effect on the lead removal from aqueous solution," *Diamond Related Materials*, 2006; **15**: 90–94.

45. Robinson, J. A., Snow, E. S., Badescu, S., Reinecke, T. L., and Perkins, F. K., "Role of defects in single-walled carbon nanotube chemical sensors," *Nano Lett.*, 2006; **6**(8): 1747–1751.

46. Hoefer, M. and Bandaru, P. R., "Determination and enhancement of the capacitance contributions in carbon nanotube based electrode systems," *Appl. Phys. Lett.*, 2009; **95**(18): 183108

47. Idea connection, Chromium as pollutant. Retrieved July 16, 2012. http://www.ideaconnection.com/solutions/491-Chromium-as-a-pollutant.html

48. Di, Z., Ding, J., Peng, X., Li, Y., Luan, Z., and Liang, J., "Chromium adsorption by aligned carbon nanotubes supported ceria nanoparticles," *Chemosphere*, 2006; **62**: 861–865.

49. Dangerous pollutants, Chromium. Retrieved July 16, 2012. http://www.dangerouspollutants.com/chromium.asp

50. Hu, J., Wang, S. W., Shao, D. D., Dong, Y. H., Li, J. X. and Wang X. K., "Adsorption and reduction of chromium(VI) from aqueous solution by multiwalled carbon nanotubes," *Open Environ. Pollut. Toxicology J.*, 2009; **1**: 66–73.

51. Pillay, K., Cukrowska, E. M., and Coville, N.J., "Multi-walled carbon nanotubes as adsorbents for the removal of parts per billion levels of hexavalent chromium from aqueous solution," *J Hazardous Mater.*, 2009; **166**: 1067–1075.

52. Hu, J., Chen, C., Zhu, X., and Wang, X., "Removal of chromium from aqueous solution by using oxidized multiwalled carbon nanotubes," *J. Hazardous Mater.*, 2009; **162**: 1542–1550.

53. Atieh, M., Bakather, O., Tawabini, B., Bukhari, A., Khaled, M., Alharthi, M., Fettouhi, M., and Abuilaiwi, F., "Removal of chromium (III) from water by using modified and nonmodified carbon nanotubes," *J. Nanomater.*, 2010; Article ID 232378.

54. Tuzen, M. and Soylak, M., "Multiwalled carbon nanotubes for speciation of chromium in environmental samples," *J. Hazardous Mater.*, 2007; **147**: 219–225.

55. Di, Z., Ding, J., Peng, X., Li, Y., Luan, Z., and Liang, J., "Chromium adsorption by aligned carbon nanotubes supported ceria nanoparticles," *Chemosphere*, 2006; **62**: 861–865.

56. Lu, C., Liu, C., and Rao, G., "Comparisons of sorbent cost for the removal of Ni^{2+} from aqueous solution by carbon nanotubes and granular activated carbon," *J. Hazardous Mater.*, 2008; **151**: 239–246.

57. Lu, C. and Liu C., "Removal of nickel(II) from aqueous solution by carbon nanotubes," *J. Chem. Technol. Biotechnol.*, 2006; **81**(12): 1932–1940.
58. Adolph, M., Xavier, Y., Kriveshini, P., and Rui, K., "Phosphine functionalised multiwalled carbon nanotubes: A new adsorbent for the removal of nickel from aqueous solution," *J. Environ. Sci.*, 2012; **24**(6): 1133–1141.
59. Kandaha, M. and Meunier, J.-L. "Removal of nickel ions from water by multi-walled carbon nanotubes," *J. Hazardous Mater.*, 2007; **146**: 283–288.
60. State of Ohio Environmental Protection Agency., "Pollution prevention opportunities for PBT chemicals, phenol." Office of pollution prevention. September 2002, number 100.
61. Lin, D. and Xing, B., "Adsorption of phenolic compounds by carbon nanotubes: Role of aromaticity and substitution of hydroxyl groups," *Environ. Sci. Technol.*, 2008; **42**: 7254–7259.
62. Vecitis, C., Schnoor, M., Rahaman, M. S., Schiffman, J., and Elimelech, M. "Electrochemical multiwalled carbon nanotube filter for viral and bacterial removal and inactivation," *Environ. Sci. Technol.*, 2011; **45**: 3672–3679.
63. Srivastava, A., Srivastava, O. N., Talapatra, S., Vajtai, R., Ajayan, P. M., "Carbon nanotube filters," *Nature Mater.*, 2004; **3**: 610–614.
64. Brady-Estevez, A., Schnoor, M., Vecitis, C., Saleh, N., and Elimelech, M., "Multiwalled carbon nanotube filter: Improving viral removal at low pressure," *Langmuir*, 2010; **26**(18): 14975–14982.
65. Brady-Estevez, A., Kang, S., and Elimelech, M., "A single-walled carbon-nanotube filter for removal of viral and bacterial pathogens," *Small*, 2008; **4**(4): 481–484.
66. Brady-Estevez, A., Nguyen, T, Gutierrez, L., and Elimelech, M., "Impact of solution chemistry on viral removal by a single-walled carbon nanotube filter," *Water Res.* 2010; **44**: 3773–3780.
67. Chungsying, L., Shih-Chieh, H., Fengsheng, S., Wanting, Z., and Wenfa, C., "Thermodynamics and regeneration studies of CO_2 adsorption on multiwalled carbon nanotubes," *Chem. Eng. Sci.*, 2010: **65**(4): 1354–1361.
68. Carbamate. Wikipedia. Retrieved July 12, 2012. http://en.wikipedia.org/wiki/Carbamates.

69. Su, F., Lu, C., Cnen, W., Bai, H., and Hwang, J., "Capture of CO_2 from flue gas via multiwalled carbon nanotubes," *Chem. Eng. Sci.*, 2010; **65**: 1354–1361.

70. Cinke, M., Li, J., Bauschlicher, Jr. C., Ricca, A., and Meyyappan, M., "CO_2 adsorption in single-walled carbon nanotubes," *Chem. Phys. Lett.*, 2003; **376**: 761–766.

71. Hu, Y. and Ruckenstein, E., "Applicability of Dubinin–Astakhov equation to CO_2 adsorption on single-walled carbon nanotubes," *Chem. Phys. Lett.*, 2006; **425**: 306–310.

72. Douwes, J., Thorne, P., Pearce, N., and Heederik, D., "Bioaerosol health effects and exposure assessment: Progress and prospects," *Ann. Occup. Hyg.*, 2003; **47**(3): 187–200.

73. Yang, S., Huang, Y., Luo, C., Li, Y., Huang, J., Chuang, C., Chen, C., Fang, W., and Chuang, C., "Inactivation efficiency of bioaerosols using carbon nanotube plasma," *CLEAN Soil, Air, Water*, 2011; **39**(3): 201–205.

74. Guan, T. and Yao, M., "Use of carbon nanotube filter in removing bioaerosols," *J. Aerosol Sci.*, 2010; **41**: 611–620.

75. Xu, Z. and Yao, M., "Effects of single-walled carbon nanotube filter on culturability and diversity of environmental bioaerosols," *J. Aerosol Sci.*, 2011; **42**: 387–396.

76. Pulskamp, K., Worle-Knirsch, J., Hennrich, F., Kern, K., and Krug, H. F., "Human lung epithelial cells show biphasic oxidative burst after single-walled carbon nanotube contact," *Carbon*, 2007; **45**: 2241–2249.

Chapter 10

Nanotechnology in Batteries

Kaufui V. Wong and Sarah Dia
Department of Mechanical and Aerospace Engineering, University of Miami, Coral Gables, FL 33124, USA
kwong@miami.edu

A relatively new field, nanotechnology has expanded to almost every scientific sector since its origin in the 1980s. This chapter focuses on the potential of nanotechnology in batteries, in particular, with a review of the current and past developments in the field. For smaller applications using lithium-ion batteries, it appears that nanotechnology has established a firm foothold. The possibilities for main streaming this advance in large batteries, for example, grid batteries, are researched, and developments to date reported. Viable grid batteries are the key to adapting wind, water, and solar (WWS) sources of energy for the power grid since none of these WWS resources are available every single hour of the day and night.

This chapter has been reprinted from the *Journal of Energy Resources Technology* (Technology Review), Vol. 139(1), Kaufui Wong and Sarah Dia, Nanotechnology in batteries, Copyright 2016, with permission from ASME.

Nanotechnology and Energy
Edited by Kaufui V. Wong

ISBN 978-981-4774-24-6 (Hardcover), 978-1-315-16357-4 (eBook)
www.panstanford.com

10.1 Introduction

At the forefront of the rechargeable portable energy industry, lithium-ion batteries have endured as the primary contender for powering up electronic devices. They are ubiquitous and have been so for more than 30 years. However, as technology advances and new portable energy requirements arise, a more sophisticated battery technology is also required. The standard Li-ion batteries have been more than sufficient for the earlier smaller devices such as cellphones, watches, and other gadgets. They have even been used for larger, more demanding applications such as electric vehicles and military devices. Yet, as we have come to know, these more demanding applications were often met with their share of shortcomings. Excessive charge times for hybrid-electric and plug-in vehicles have prevented them from becoming as mainstream as their more practical, though be it, less clean-energy-friendly counterparts. Combustible properties have made them a safety concern, limiting their use for grander endeavors. Essentially, the advances made in technology soon found a major drawback in the performance offered by standard Li-ion batteries, but as the expansion of the nanotechnological era came about, so too came the possibility that more might be expected of a battery [1–7].

Providing a sustainable and truly practical solution to all more demanding needs has been a topic of extensive research for several years. The application of nanotechnology in batteries has been repeatedly studied and proclaimed as a very probable solution to what might have been missing.

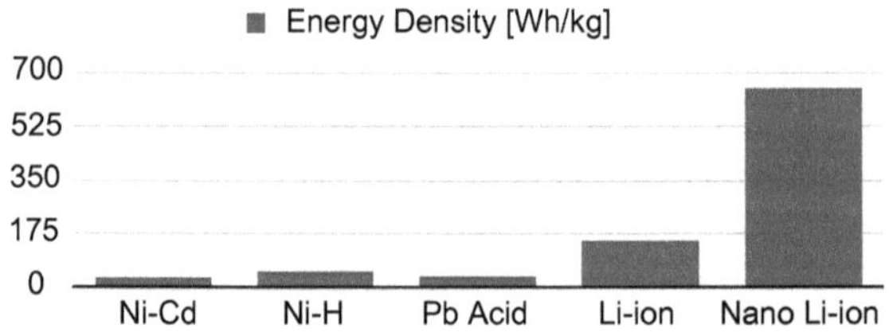

Figure 10.1 Visible improvements offered from Li-ion batteries, even before nanotechnology, as compared to older models [5, 7].

Table 10.1 presents a general overview of some of these approaches in using nanotechnology to fill that gap. Among these included herein are numerous others, as a number of chemical and material combinations are possible and will continue to be explored for several years to come [6–24].

Table 10.1 Current and past approaches to using nanotechnology in batteries

S. No.	Developers	Approach/description
1.	A123Systems	Nanophosphate® battery with lithium iron phosphate cathode [8].
2.	mPhase Technologies	Electrical control of wetting behavior of liquids on nanostructured surfaces [9].
3.	HE3DA Company in Prague	Expansion to 3D electrode using lithium nanomaterials [10].
4.	Hanyang University, and others	KB-600JD, OMC, rGO, and CS, as nanostructured carbon materials for cathode used in rechargeable Li-SO_2 batteries [7].
5.	La Sapienza University, University G. D'Annunzio	Anode: LiAl, $Li_{21}Si_5$, $Li_{22}Sn_5$, Li_3Sb, and LiC_6 with lithium particle size reduced to nanoscale. Cathode: $LiFePO_4$, with nano-morphological modifications. Electrolyte: PEO-LiX with ceramic powders dispersed to PEO-LiX matrix at nanoparticle size [6].
6.	North Caroline State University	Aligned silicon-coated carbon nanotubes in anodes of Li-ion batteries [11].
7.	MIT	Functionalized multiwalled carbon nanotubes for positive electrode of Li-ion battery [12].
8.	Rice University	Coaxial manganese oxide/carbon nanotube arrays in porous alumina templates for cathode [13].
9.	Stanford University	Nanostructured silicon anodes: double-walled nanotubes, yolk–shell nanoparticles [14].
10.	Stanford University, SLAC, UNIST	Silicon nanolayer–embedded graphite [15].
11.	Stanford University	Silicon nanoparticles encapsulated by conductive carbons layers [16].
12.	USC	Li-ion battery made with silicon nanoparticles in anode [17].
13.	LMU Munich and Waterloo University	Mesoporous carbon nanoparticle cathode, with sulfur in nanopores [18].

(*Continued*)

Table 10.1 (*Continued*)

S. No.	Developers	Approach/description
14.	Stanford University	Crystalline copper hexacyanoferrate nanoparticles [19].
15.	University of California	Gold nanowire in manganese dioxide shell, coated in electrolyte gel [20].
16.	Stanford University	Silicon nanowires grown on stainless steel substrate for anode of Li-ion battery [21].
17.	Stanford University	Lithium sulfur batteries with carbon nanofiber covered sulfur cathode [22].
18.	Rensselaer Polytechnic Institute	Photothermally reduced graphene for anode of Li-ion batteries [23].
19.	Stanford University and SLAC	Nickel hexacyanoferrate (NiHCF) nanoparticle electrodes, with sodium and potassium ions [24].

While a majority of the many approaches dedicated to improving batteries through nanotechnology are still in the research and development phase, some have already been assembled into products, tested, and some even sold. Among these are A123Systems, mPhase Technologies, and HE3DA (#1–3 in Table 10.1). The nanoenabled $LiFePO_4$ cathode of A123Systems and the 3D electrode of HE3DA are later discussed, as they exploit popular approaches studied by several research groups.

Others, still in the R&D phase, should not be overlooked, as many offer promising, if not superior, possibilities, and may one day make it to the market as well. Silicon, for instance, has been a widely attractive material for use in batteries. Research teams pursuing that use have been particularly popular in Stanford University, among others (#9–12, 16 in Table 10.1). Not only is Si one of the most abundant elements in the earth's crust, it offers the highest volumetric capacity (9786 mAh/cm^3) and gravimetric capacity (4200 mAh/g , $Li_{22}Si_5$) among all other candidates, besides lithium metal itself [14, 25].

Carbonaceous materials, though deemed unpromising by some due to cycling and safety concerns, are actually still very commonly studied and ventured to improve [3]. Potential in these materials has been acknowledged, so great effort has been made to exploit the potential, while controlling the unwanted properties associated with them, with the aid of nanotechnology. Most commonly studied appear to be carbon nanotubes; however, other forms of nanostructured carbon materials have been extensively studied as well (#4, 6–8, 13, 17). Some varieties of these carbonaceous materials for batteries are later discussed.

A large variety of other approaches exist and will continue to be explored, whether similar to these here mentioned or not, yet bound still by one key point in common—the adoption of nanotechnology for the vast advantageous attributes offered in none but the nanoscale. These attributes, along with key approaches in utilizing them, will be discussed in this chapter.

10.2 Energy Density, Power Density, and Cycling Challenge

Energy and power densities are considered two of the main factors in truly gauging a battery's potential, along with cycling ability. They reflect not just what a battery may offer in terms of stored electric potential, but for how long, and how well. Although lithium-ion batteries have revolutionized the industry with their comparatively high energy and power densities in the past, they reveal serious limitations for more demanding applications. For the higher densities required, there is a need of significantly higher capacities, which standard Li-ion batteries alone cannot fulfill. Reasons for their limitations are plentiful, including small surface areas, caustic structural changes during cycling, slow solid state Li-diffusion rates, and others [2].

10.2.1 Energy Density

Energy density is defined as the amount of energy stored in a system, per unit volume or mass. In a battery, the energy (watt-hour) is found as a function of the capacity or charge (ampere-hour) and electric potential (voltage) offered in the battery chemistry, which would render the amount of power expanded for 1 h, i.e., energy [1].

Capacity × Voltage = Energy

$$\text{Amp-hour} \times \text{Volt} = \text{Wh}$$

$$\text{Amp-hour} \times \text{Joule}/(\text{s} \times \text{Amp}) \times \text{s/h} = \text{Ws}$$

The energy density would then be watt-hour per unit volume or mass (Wh/kg or Wh/L), which offers a ratio applicable to a variety of battery sizes. For larger energy needs, a bigger battery is often used. However, with the rise of new technological pursuits comes a new challenge—meeting greater energy needs, in not so large a battery, i.e., increasing the energy density of batteries.

Given that the energy density is a product of a battery's specific charge capacity (Ah/g) and voltage, the primary objective scientists pursue is to increase either or both of these components. A higher capacity would render an increased energy density, as would a higher voltage, although voltage tends to be a more complex property to significantly alter without compromising battery safety, so more efforts are placed in capacity [3, 25, 26]. Extensive scientific efforts have thus been made in this particular direction: increasing capacity to increase energy density.

A battery is made of three key components: the two electrodes, anode and cathode, and the electrolyte, which separates the two electrodes. One of the most commonly used anode materials for Li-ion batteries, graphite, provides good cyclability and safety, but a relatively low specific capacity. In graphite, there could be the intercalation of only one lithium ion for every six carbon atoms, as shown, rendering the low 372 mAh/g capacity.

$$\text{C} + x\text{Li} + xe^- \Leftrightarrow \text{Li}_x\text{C}_6$$

Better performing batteries were needed, however, and in attempts to increase capacity, a number of materials have been studied over the years. Some of the particularly favorable alternative candidates for the negative electrode have been the metals capable of reversibly reacting with lithium, such as Al, Si, and Sn, to form lithium alloy compounds, according to the following reaction, where M is the given metal [6]:

$$x\text{Li}^+ + xe^- + \text{M} \Leftrightarrow \text{Li}_x\text{M}$$

For example, Li alloyed with Sn, to form an $Li_{4.4}Sn$ compound, can render a theoretical capacity of 993 mAh/g.

$$\text{Li}_{4.4}\text{Sn} \Rightarrow \text{Sn} + 4.4\text{Li}^+ + 4.4e^- \quad \text{(charge)}$$

$$\text{Sn} + 4.4\text{Li}^+ + 4.4e^- \Rightarrow \text{Li}_{4.4}\text{Sn} \quad \text{(discharge)}$$

These combinations offer significantly higher specific capacity and even voltage magnitudes, compared to graphite. However, limitations still plague such Li-ion batteries, as the necessity of large amounts of lithium eventually results in severe changes in volume in the accompanying metal, as listed in Table 10.2.

Table 10.2 Volume changes of some anode materials, upon cycling

Material	Volume change
Al	196%
Si	413%
Sn	258%
Sb	236%
Graphite	114%

Source: Ref. [27].

Upon Li insertion/extraction or Li alloying/de-alloying, volume is shown to undergo substantial expansion–contraction, almost linearly dependent on the concentration of lithium. Mechanical strains ensue, leading to electrode cracking, and eventual disintegration after a number of charge–discharge cycles [4–6, 25].

The problem analyzed in these metal alloy electrodes is specifically a result of the particle size. The magnitude of the particle size appears to be in direct relationship to the magnitude of the mechanical strain. Large particle size, as present in Li-ion batteries pre-nanotechnology, results in poorly accommodated, large mechanical strains and all that ensues. However, in order to not waste the high capacity offered by these lithium alloys, nanotechnology can be employed—morphological particle size reduction to a few nanometers and/or creation of specially designed nanostructures. This, in turn, dramatically stabilizes the cycling reactions due to the improved ability of smaller particles to accommodate the mechanical strains, and enhances power as the lithium-ion diffusion length diminishes, as discussed later. These modified properties and subsequent reactions essentially allow the high capacity factor of these lithium alloy materials to bring the world one step closer to increasing the energy density of a battery without some of the inadvertent reactions that once impeded them [6].

In Table 10.3, theoretical specific capacities of some lithium storage metals are shown, as compared to graphite. Visibly, lithium alloy combinations prove remarkably beneficial to increasing capacity, and hence, energy density.

Table 10.3 Theoretical specific capacities of some lithium storage metals, compared to graphite

Lithium alloy combination	Capacity [mAh/g]
LiAl	993
$Li_{21}Si_5$	4008
$Li_{22}Sn_5$	994
Li_3Sb	660
LiC_6	372

Source: Ref. [6] (#5 in Table 10.1).

10.2.2 Cycling

As one of the many issues with standard Li-ion batteries being the large volume expansion–contraction and subsequent electrode cracking and disintegration, cycling has become a major concern. Particularly for more demanding applications, such as electronic vehicles or grid-scale systems, batteries with the ability to undergo at least several thousand cycles of charge and discharge without losing significant capacity is crucial. In standard, bulk materials of micro-sized particles, the repeated straining caused by the volume changes results in a continuous and rapid decrease in discharge capacity with number of cycles. Si, a popular choice for anode material, demonstrated a 70% drop in reversible capacity after only five cycles. During lithiation, particles expand and repeatedly crack into one another, and during delithiation, particles contract and detach, cutting important electrical connections. This all in all contributes to the drastic capacity fading that occurs with each ensuing cycle [14, 28, 29].

Another key factor affecting cycle life is the solid electrolyte interphase (SEI). This layer forms at the surface between the electrode and organic electrolyte as the decomposition of the latter takes place. It is a spontaneous occurrence, and favorable when the potential of anode is below ~1 V vs Li/Li^+. In order to prevent further side chemical reactions and ensure a long cycle life, the layer must be stable, dense, ionically conducting, and electronically insulating. Yet, again, a difficulty is presented in forming a stable SEI layer due to the inevitable large changes in volume, as occur on the micro-scale. Just as the SEI is formed upon expansion, it can crack and break as the particles shrink from each other, causing the SEI to be unstable as it re-forms and becomes thicker and thicker with the number of cycles.

The thick SEI layer formation is a challenge that may be present even in some nanomaterials, as a side effect of high surface energy and increased electrolyte/electrode surface area, which induces undesirable surface reactions. Particularly with nano-sized $LiFePO_4$, studies demonstrate instability in air atmosphere increases as the particle size is reduced; the smaller the particle is, the more area is exposed to air. The thick SEI films that form on the surface of the anode consume a lot of lithium ions that are supplied by the cathode, and essentially affect the cycle life of the battery. Alternative approaches and materials in studies conducted by a number of researchers have provided solutions to this problem and, in contrast, have been aided by the use of nanostructures [11, 14, 28].

Essentially, the improved cycling properties of nanostructured materials offer a significant benefit to batteries, as they can absorb and accommodate the effects of the volume changes, overcome critical SEI challenges, and internally stabilize the cycle performance of the electrode, and hence the battery.

In Table 10.4, some examples of high cycle numbers attained upon the use of nanotechnology are shown.

Table 10.4 Some approaches and respective cycling values

Approach/material	Discharge cycles	Capacity retention
Double-walled Si nanotubes [14]	6,000	88%
CuHCF [19]	40,000	83%
NiHCF (with Na) [25]	5,000	100%
NiHCF (with K) [25]	1,000	100%
Carbon-covered Si nanoparticle [16]	1,000	97%
$LiFePO_4$/Carbon coating [29]	1,100	95%
Nanophosphate $LiFePO_4$ [8]	8,000	~100%

10.2.3 Power Density

Another important consideration in the viability and success of a battery, especially for electronic vehicles and grid-energy storage systems, is the matter of its power density. With power density, the importance of time emerges. Power is a value of energy discharged per second (J/s = W), with the amount of power per unit volume (W/m^3) being the power density. In a speed-driven market, the rate at which energy can be drawn is an important factor in making batteries a comparable alternative to existing fuels. Currently, most battery alternatives are still relatively far from performing at the same rates made possible with fuels, such as coal or petrol, yet, in recent years, many advances have been, and continue to be, made.

While several factors may affect a battery's power density, the main variables are the properties associated with the discharge process, specifically, the lithium diffusion rates during insertion/extraction. In order to increase the power density, or the rate at which energy can be drawn of a battery, higher Li-diffusion rates would be necessary [25, 28].

The time associated with Li-ion diffusion in electrode materials is a factor of two properties: the Li-ion diffusion coefficient (D_{Li}), an intrinsic property of the material, and diffusion length (L), related to the particle size, as shown in the following relationship: $t = L^2/D_{Li}$.

As is evident, the longer the length of diffusion, the more the time required. Similarly, as the diffusion length diminishes, say to the nanoscale, the shorter the intercalation time, and the higher the resulting power density. Therefore, for a future where higher power density is possible for batteries, better electrode materials with nanoscale particles are a necessity [4, 25, 28].

10.3 Nano-enhanced Approaches

After evaluating some of the key factors determining a battery's chemical significance, it is possible to evaluate some of the approaches, as listed in Table 10.1.

10.3.1 Nanostructured Carbon Cathodes

Several approaches have been pursued to improve batteries with nanotechnology. Though carbonaceous materials have been deemed less electrochemically effective than some other materials, such as the aforementioned metal alloys, they have still been experimented with and returned with some rather favorable results when nanotechnology was employed. In research concentrating on Li-SO_2 batteries, it was demonstrated that nanostructured carbon cathode materials could potentially provide reversible capacities of higher than 1000 mAh/g, with working potential of 3 V, and even provide a theoretical energy density of about 70% higher than that of currently used Li-ion batteries (651 Wh/kg) [7].

Table 10.5 lists the discharge capacities of some of the higher performing carbon materials of the study, visibly comparable to those of the lithium alloy combinations. Though KB-600JD (Ketjenblack EC-600JD carbon black, Akzo Nobel, Japan) and rGO (reduced graphene oxide) cathodes produced higher capacities, it should be noted that they showed continuous capacity fading during cycling, whereas the OMC (ordered mesoporous carbon) cathode proved to be highly stable with excellent capacity retention over 150 cycles.

Table 10.5 Carbon materials and respective capacities

Carbon	Capacity (mAh/g)
KB	1654
OMC	1050
rGO	1190

Source: Ref. [7] (#4 in Table 10.1).

The advantage of these nanostructured materials lies particularly in the pore size of the particles. Although carbon materials with higher active surface areas allow for higher discharge capacities, surprisingly, no relationship was found between surface area and reversible capacity for the carbon materials tested. On the other hand, the nano-carbon cathodes proved to have a direct correlation to the pore size of the structure, with the best electrochemical performance belonging to carbon materials of larger mesopores or macropores. However, the best sample was not that with the largest pore size, but the best balance of meso- and macroporous structure for lasting cycle performance. The large pore size in the carbon electrode also proved favorable as a buffering space for when the lithium-related volume changes and subsequent mechanical and electrical degradations occur, as they did in lithium alloy storage materials. The large pore size offered in the OMC cathode can effectively accommodate the insulating discharge products and maintain the consistent mass transport of the catholyte due to reduced pore blockage. Essentially, again, the nano-modifications of previously sub-par performing carbonaceous materials proved greatly beneficial to improving the capacity and overall cycle performance of batteries, particularly Li-SO_2 batteries [6, 7].

10.3.2 Lithium Iron Phosphate ($LiFePO_4$) Cathode

A particularly promising and commonly used cathode material is lithium iron phosphate. Even in its pre-nanoscale condition, $LiFePO_4$ is an attractive choice, due to its chemical and thermal stability, low toxicity, and tenably high capacity (170 mAh/g). It is even the material of choice for an already existing Li-ion battery company, A123Systems [4, 6, 8].

$LiFePO_4$ has an interesting quality, however. While nanotechnology introduces several new benefits to this material, one of its main inherent attributes, stability, is actually known to be significantly compromised when reduced to the nanoscale alone. As previously mentioned, the increased exposure to air atmosphere, due to the higher surface area present in nanoparticles, has a direct effect on the stability of $LiFePO_4$. Yet, as also mentioned, many approaches have addressed these stability issues as well as enabled the use of all the benefits that come with $LiFePO_4$ nanoparticles. One such example comes in the form of surface coating. Full carbon surface coating of $LiFePO_4$ nanoparticles has been shown to be a promising way of increasing stability in air atmosphere, as they shield the nanoparticles from direct contact with air. Essentially, particular nanostructures and nanoscale morphological modifications enable the ability to control the undesired side phenomena related to LiFePO4. Even when, in some instances, there may be side effects associated with the nanoscale, further adjustments and alternative nanostructures have been shown to resolve many issues [4, 6, 28].

Another use of nanotechnology was in correcting a problem that was inherent to the larger-scale form of $LiFePO_4$. While the intercalation/deintercalation process of $LiFePO_4$ is usually slow, causing a resulting low power density, it was determined that by reducing the particle size to the nanoscale, rate capabilities were dramatically increased, with good cycling capacity and high rates. In the instance of A123System's Nanophosphate $LiFePO_4$ battery, more than 8000 cycles were achieved with full depth of discharge and little impedance growth [8]. These performance results are particularly desirable for electronic vehicles and grid-energy storage systems [4, 6].

10.3.3 3D Electrodes

Another company that is already working on a battery prototype and with a planned market launch at the end of 2016 is HE3DA, of Czech Republic. The battery developed here uses a less discussed, though growing in popularity, approach to batteries: 3D electrodes. An alternative to thin film (2D) technology, 3D electrodes have shown to significantly increase volumetric capacity, density, safety, and other key components of a battery. Besides HE3DA's 3D battery, which boasts of a 500 Wh/L battery and 5000 cycle lifespan, several other variations exist and continue to be explored in the area of 3D batteries with nanostructured electrodes for the various benefits they seem to offer [4, 10].

10.3.4 Si Nanostructures

A most promising negative electrode material, garnering special attention by a large group of researchers, is silicon. Not surprisingly, it offers some of the best characteristics necessary in a battery, when compared to some carbonaceous or other metal alternatives. At 4200 mAh/g, it has one of the highest theoretical capacities of any material, with nearly 12 times that of the commonly used graphite anode. It is also an abundant natural resource and not environmentally harmful [11, 14].

Yet, as has been the case with most materials, even inherent benefits can sometimes be lost due to certain challenges associated with their intercalation process with lithium. In Si anodes, the main challenges relate to the severe volume changes and subsequent SEI instability upon cycling. When in the nanoscale, some other challenges are presented in the form of side reactions with the electrolyte, as well as low volumetric capacity. However, as has been done for other materials previously mentioned, specific nanostructures designed to overcome these specific challenges have been, and continue to be, developed for ever-increasing improvements in battery performance.

In a study comparing some of the many different approaches to using nanostructured silicon in batteries, summarized in Table 10.6, there were some rather promising findings regarding, especially, the surface-clamped hollow Si nanostructured, sometimes termed the double-walled Si nanotubes, and the yolk–shell design (#9 in Table 10.1). These particular approaches have been explored in attempts to solve the stability issues related to the SEI, as well as other challenges. The double-walled Si nanotube anode serves as an example, in which the unstable SEI problem was overcome using a strong coating layer, SiO*x*, able to prevent the Si particles from expanding out and cracking into each other upon cycling. It provides a layer to preserve the anode while still allowing the lithium ions to pass through, creating a static and, therefore, more stable outer surface at the electrode–electrolyte interface. These structures also benefit from the mass other advantages made possible with the nanoscale, such as short Li-ion diffusion length and subsequent high power density [14].

Table 10.6 Summary of some Si nanostructure cycle and capacity results

Si nanostructure	Capacity	Cycles	Notes
Carbon-coated Si nanowire	1326 mAh/g	40	Stable
Core–shell Si nanowire	1000 mAh/g	100	90% capacity retention
Si nanoparticles	1200 mAh/g	130	Stable
Hollow Si nanostructures	2725 mAh/g	700	Capacity down 8% per 100 cycles
Si nanotubes	~3200 mAh/g	200	89% capacity retention
Double-walled Si nanotube	2970 mAh/g	6000	88% capacity retention
Yolk–shell Si nanostructure	2800 mAh/g	1000	74% capacity retention

Source: Ref. [14].

10.4 Discussion

While Li-ion batteries have seen dramatic improvements throughout the years, a lot of work has yet to be done in order to make batteries a comparable alternative to existing fuels. Nanotechnology is not a panacea. Even with its use, so far, energy density and other values fall far behind other fuels, as seen in Fig. 10.2 [30]. Yet, the suitability of a battery combination as an energy source is dependent not only on its energy density, but also on its mass. This is the rationale for considering the nano Li-ion battery for a grid battery, since the mass of the battery can be made rather large. This mass keeps renewing itself upon recharge because it is a battery. All other competitive fuels (shown in Fig. 10.2) such as coal, the mass diminishes to nothing with electricity production and needs to be replenished by trucking/shipping it in.

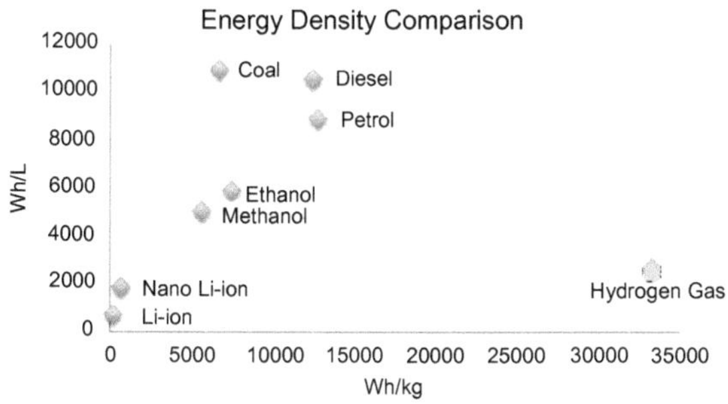

Figure 10.2 Energy density comparison [30].

In the direction green-energy technology appears to be taking, energy storage systems for electric vehicles, alternative grid-energy systems, and others will be a necessity. Nanotechnology simply offers a drastic improvement from what is already there, to what will have to be relied upon. With its various special properties not known to be possible in bulk materials [31], nanotechnology brings batteries closer to that appropriate suitability, and may very well change the technological landscape of the future, despite its comparative limitations.

Some developers, such as Exergonix, have already proven the viability of nano-based grid batteries, launching a 1 MW nanotech Battery Energy Storage System (BESS), in Kansas City in 2012. It is an adaptable system, with the ability to scale up or down in size according to demand, for the accommodation to different energy storage needs [32]. In general, it may be impractical to scale up a battery from the nanoscale to a large grid-scale battery, so it must be noted that nanotechnology can be applied to a battery in different ways. A nano-enhanced battery is not a wholly nano-sized battery. While the dimensions of some particles may be on the nanoscale, a nano-battery may not always be entirely built on the nanoscale, as is particularly the case for grid batteries. A battery can be built up from nanoparticles or nanostructures, considered primary nanomaterials, or simply be a combination built from micro-particles with internally nanoscale regions, considered secondary nanomaterials [33]. In the case of the Exergonix BESS and other such grid systems, secondary nanomaterials are likely employed.

10.5 Conclusion

Batteries of all sizes can benefit from nanotechnology; this is true whether the batteries are intended for devices as small as hearing aids or as large as grid-energy storage applications. While improving energy density might be a primary objective for applications requiring smaller, lighter batteries, objectives vary when it comes to fulfilling larger-sized demands. No matter the particular necessity, it has been repeatedly studied and shown that nanostructured materials offer special properties not known to be possible in bulk materials and can, therefore, be widely applied for a variety of uses in which energy storage is required.

The development of large batteries for the electric power industry would be a big leap forward. If overheating problems with solid electrolytes can be overcome, among other problems, these large batteries will be the fulcrum for renewable energies to leapfrog to center stage with regards to power generation. Electric power via WWS sources of energy will be more easily adapted as grid power since these sources cannot be dependent on 24 h a day. This, in turn, will cut carbon dioxide emissions quite a lot and reduce the problems of the energy–water–food nexus [34–36] occurring in many places around the world because of climate change, since climate change [37–39] can be mitigated with carbon dioxide reduction. Carbon dioxide reduction, in turn, is achieved if the combustion of fossil fuels is reduced worldwide.

References

1. Tarascon, J. M. and Armand, M. Issues and challenges facing rechargeable lithium batteries. *Nature*, 2001; **414**(6861): 359–367.
2. Ji, L., Lin, Z., Alcoutlabi, M., and Zhang, X. Recent developments in nanostructured anode materials for rechargeable lithium-ion batteries. *Energy & Environmental Science*, 2011; **4**(8): 2682–2699.
3. Manthiram, A. Materials challenges and opportunities of lithium ion batteries. *Journal of Physical Chemistry Letters*, 2011; **2**(3): 176–184.
4. Bruce, P. G., Scrosati, B., and Tarascon, J. M. Nanomaterials for rechargeable lithium batteries. *Angewandte Chemie International Edition*, 2008; 47(16): 2930–2946.
5. Stura, E. and Nicolini, C. New nanomaterials for light weight lithium batteries. *Analytica Chimica Acta*, 2006; **568**(1): 57–64.
6. Panero, S., Scrosati, B., Wachtler, M., and Croce, F. Nanotechnology for the progress of lithium batteries R&D. *Journal of Power Sources*, 2004; **129**(1): 90–95.
7. Jeong, G., Kim, H., Park, J. H., Jeon, J., Jin, X., Song, J., Kim, B. R., Park, M. S., Kim, J. M., and Kim, Y. J. Nanotechnology enabled rechargeable Li–SO_2 batteries: Another approach towards post-lithium-ion battery systems. *Energy & Environmental Science*, 2015; **8**(11): 3173–3180.
8. Vartanian, C. and Bentley, N. A123 Systems' advanced battery energy storage for renewable integration. In *Power Systems Conference and Exposition (PSCE), 2011 IEEE/PES 2011* Mar 20 (pp. 1–6). IEEE.
9. Krupenkin, T. N., Taylor, J. A., Schneider, T. M., and Yang, S. From rolling ball to complete wetting: The dynamic tuning of liquids on nanostructured surfaces. *Langmuir*, 2004; **20**(10): 3824–3827.
10. HE3DA S.R.O., "Lithium Accumulator," 2015: n. pag. Print.
11. Fu, K., Yildiz, O., Bhanushali, H., Wang, Y., Stano, K., Xue, L., Zhang, X., and Bradford, P. D. Aligned carbon nanotube–silicon sheets: A novel nano-architecture for flexible lithium ion battery electrodes. *Advanced Materials*, 2013; **25**(36): 5109–5114.
12. Lee, S. W., Yabuuchi, N., Gallant, B. M., Chen, S., Kim, B. S., Hammond, P. T., and Shao-Horn, Y. High-power lithium batteries from functionalized carbon-nanotube electrodes. *Nature Nanotechnology*, 2010; **5**(7): 531–537.

13. Reddy, A. L., Shaijumon, M. M., Gowda, S. R., and Ajayan, P. M. Coaxial MnO_2/carbon nanotube array electrodes for high-performance lithium batteries. *Nano Letters*, 2009; **9**(3): 1002–1006.
14. Wu, H. and Cui, Y. Designing nanostructured Si anodes for high energy lithium ion batteries. *Nano Today*, 2012; **7**(5): 414–429.
15. Ko, M., Chae, S., Ma, J., Kim, N., Lee, H. W., Cui, Y., and Cho, J. Scalable synthesis of silicon-nanolayer-embedded graphite for high-energy lithium-ion batteries. *Nature Energy*, 2016; **1**: 16113.
16. Liu, N., Lu, Z., Zhao, J., McDowell, M. T., Lee, H. W., Zhao, W., and Cui, Y. A pomegranate-inspired nanoscale design for large-volume-change lithium battery anodes. *Nature Nanotechnology*, 2014; **9**(3): 187–192.
17. Ge, M., Rong, J., Fang, X., Zhang, A., Lu, Y., and Zhou, C. Scalable preparation of porous silicon nanoparticles and their application for lithium-ion battery anodes. *Nano Research*, 2013; **6**(3): 174–181.
18. Schuster, J., He, G., Mandlmeier, B., Yim, T., Lee, K. T., Bein, T., and Nazar, L. F. Spherical ordered mesoporous carbon nanoparticles with high porosity for lithium–sulfur batteries. *Angewandte Chemie*, 2012; **124**(15): 3651–3655.
19. Wessells, C. D., Huggins, R. A., and Cui, Y. Copper hexacyanoferrate battery electrodes with long cycle life and high power. *Nature Communications*, 2011; **2**: 550.
20. Le Thai, M., Chandran, G. T., Dutta, R. K., Li, X., and Penner, R. M. 100k cycles and beyond: Extraordinary cycle stability for MnO_2 nanowires imparted by a gel electrolyte. *ACS Energy Letters*, 2016; **1**: 57–63.
21. Chan, C. K., Peng, H., Liu, G., McIlwrath, K., Zhang, X. F., Huggins, R. A., and Cui, Y. High-performance lithium battery anodes using silicon nanowires. *Nature Nanotechnology*, 2008; **3**(1): 31–35.
22. Zheng, G., Yang, Y., Cha, J. J., Hong, S. S., and Cui, Y. Hollow carbon nanofiber-encapsulated sulfur cathodes for high specific capacity rechargeable lithium batteries. *Nano Letters*, 2011; **11**(10): 4462–4467.
23. Mukherjee, R., Thomas, A. V., Krishnamurthy, A., and Koratkar, N. Photothermally reduced graphene as high-power anodes for lithium-ion batteries. *Acs Nano*, 2012; **6**(9): 7867–7878.
24. Wessells, C. D., Peddada, S. V., Huggins, R. A., and Cui, Y. Nickel hexacyanoferrate nanoparticle electrodes for aqueous sodium and potassium ion batteries. *Nano Letters*, 2011; **11**(12): 5421–5425.
25. Goriparti, S., Miele, E., De Angelis, F., Di Fabrizio, E., Zaccaria, R. P., and Capiglia, C. Review on recent progress of nanostructured anode

materials for Li-ion batteries. *Journal of Power Sources*, 2014; **257**: 421–443.

26. Ortiz, G. F., López, M. C., Li, Y., McDonald, M. J., Cabello, M., Tirado, J. L., and Yang, Y. Enhancing the energy density of safer Li-ion batteries by combining high-voltage lithium cobalt fluorophosphate cathodes and nanostructured titania anodes. *Scientific Reports*, 2016; **6**.
27. Carnegie Mellon University. Self-Contained, Alloy Type, Thin Film Anodes For Lithium-Ion Batteries. 2016: n. pag. Print.
28. Wang, Y., Li, H., He, P., Hosono, E., and Zhou, H. Nano active materials for lithium-ion batteries. *Nanoscale*, 2010; **2**(8): 1294–1305.
29. Wu, H. B., Zhang, G., Yu, L., and Lou, X. W. One-dimensional metal oxide–carbon hybrid nanostructures for electrochemical energy storage. *Nanoscale Horizons*, 2016; **1**(1): 27–40.
30. Wikipedia, "Energy Density". Retrieved 9/05/16. https://en.wikipedia.org/wiki/Energy_density.
31. Dunn, B., Kamath, H., and Tarascon, J. M. Electrical energy storage for the grid: A battery of choices. *Science*, 2011; **334**(6058): 928–935.
32. New Nanotech Battery Energy Storage System Debuts in Kansas City [Internet]. *CleanTechnica*, 2012.
33. Arico, A. S., Bruce, P., Scrosati, B., Tarascon, J. M., and Van Schalkwijk, W. Nanostructured materials for advanced energy conversion and storage devices. *Nature Materials*, 2005; **4**(5): 366–377.
34. Wong, K. V. (Expert View) Recommendations for energy water nexus problems. *Journal of Energy Resources Technology*, 2014; **136**(3): 034701.
35. Wong, K. V. and Pecora, C. Recommendations for energy-water-food nexus problems, *Journal of Energy Resources Technology*, 2014; **137**(3): 032002.
36. Wong, K. V. (Expert View) Energy-water-food nexus and recommendations for security. *Journal of Energy Resources Technology*, 2015; **137**: 034701-1.
37. Wong, K. V. The second law of thermodynamics and heat release to the global environment by human activities. ASME International Mechanical Engineering Congress and Exposition, *Energy Systems Analysis, Thermodynamics and Sustainability; NanoEngineering for Energy; Engineering to Address Climate Change, Parts A and B*, 2010; **5**: 469–472. doi:10.1115/IMECE2010-38201.

38. Wong, K. V., Dai, Y., and Paul, B. Anthropogenic heat release into the environment. *Journal of Energy Resources Technology*, 2012; **134**(4): 041602.

39. Wong, K. V. *Climate Change*, Momentum Press, NY, 2016, 195 pp., ISBN-13:978-1-60650-848-0.

Chapter 11

An Investigation of the Thermal Performance of Cylindrical Heat Pipes using Nanofluids

Maryam Shafahi,[a] Vincenzo Bianco,[b] Kambiz Vafai,[a] and Oronzio Manca[b]

[a]*Department of Mechanical Engineering, University of California Riverside, Riverside, CA 92521, USA*

[b]*Dipartimento di Ingegneria Aerospaziale e Meccanica, Seconda Università degli Studi di Napoli, Via Roma 29, 81031 Aversa (CE), Italy*

vafai@engr.ucr.edu

In this chapter, a two-dimensional analysis is used to study the thermal performance of a cylindrical heat pipe utilizing nanofluids. Three of the most common nanoparticles, namely Al_2O_3, CuO, and TiO_2, are considered the working fluid. A substantial change in the heat pipe thermal resistance, temperature distribution, and maximum capillary heat transfer of the heat pipe is observed when using a nanofluid. The nanoparticles within the liquid enhance

This chapter has been reprinted from the *International Journal of Heat and Mass Transfer*, Vol. 53, Maryam Shafahi, Vincenzo Bianco, Kambiz Vafai, and Oronzio Manca, An investigation of the thermal performance of cylindrical heat pipes using nanofluids, 376–383, Copyright 2009, with permission from Elsevier.

Nanotechnology and Energy
Edited by Kaufui V. Wong

ISBN 978-981-4774-24-6 (Hardcover), 978-1-315-16357-4 (eBook)
www.panstanford.com

the thermal performance of the heat pipe by reducing the thermal resistance while enhancing the maximum heat load it can carry. The existence of an optimum mass concentration for nanoparticles in maximizing the heat transfer limit is established. The effect of particle size on the thermal performance of the heat pipe is also investigated. It is found that smaller particles have a more pronounced effect on the temperature gradient along the heat pipe.

11.1 Introduction

Heat pipes and their applications in thermal management have been studied for decades. They constitute an efficient, compact tool to dissipate substantial amount of heat from various engineering systems, including electronic components. Heat pipes are able to dissipate substantial amount of heat with a relatively small temperature drop along the heat pipe while providing a self-pumping ability due to an embedded porous material in their structure. A limiting factor for the heat transfer capability of a heat pipe is related to the working fluid transport properties. In order to overcome this limitation, the thermophysical properties of the fluid can be improved. An innovative way to enhance liquid thermal conductivity is the dispersion of highly conductive solid nanoparticles within the base fluid. This new generation of conductive fluids with nanoparticles is referred to as nanofluids [1]. The nanoparticles within the fluid change the other thermophysical properties such as density and viscosity. A number of investigations have been conducted to study boiling and natural and forced convection using nanofluids [2].

Due to the useful features of a nanofluid, various research groups [3–14] have tried to engage it within a heat pipe and study the subsequent thermal enhancement experimentally. Different nanoparticles such as silver [3, 8, 11, 13], CuO [5, 9], diamond [6, 12], titanium [4, 7], nickel oxide [10], and gold [14] have been utilized within the heat pipe working fluid. The improved thermal performance is observed through a reduction in thermal resistance [3, 6–9, 11, 13, 14], a drop in the temperature gradient along the heat pipe [3, 12, 13], an increase in the heat pipe efficiency [4], and an enhancement in the overall heat transfer coefficient [10]. In some studies [5, 9], the existence of an optimum amount of nanoparticle mass concentration providing the highest thermal performance has been established. Liu et al. [9] have shown that the heat pipe operating pressure has a significant effect on thermal performance.

In another study, Riehl [10] has observed that a higher heat transfer coefficient can be seen when using nanoparticles in water under low heat input conditions. Tsaia [14] investigated the influence of particle size on the heat pipe thermal performance.

Almost all of the research work on the use of nanoparticles in heat pipes is experimental. To the best of authors' knowledge, there is a lack of information in modeling heat pipe characteristics in the presence of a nanofluid. The focus of the present work is to model the influence of a nanofluid on the thermal performance of a heat pipe. A comprehensive analytical model proposed by Zhu and Vafai [15] is modified to include the effects of nanofluids within a cylindrical heat pipe. A common range of concentration for different nanoparticles, namely Al_2O_3, CuO, and TiO_2, in water is considered the operational fluid within the heat pipe under various heat inputs. The temperature profile, thermal resistance, and the maximum heat transfer limits are investigated. A significantly higher thermal performance is observed due to a reduction in the thermal resistance as well as the end to end temperature gradient. Furthermore, the maximum capillary limit heat transfer increases when using a nanofluid as the working fluid. To the best of our knowledge, there is no study on the effect of a nanofluid on the maximum heat transfer in heat pipes. Our results establish the existence of an optimum concentration of nanoparticles within the working fluid with respect to attaining the maximum heat transfer. This optimum concentration is established for different nanofluids under different operational conditions.

Nomenclature

K	permeability of the wick [m^2]
k	thermal conductivity [W/m K]
k_{eff}	effective thermal conductivity of the liquid-saturated wick [W/m K]
k_{layer}	nanolayer thermal conductivity [W/m K]
k_{wall}	thermal conductivity of the heat pipe wall [W/m K]
h	convective heat transfer coefficient [W/m^2 K]
h_{fg}	latent heat of the working fluid [KJ/kg]
L	length of the heat pipe [m]
L_a	length of the adiabatic section [m]
L_c	length of the condenser section [m]
L_e	length of the evaporator section [m]
P	Pressure [Pa]
Q	input heat [W]

Q_c	heat transfer rate at the condenser section [W]
Q_e	heat transfer rate at the evaporator section [W]
r	cylindrical coordinates [m]
r_p	nanoparticles radius [nm]
r_c	effective pore radius of the wick [m]
R_o	heat pipe's outer radius [m]
R_t	heat pipe thermal resistance [K/W]
R_v	vapor core radius [m]
R_w	heat pipe's inner radius [m]
T	temperature [K]
T_b	bulk temperature of the coolant in cooling jack [K]
u	horizontal velocity component [m/s]
U	velocity along the axis [m/s]
v	vertical velocity component [m/s]
v_1	vapor injection velocity [m/s]
v_2	vapor suction velocity [m/s]
w	nanolayer thickness [nm]
x	cylindrical coordinates [m]
Greek symbols	
ϕ	particle concentration
ε	porosity of the wick
γ	porous wick shape parameter $\left(\sqrt{\varepsilon / K}\right)$
μ	dynamic viscosity [N s/m^2]
μ^+	dimensionless dynamic viscosity $\left(\frac{\mu_l}{\mu_v}\right)$
ρ	density [kg/m^3]
ρ^+	dimensionless density $\left(\frac{\rho_l}{\rho_v}\right)$
σ_1	surface tension of the working fluid [N/m]
Subscripts	
0	heat pipe with pure fluid
bf	base fluid
i	liquid–vapor interface
l	liquid phase
max	maximum
nf	nanofluid

np	nanoparticle
p	particle
v	vapor
Superscript	
+	dimensionless quantity

11.2 Analysis

The schematic of a cylindrical heat pipe used in the present analysis is given in Fig. 11.1. The heat applied at the evaporator section causes vaporization and, consequently, a pressure increase in the working fluid. The vapor formed in the evaporator flows toward the condenser as a result of the pressure difference between the evaporator and condenser regions. Vapor condenses releasing its latent heat in the condenser, and the condensate flows back to the evaporator through the porous wick, thus completing the cycle. The mathematical model adopted in this work is based on the following assumptions: The process is steady state; radiative and gravitational effects are negligible; and the fluid is considered Newtonian and incompressible. Moreover, the injection and suction velocities at the liquid–vapor interface are considered to be uniform. Also, the wick is assumed to be isotropic and saturated with the working fluid. The liquid flow within the porous wick is modeled using the generalized momentum equation [16].

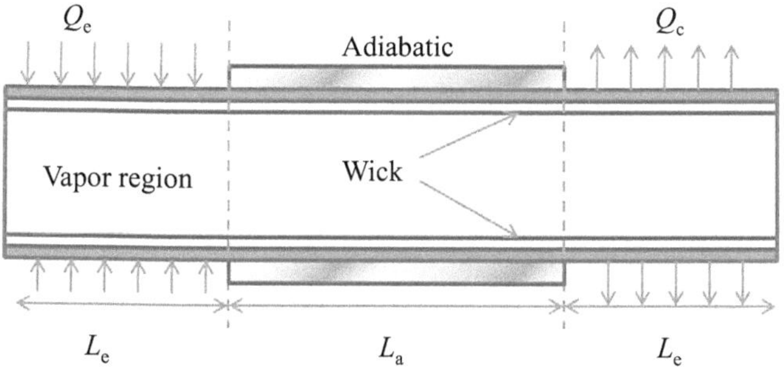

Figure 11.1 Schematic of a cylindrical heat pipe under consideration.

11.2.1 Governing Equations

The mass and momentum conservation equations for liquid and vapor regions are [15]:

Vapor regions

$$\frac{\partial u_v}{\partial x}+\frac{\partial v_v}{\partial r}+\frac{v_v}{r}=0 \tag{11.1}$$

$$\rho_v\left(u_v\frac{\partial u_v}{\partial x}+\frac{\partial u_v}{\partial r}\right)=-\frac{\partial p_v}{\partial x}+\mu_v\left(\frac{\partial^2 u_v}{\partial r^2}+\frac{1}{r}\frac{\partial u_v}{\partial r}\right) \tag{11.2}$$

$$\frac{\partial p_v}{\partial r}=0 \tag{11.3}$$

Liquid region

$$\frac{\partial u_l}{\partial x}+\frac{1}{r}\frac{\partial}{\partial r}(ru_l)=0 \tag{11.4}$$

$$\frac{\mu_l}{\varepsilon}\left(\frac{\partial^2 u_l}{\partial r^2}+\frac{1}{r}\frac{\partial u_l}{\partial r}\right)-\frac{\mu_l}{K}u_l-\frac{\rho_l F\varepsilon}{K^{1/2}}|u_l|u_l-\frac{\partial p_l}{\partial x}=0 \tag{11.5}$$

where ε, K, and F, are porosity, permeability, and a geometric function based on the porous wick structure described by Vafai [17].

11.2.2 Analytical Solution

The analysis of Zhu and Vafai [15] is modified to incorporate the presence of a nanofluid within the heat pipe. The analysis is carried on to obtain the velocity, pressure, temperature, and the maximum heat transfer limit for the heat pipe for various operational conditions using nanofluids.

11.2.2.1 Liquid velocity profile

The maximum liquid velocity profile is determined by integrating the liquid continuity and applying proper boundary conditions [15]. The liquid velocity can be presented as:

$$U_l(x)=\begin{cases}-Bv_1x & 0\le x\le L_e\\ -Bv_1L_e & L_e\le x\le L_e+L_a\\ -Bv_2(L-x) & L_e+L_a\le x\le L_e+L_a+L_c\end{cases} \tag{11.6}$$

$$L = L_e + L_a + L_c$$

where B is given in Table 11.1 and can be seen that it is dependent on the liquid density.

11.2.2.2 Liquid and vapor pressure distributions

Liquid and vapor pressure distributions are obtained by integrating the momentum equations in the radial direction. Applying the appropriate boundary conditions [15], the pressure distributions can be presented as:

Vapor pressure

$$p_v(x) = p_v(0) + \begin{cases} (Gv_1^2 + Mv_1)x^2 & 0 \le x \le L_e \\ (Gv_1^2 - Mv_1)L_e^2 + 2Mv_1 L_e x & L_e \le x \le L_e + L_a \\ (Gv_2^2 - Mv_2)(x - L)^2 + Mv_2(L + L_a)L_c & \\ \qquad L_e \le L_a \le x \le L_e + L_a + L_c \end{cases} \tag{11.7}$$

Liquid pressure

$$p_l(x) = p_l(0) + \begin{cases} (G_2 v_1 x^2 & 0 \le x \le L_e \\ (G_2 v_1 L_e (2x - L_e) & L_e \le x \le L_e + L_a \\ -(G_2 v_2 [(x - L)^2 - (L + L_a)L_c] & L_e + L_a \le x \le L_e + L_a + L_c \end{cases} \tag{11.8}$$

where v_1 and v_2 are vapor injection and suction velocities, and G, M, and G_2 are given in Table 11.1. It can be seen that the liquid pressure explicitly depends on the density and viscosity of the working fluid. The nanofluid's density is calculated based on a simple representation of the particle and fluid phases, which has been shown to provide a reasonable result based on the experimental work of Pak and Cho [20], while a classical model proposed by Brinkman [21] is used to obtain the viscosity. As such, the density and viscosity of the nanofluid can be presented as:

$$\rho_{nf} = \rho_p \phi + (1 - \phi)\rho_{bf} \tag{11.9}$$

$$\mu_{nf} = \frac{\mu_{bf}}{(1-\phi)^{2.5}} \tag{11.10}$$

Table 11.1 Analytical solution parameters

$A=\frac{(G_2-2M_1)(L+L_a)}{4\pi R_v\rho_v h_{fg}}$	$B=\frac{2R_v}{\rho_{nf}^{+}(R_w^2-R_v^2)}\left(1+\frac{8\rho^{+}}{\mu^{+}\gamma^2R_v^2}\right)$
$C=\frac{8}{\mu_{nf}^{+}\gamma R_v^2}-B$	$D=CR_v$
$G_2=\frac{\mu_{nf}}{2K}\left(B+\frac{1}{\gamma R_v}\frac{2C}{1-\left(\frac{R_W}{R_v}\right)^2}\right)$	$M=\frac{4\mu_v}{R_v^3}(D-2)$
	$G=-\frac{\rho_v}{3R_v^2}(D^2-7D+16)$

11.2.2.3 Temperature distribution

The wall temperature profile is obtained assuming uniform wall temperature along the condenser and evaporator sections. A heat conduction model is used for the wall and liquid–wick region. As such, the heat pipe temperature profile can be presented in the following form [15]:

$$T_{wall}(x)=\begin{cases} T_b+\frac{Q}{2\pi L_c}\left[\left(\frac{\ln\left(\frac{R_0}{R_w}\right)}{k_{wall}}+\frac{\ln\left(\frac{R_w}{R_v}\right)}{k_{eff}}\right)\left(1+\frac{L_c}{L_e}\right)+\frac{1}{hR_0}\right] & 0\le x\le L_e \\ T_b+\frac{Q}{2\pi L_c}\left(\frac{\ln\left(\frac{R_0}{R_w}\right)}{k_{wall}}+\frac{\ln\left(\frac{R_w}{R_v}\right)}{k_{eff}}+\frac{1}{hR_0}\right) & L_e\le x\le L_e+L_a \\ T_b+\frac{Q}{2\pi L_c hR_0} & L_e+L_a\le x\le L \end{cases} \tag{11.11}$$

where k_{wall} is the thermal conductivity of the heat pipe wall and k_{eff} is the effective thermal conductivity of the liquid-saturated wick. The steady-state operation in the condenser can be presented as:

$$Q_c = 2\pi R_0 L_c h(T_{wall,c} - T_b) \tag{11.12}$$

where R_0, L_c, h, $T_{wall,c}$, and T_b are the heat pipe's outer radius, length of the condenser section, outside convective heat transfer coefficient, wall temperature at the condenser section, and the bulk temperature of the coolant.

Temperature distribution results are compared with the experimental results of Huang et al. [22] and were found to be in very good agreement. It can be seen that the heat pipe performance is closely linked to the employed working fluid properties. Particularly, the temperature distribution is dependent on the effective thermal conductivity of the porous wick, which can be presented as [18]:

$$k_{eff} = \frac{k_{nf}[(k_{nf}+k_s)-(1-\varepsilon)(k_{nf}-k_s)]}{[(k_{nf}+k_s)+(1-\varepsilon)(k_{nf}-k_s)]} \tag{11.13}$$

As seen in Eq. (11.13), k_{eff} is a function of nanofluid and solid matrix conductivities and porosity of the wick. Using a nanofluid within the heat pipe improves its thermal performance primarily through an increase in the thermal conductivity of the working fluid. k_{eff} increases when thermal conductivity of the nanofluid, k_{nf}, is increased, and it leads to an enhancement of the thermal performance of the heat pipe. Nanofluid's thermal conductivity is obtained using the model proposed by Yu and Choi [19, 23], which can be presented as:

$$k_{nf} = \frac{k_{pe}+2k_l+2(k_{pe}-k_l)(1+\beta)^3\phi}{k_{pe}+2k_l-(k_{pe}-k_l)(1+\beta)^3\phi}k_l \tag{11.14}$$

$$k_{pe} = \frac{[2(1-\alpha)+(1+\beta)^3(1+2\alpha)]\alpha}{-(1-\alpha)+(1+\beta)^3(1+2\alpha)}k_p \tag{11.15}$$

$$\alpha = \frac{k_{layer}}{k_p};\ \beta = \frac{w}{r_p} \tag{11.16}$$

where k_p and k_{layer} are the thermal conductivities of the nanoparticle and nanolayer, respectively. They proposed a modified Maxwell model to account for the effect of the nanolayer, by replacing the thermal conductivity of solid particles in the Maxwell model with the modified thermal conductivity of particles k_{pe} as seen in Eq. (11.14a). A nanoparticle with the radius r_p is assumed to be surrounded by a nanolayer of thickness w. This model has the advantage of linking the nanofluid's thermal conductivity to the nanoparticle's diameter, taking into account the effect of diameter variation on the thermal behavior of the fluid. The base fluid physical properties change versus temperature is taken into account for the applied range of 297 K < T < 370 K.

11.2.2.4 Maximum heat transfer limit

The capillary pressure of the wick provides a steady-state operational pressure for the heat pipe. The maximum heat input that a heat pipe can remove can be obtained based on the capillary pressure limit as:

$$\Delta p_v(X_{max} - X_{min}) + \Delta p_l(X_{max} - X_{min}) + p_c(X_{min}) = \frac{2\sigma_l}{r_c} \quad (11.15)$$

To solve Eq. (11.15), it is necessary to know vapor and liquid pressure distributions, which are obtained by integrating the vapor and liquid momentum equations in the radial direction. The maximum heat transport by the heat pipe when boundary and inertial effects are neglected can be presented as [15]:

$$Q_{max} = \frac{2\sigma_1}{Ar_c} \quad (11.16)$$

The maximum heat transport capillary limit is affected by the density and viscosity of the nanofluid as seen in Eqs. (11.15) and (11.16) and Table 11.1.

11.3 Results and Discussion

The analysis was carried out by incorporating the effect of nanofluids in the analytical model given by Zhu and Vafai [15]. The heat pipe considered in the present chapter has a total length of 89 cm. The condenser has a length of 20 cm, while the evaporator and adiabatic sections are 60 and 9 cm in length, respectively. The outer radius of the heat pipe is taken as 9.55 mm, the inner radius is 9.4 mm, and the vapor core radius is 8.65 mm. It should be noted that the dimensions used here are just nominal. We have established that the presented results hold for a substantial variation in the nominal geometrical dimensions given in here.

In the present analysis, three different types of water-based nanofluids, namely Al_2O_3, CuO, and TiO_2, were considered. In this work, a reasonable concentration range is chosen for different particle sizes, and the performance of the cylindrical heat pipe is investigated for different heat inputs. In what follows, the influence of nanofluid on liquid velocity, pressure profile, temperature profile, thermal resistance, and maximum heat transport capability is investigated.

As seen in Fig. 11.2, the maximum liquid velocity decreases when increasing the nanoparticle concentration within the working fluid. This is due to an increase in the liquid density in the presence of more nanoparticles, as can be seen in Eq. (11.9). As a result of this increase in the nanofluid density, a slower liquid flow is observed. Figure 11.3 displays the effect of nanoparticles on liquid and vapor pressure distributions. With respect to the liquid pressure, it can be seen that initially the pressure gradient decreases as the concentration of the nanoparticles increases. This trend continues up to a specific concentration of nanoparticles within the working fluid. However, this trend reverses once the concentration increases beyond this specific value. It is observed that pressure distribution along the heat pipe changes slightly between 5% and 10% of nanofluid concentration and there is a reversal effect on the pressure distribution changes when the concentration reaches 20%. This behavior is due to the opposite roles played by density and viscosity, both growing with particle concentration in the liquid. At the beginning, density effect prevails leading to a smaller pressure drop. After reaching a critical concentration level, the increase in viscosity overcomes the density effect resulting in a larger liquid pressure drop. Physically, an increase in density reduces the liquid velocity, which in turn results in a lower shear stress. In contrast, an increase in viscosity increases the shear stress. These two opposite effects cause the presence of a critical concentration for nanoparticles. Figure 11.3 also shows that the vapor pressure drop increases when particle concentration increases. This is due to the changes in the physical properties of vapor and liquid as seen in Eq. (11.7) and Table 11.1. Figure 11.4 shows the wall temperature distribution for different nanofluids with different concentration levels. The figure clearly highlights the positive impact of the nanoparticles on the thermal performance of the heat pipe. A lower temperature difference between evaporator and condenser can be observed when nanofluids are used. For a 4% particle concentration, the temperature gradient is reduced by 5%, 3%, and 5% for Al_2O_3, TiO_2, and CuO, respectively.

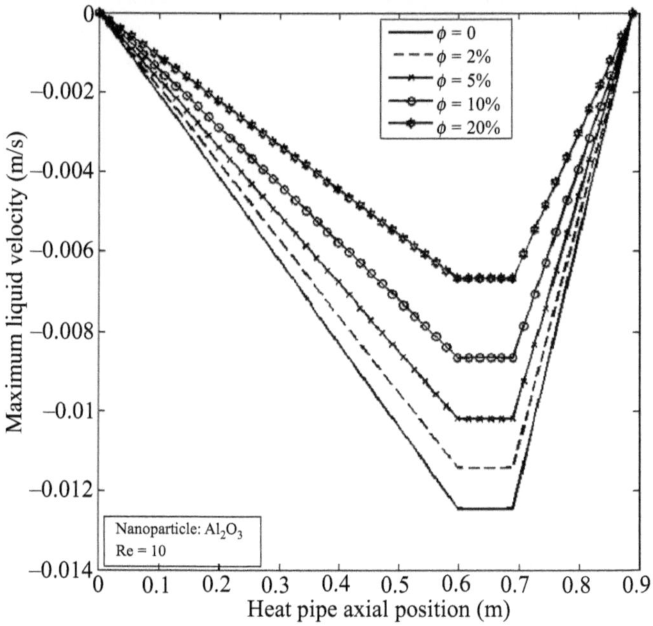

Figure 11.2 The effect of nanoparticle concentration levels on the maximum liquid velocity profile.

The effect of particle diameter and concentration levels on the temperature difference between evaporator and condenser for various heat loads is shown in Fig. 11.5. The figure shows that increasing the particle concentration decreases the temperature difference between the evaporator and condenser. Moreover, the temperature difference decreases with a reduction in particle diameter. It can be seen that for a constant temperature difference between condenser and evaporator, the use of a nanofluid as the working fluid allows the heat pipe to operate under a larger heat load. A nanofluid-based heat pipe is able to dissipate up to 26% more heat without experiencing an increase in the wall temperature.

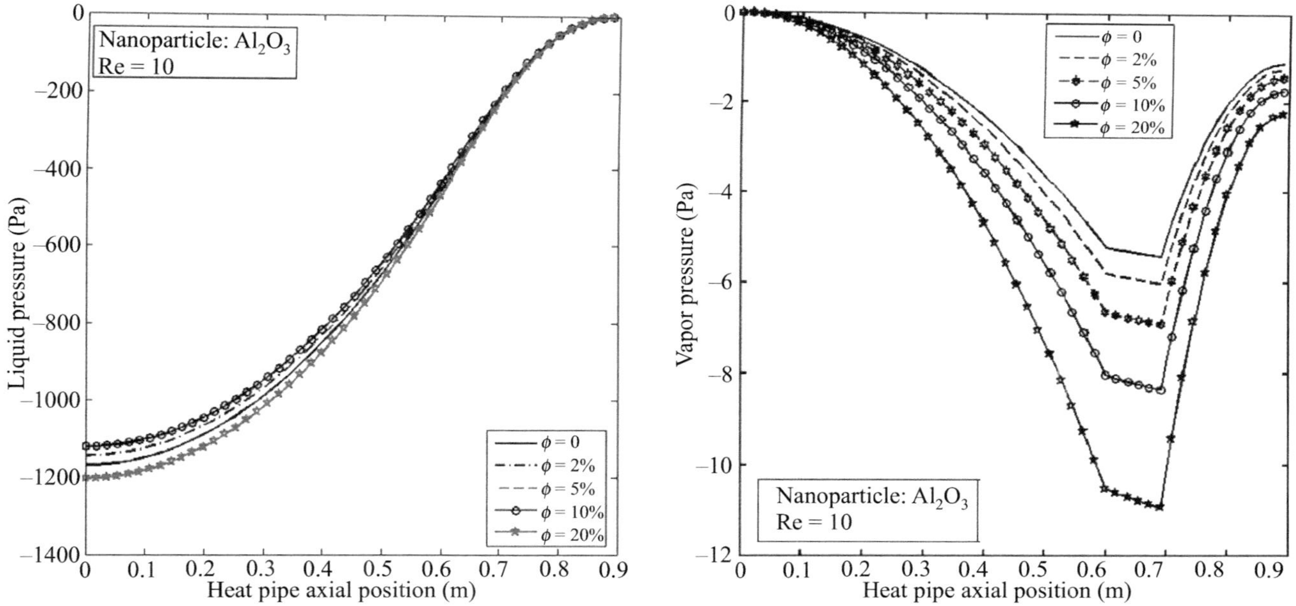

Figure 11.3 The effect of nanoparticle concentration levels on the liquid and vapor pressure distributions.

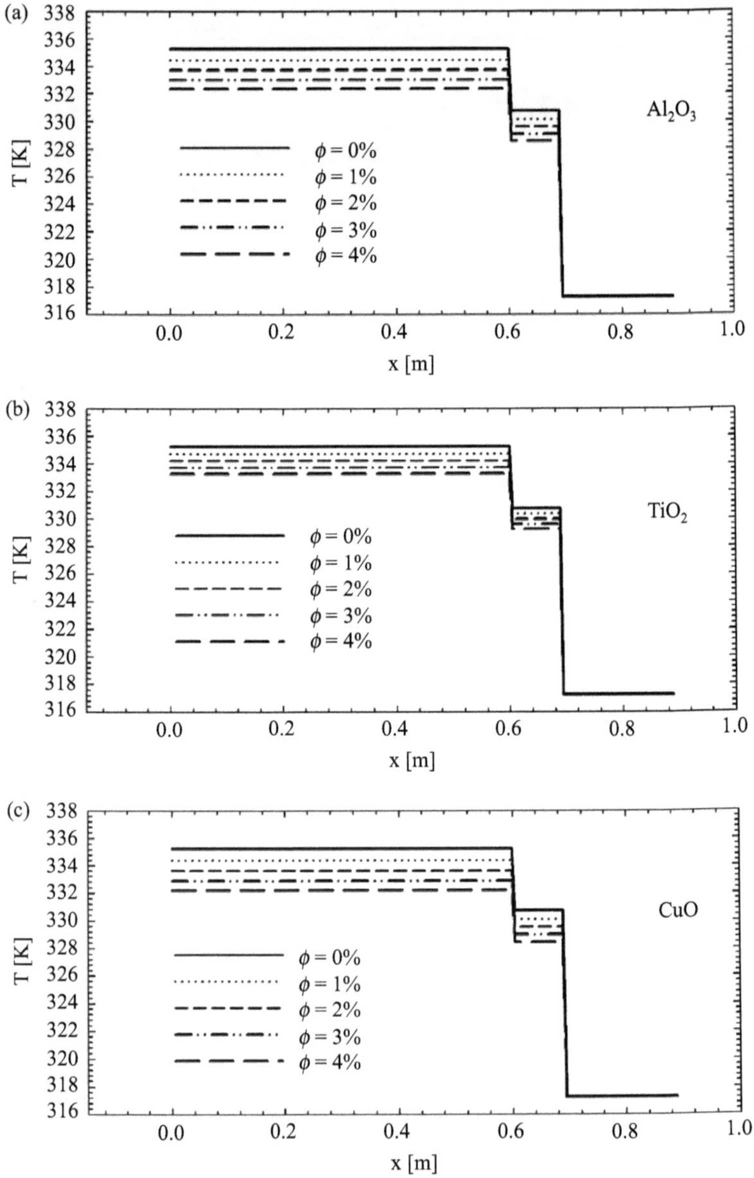

Figure 11.4 Heat pipe temperature distribution for different particle concentration levels analyzing different nanoparticles: (a) Al_2O_3, (b) TiO_2, and (c) CuO; d_p = 20 nm.

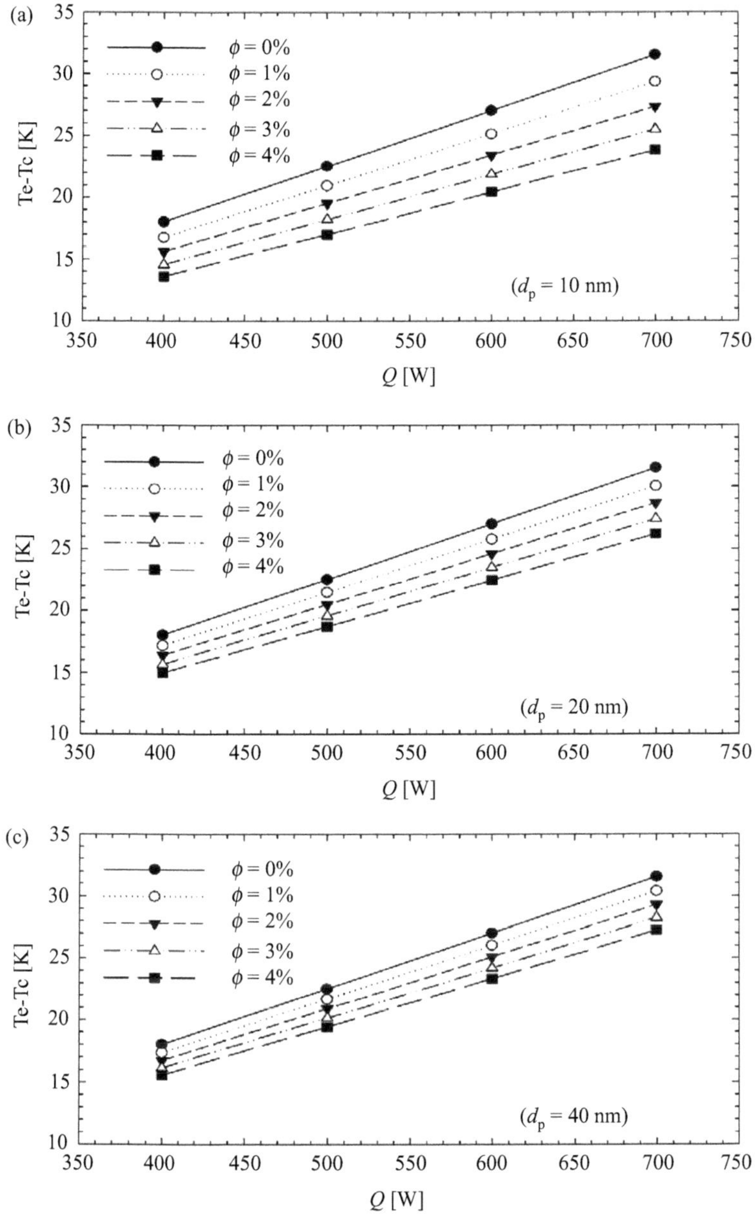

Figure 11.5 Effect of CuO particle concentration levels on the thermal performance of a heat pipe under various heat input and for different particle diameters: (a) $d_p = 10$ nm, (b) $d_p = 20$ nm, and (c) $d_p = 40$ nm.

Figure 11.6 shows that it is possible to down size a heat pipe's dimension when using nanofluid as the working fluid. In this figure, L_0 is the nominal length of a water-based heat pipe and L designates the length for the nanofluid-based heat pipe. As can be seen in Fig. 11.6, there is a significant reduction in the size of a heat pipe when using a nanofluid. For example, the presence of 10 nm CuO particles in water results in 78% reduction in the nominal size.

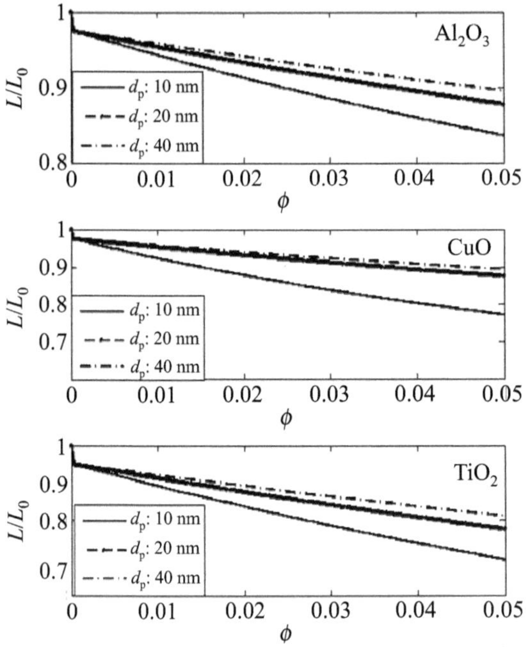

Figure 11.6 Effect of different particle diameter and concentration levels on the size of the heat pipe: (a) Al_2O_3, (b) TiO_2, and (c) CuO ($T_{wall} \leq 90°C$).

The influence of different nanoparticle concentration levels on the heat pipe thermal resistance is shown in Fig. 11.7. The thermal resistance of the heat pipe is defined as $[Q/(T_e - T_c)]^{-1}$.

In this figure, the thermal resistance ratio of the heat pipe with nanofluid over one with pure water is shown. A nominal particle diameter of 10 nm is utilized, and the heat load is varied from 200 to 800 W. It can be seen that increasing the nanoparticle concentration decreases the heat pipe thermal resistance and provides a better performance. For example, a 75% reduction in the thermal resistance ratio is obtained for a 4% CuO nanoparticle concentration level.

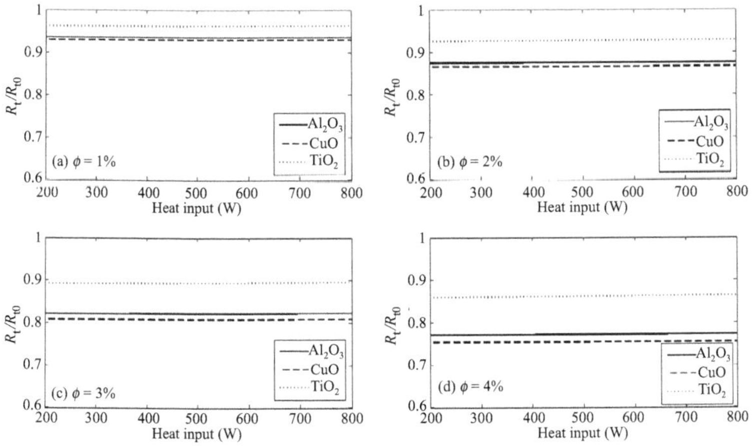

Figure 11.7 The effect of different nanoparticle concentration levels on the heat pipe thermal resistance: (a) $\phi = 1\%$, (b) $\phi = 2\%$, (c) $\phi = 3\%$, (d) $\phi = 4\%$.

Figure 11.8 represents the maximum heat transport capillary limit as a function of particles concentration for all the investigated nanofluids. The figure clearly establishes the existence of an optimum nanoparticle concentration level for each nanofluid in maximizing the thermal performance of the heat pipe. Increasing the nanoparticle concentration enhances the maximum heat transfer until a critical concentration level is reached. After this critical level, an increase in the concentration decreases the maximum heat transfer. It is worth mentioning that both density and viscosity changes have a substantial effect on the thermal performance of the heat pipe. This is similar to what was experienced with the liquid pressure and once again is due to the opposite roles that density and viscosity have in affecting the Q_{max}. From a physical point of view, this occurs due to an increase in density and viscosity with an increase in the concentration level, which results in a larger mass flow and higher pressure loss, respectively. These two factors have positive and negative effects on Q_{max}, resulting in the existence of an optimum particle concentration level. As seen in Fig. 11.8, the optimum concentration for Al_2O_3, CuO, and TiO_2 are about 5%, 15%, and 7%, respectively.

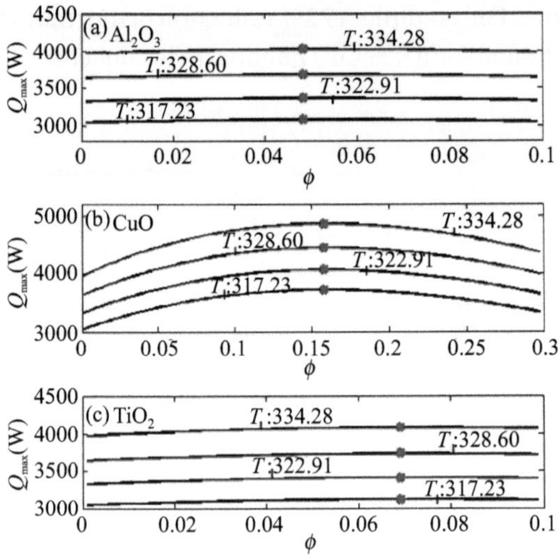

Figure 11.8 The effect of nanoparticles concentration levels on the maximum heat load carrying capability of the heat pipe for different condenser temperatures: (a) Al_2O_3, (b) CuO, (c) TiO_2.

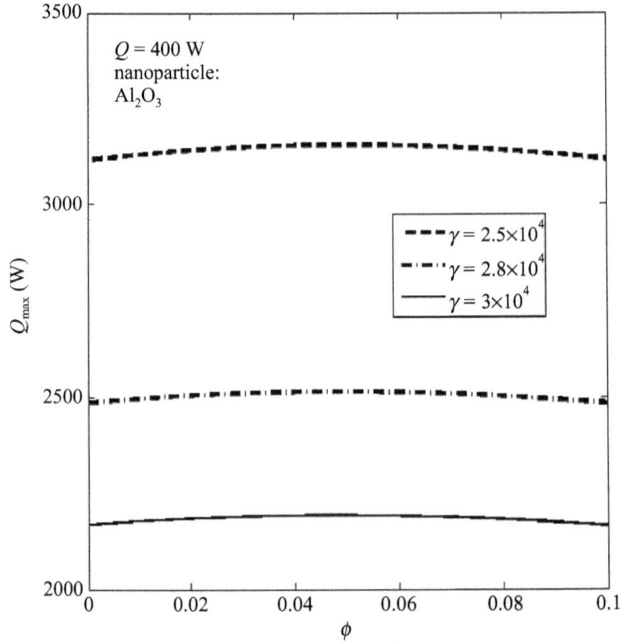

Figure 11.9 The effect of porous media shape parameters (γ) and nanoparticle concentration levels on the maximum heat transfer capability of the heat pipe.

The effect of porous media characteristics, shown in terms of γ, on the maximum heat transfer can be seen in Figs. 11.9 and 11.10. Figure 11.9 represents the Q_{max} variations versus changes in nanoparticle concentration level and γ. The smaller γ corresponds to the larger permeability of the porous wick and bigger maximum heat load of the heat pipe.

Figure 11.10 shows the effect of variation in γ on Q_{max}. It can be seen that increasing γ reduces the maximum heat load capacity of the heat pipe. The three-dimensional representation for the maximum heat transfer in terms of nanoparticle concentration and γ is shown in Fig. 11.10. The existence of an optimum value of nanoparticle concentration level in maximizing the heat load capacity can be seen in this figure.

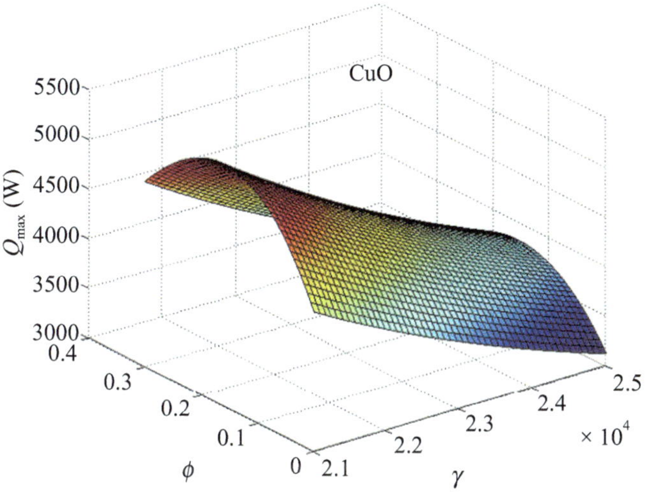

Figure 11.10 Surface representation of the heat pipe maximum heat transfer in terms of nanoparticle concentration level and the porous media shape parameters (γ).

11.4 Conclusion

The thermal performance of cylindrical heat pipes utilizing a nanofluid as the working fluid has been investigated. Three of the most common nanoparticles, namely Al_2O_3, CuO, and TiO_2, are considered. The heat pipe velocity, pressure, temperature, and maximum heat transfer limit are obtained for different nanoparticle concentration levels and sizes. Moreover, the possibility of reducing the size of the cylindrical heat pipe by utilizing nanofluids is studied. The results from our model confirm previous experimental results, i.e., the thermal performance of a heat pipe is improved and temperature gradient along the heat pipe and thermal resistance across the heat pipe are reduced when nanofluids are utilized as the working fluid. It is shown that the thermal resistance decreases as the concentration increases or as the particle diameter decreases. The influence of nanofluid and the geometrical characteristics of the wick on the maximum heat load carrying capability of the cylindrical heat pipe is investigated. The existence of an optimum concentration level in producing the maximum heat transfer is established.

References

1. S. U. S. Choi, "Enhancing thermal conductivity of fluids with nano-particles," *ASME FED*, 1995; **231**: 99–105.
2. S. Kakac and A. Pramuanjaroenkij, "Review of convective heat transfer enhancement with nanofluids," *Int. J. Heat Mass Transfer*, 2009; **52**: 3187–3196.
3. S. W. Kang, W. C. Wei, S. H. Tsai, and S. Y. Yang, "Experimental investigation of silver nano-fluid on heat pipe thermal performance," *Appl. Therm. Eng.*, 2006; **26**: 2377–2382.
4. P. Naphon, D. Thongkum, and P. Assadamongkol, "Heat pipe efficiency enhancement with refrigerant–nanoparticles mixtures," *Energy Convers. Manage.*, 2009; **50**: 772–776.
5. X. F. Yang, Z. H. Liu, and J. Zhao, "Heat transfer performance of a horizontal micro-grooved heat pipe using CuO nanofluid," *J. Micromech. Microeng.*, 2008; **18**: 035038.
6. H. B. Ma, C. Wilson, Q. Yu, K. Park, U. S. Choi, and M. Tirumala, "An experimental investigation of heat transport capability in a nanofluid oscillating heat pipe," *J. Heat Transfer*, 2006; **128**: 1213–1216.
7. P. Naphon, P. Assadamongkol, and T. Borirak, "Experimental investigation of titanium nanofluids on the heat pipe thermal efficiency," *Int. Commun. Heat Mass Transfer*, 2008; **35**: 1316–1319.
8. S. W. Kang, W. C. Wei, S. H. Tsai, and C. C. Huang, "Experimental investigation of nanofluids on sintered heat pipe thermal performance," *Appl. Therm. Eng.*, 2009; **29**: 973–979.

9. Z. Liu, J. Xiong, and R. Bao, "Boiling heat transfer characteristics of nanofluids in a flat heat pipe evaporator with micro-grooved heating surface," *Int. J. Multiphase Flow*, 2007; **33**: 1284–1295.

10. R. R. Riehl, "Analysis of loop heat pipe behavior using nanofluid," in: *Heat Powered Cycles International Conference (HPC)*, New Castle, UK, Paper 06102, 2006.

11. Y. Chen, W. Wei, S. Kang, and C. Yu, "Effect of nanofluid on flat heat pipe thermal performance," in: *24th IEEE SEMI-THERM Symposium*, pp. 16–19.

12. H. B. Ma, C. Wilson, B. Borgmeyer, K. Park, Q. Yu, S. U. S. Choi, and M. Tirumala, "Effect of nanofluid on the heat transport capability in an oscillating heat pipe," *Appl. Phys. Lett.*, 2006; **88**: 143116.

13. Y. Lin, S. Kang, and H. Chen, "Effect of silver nano-fluid on pulsating heat pipe thermal performance," *Appl. Therm. Eng.*, 2008; **28**: 1312–1317.

14. C. Y. Tsaia, H. T. Chiena, P. P. Dingb, B. Chanc, T. Y. Luhd, and P. H. Chena, "Effect of structural character of gold nanoparticles in nanofluid on heat pipe thermal performance," *Mater. Lett.*, 2004; **58**: 1461–1465.

15. N. Zhu and K. Vafai, "Analysis of cylindrical heat pipes incorporating the effects of liquid–vapor coupling and non-Darcian transport: A closed form solution," *Int. J. Heat Mass Transfer*, 1999; **42**: 3405–3418.

16. K. Vafai and C. L. Tien, "Boundary and inertia effects on flow and heat transfer in porous media," *Int. J. Heat Mass Transfer*, 1981; **24**: 195–203.

17. K. Vafai, "Convective flow and heat transfer in variable porosity media," *J. Fluid Mech.*, 1984; **147**: 233–259.

18. S. W. Chi, *Heat Pipe Theory and Practice*, Hemisphere, Washington, DC, 1976.

19. S. K. Das, S. U. S. Choi, W. Yu, and T. Pradeep, *Nanofluids Science and Technology*, John Wiley & Sons, Hoboken, 2008.

20. B. C. Pak and Y. I. Cho, "Hydrodynamic and heat transfer study of dispersed fluids with submicron metallic oxide particles," *Exp. Heat Transfer*, 1998; **11**: 151–170.

21. H. C. Brinkman, "The viscosity of concentrated suspensions and solution," *J. Chem. Phys.*, 1952; **20**: 571–581.

22. L. Huang, M. S. El-Genk, and J.-M. Tournier, "Transient performance of an inclined water heat pipe with a screen wick," in: *ASME National Heat Transfer Conference, Atlanta, Heat Pipes and Capillary Pumped Loops*, 1993; **236**: 87–92.

23. W. Yu and S. U. S. Choi, "The role of interfacial layers in the enhanced thermal conductivity of nanofluids: A renovated Maxwell model," *J. Nanoparticle Res.*, 2003; **5**: 167–171.

Chapter 12

A Critical Synthesis of Thermophysical Characteristics of Nanofluids

Khalil Khanafer[a,b] and Kambiz Vafai[c]
[a]*Vascular Mechanics Laboratory, Department of Biomedical Engineering, University of Michigan, Ann Arbor, MI 48109, USA*
[b]*Vascular Mechanics Laboratory, Section of Vascular Surgery, University of Michigan, Ann Arbor, MI 48109, USA*
[c]*Department of Mechanical Engineering, University of California Riverside, Riverside, CA 92521, USA*
vafai@engr.ucr.edu

12.1 Introduction

Recent advances in nanotechnology have led to the development of a new, innovative class of heat transfer fluids (nanofluids) created by dispersing nanoparticles (10–50 nm) in traditional heat transfer fluids [1]. Nanofluids appear to have the potential to significantly increase heat transfer rates in a variety of areas such as industrial cooling applications, nuclear reactors, transportation industry

---

This chapter has been reprinted from the *International Journal of Heat and Mass Transfer*, Vol. 54, Khalil Khanafer and Kambiz Vafai, A critical synthesis of thermophysical characteristics of nanofluids, 4410–4428, Copyright 2011, with permission from Elsevier.

Nanotechnology and Energy
Edited by Kaufui V. Wong

ISBN 978-981-4774-24-6 (Hardcover), 978-1-315-16357-4 (eBook)
www.panstanford.com

(automobiles, trucks, and airplanes), micro-electromechanical systems (MEMS), electronics and instrumentation, and biomedical applications (nano-drug delivery, cancer therapeutics, cryopreservation) [2]. Possible improved thermal conductivity translates into higher energy efficiency, better performance, and lower operating costs.

A large number of studies related to the heat transfer enhancement using nanofluids, both experimentally and theoretically, were conducted by a number of investigators [1–15]. Although various studies have shown that nanofluids illustrate higher heat transfer enhancement than those of base fluids, contradictory results on nanofluids behavior were also reported. For example, Pak and Cho [16] demonstrated that the Nusselt number for Al_2O_3–water and TiO_2–water nanofluids increased with increasing volume fraction as well as the Reynolds number. However, the convective heat transfer coefficient for nanofluids at a volume fraction of 3% was found to be 12% smaller than that of pure water when considering a constant average velocity [16]. Yang et al. [17] reported a similar conclusion for graphite nanofluids in the laminar flow regime. On the contrary, many other researchers have reported heat transfer enhancement using nanofluids [6, 11, 15, 18–22]. For instance, Xuan and Li [15] conducted an experimental study to investigate convective heat transfer and flow features of nanofluids. Their results show that convective heat transfer coefficient of the nanofluid increased with the flow velocity as well as the volume fraction of nanoparticles, and it was larger than that of the base water under the same flow velocity. Das et al. [22] had shown experimentally that the thermal conductivity of nanofluid increases with an increase in temperature. They observed that a two- to fourfold increase in the thermal conductivity can be achieved over the temperature range of 21–52°C.

In contrast to research investigations related to forced convective heat transfer using nanofluids, few studies are found in the literature on the use of nanofluids in natural convective heat transfer. Khanafer et al. [6] conducted a numerical study to determine natural convection heat transfer of nanofluids in an enclosure under various physical constraints. Their results illustrated that the average Nusselt number increases with an increase in the particle volume fraction for different Grashof numbers. Kim et al. [23] proposed a factor that described the effect of nanoparticle addition on the convective instability and heat transfer characteristics of a base fluid. The new factor included the effect of the ratio of the thermal conductivity of nanoparticles to that of the base fluid, shape factor of the nanoparticles, volume fraction of nanoparticles, and the heat capacity ratio. Their results indicate that the heat transfer coefficients in the presence of nanofluids increase with an increase in the volume fraction of nanoparticles. Convective heat transfer enhancement using nanofluids was also observed experimentally by Nanna et al. [24] and Nanna and Routhu [25].

Nomenclature

c_{eff}	heat capacity of nanofluids
c_{f}	heat capacity of the base fluid
c_{p}	heat capacity of nanoparticles
d_{p}	nanoparticles diameter
h	interparticle spacing
k	thermal conductivity
k_{B}	Stefan–Boltzmann constant
k_{eff}	thermal conductivity of nanofluids
k_{H}	huggins coefficient
k_{layer}	thermal conductivity of the nanolayer
m	mass
n	empirical shape factor
Pr	Prandtl number
Re	Reynolds number
t	thickness of the nanolayer
T	temperature
V	volume
Greek symbols	
β_{eff}	thermal expansion coefficient of nanofluids
β_{f}	thermal expansion coefficient of the base fluid
β_{p}	thermal expansion coefficient of nanoparticle
ρ	density
ρ_{eff}	density of nanofluids
ψ	particle sphericity
ϕ_{p}	volume fraction of nanoparticles

$\phi_{p,max}$	maximum volume fraction of nanoparticles
μ_{eff}	dynamic viscosity of nanofluids
μ_f	dynamic viscosity of the base fluid
$\mu_{Brownian}$	dynamic viscosity due to Brownian motion
Subscripts	
f	fluid
p	nanoparticle

On the other hand, Putra et al. [13] found experimentally that the presence of nanoparticles (Al_2O_3 and CuO) in water-based nanofluids inside a horizontal cylinder decreased natural convective heat transfer coefficient with an increase in the volume fraction of nanoparticles, particle density as well as the aspect ratio of the cylinder. Ding et al. [26] have also reported experimentally that natural convective heat transfer coefficient decreases systematically with an increase in nanoparticle concentration, and the deterioration was partially attributed to the higher viscosity of nanofluids. Chang et al. [27] performed natural convection experiments with Al_2O_3 microparticle (250 nm) aqueous suspensions in thin enclosures. Their results seem to indicate that the particles have a negligible effect on the Nusselt number values for a vertical enclosure. However, for horizontal enclosure, there was a decrease in Nusselt number compared to the presence of pure water at lower Rayleigh numbers and higher particle concentrations. The authors attributed this anomalous behavior to sedimentation.

Ho et al. [28] had experimentally investigated natural convection heat transfer of a nanofluid in a vertical enclosure for different particle sizes and various volume fractions of nanoparticles (Al_2O_3) ranging from 0.1% to 4% and Rayleigh number variations in the range of 10^5–10^8. Systematic heat transfer degradation was observed in their measurements for nanofluids containing nanoparticles with volume fractions greater than 2% over the entire range of Rayleigh numbers. However, heat transfer enhancement around 18% compared with pure water was exhibited for nanofluid containing lower nanoparticle concentrations of 0.1% at high Rayleigh numbers.

Currently, there are no reliable theoretical models to determine the anomalous thermal conductivity of nanofluids. Many researchers have based the thermal conductivity of nanofluids on thermal conductivities of fluid and nanoparticles, shape and surface area of nanoparticles, volume fraction, and temperature [29]. Keblinski et al. [30] and Eastman et al. [31] proposed four main mechanisms for thermal conductivity enhancement of nanofluids. These include Brownian motion of nanoparticles, molecular-level layering of the liquid at the liquid–particle interface, heat transport within the nanoparticles, and the effects of nanoparticle clustering. They had reasoned that the effect of Brownian motion can be neglected due to the greater input of the thermal diffusion as compared to the Brownian diffusion. Evans et al. [32] had also demonstrated that the hydrodynamics effects associated with Brownian motion have a minor effect on the thermal conductivity of the nanofluid using the molecular dynamics simulations and the simple kinetic theory.

The effect of the solid–liquid interfacial layer on the thermal conductivity enhancement of nanofluids was studied theoretically by many researchers [33–38]. For example, Yu and Choi [33, 34] proposed a theoretical model for the effect of a solid–liquid interface based on the Hamilton–Crosser model for suspensions of non-spherical particles. They attempt to demonstrate that the solid-liquid interfacial layers play an important role in augmenting the thermal conductivity of nanofluids through the use of their Hamilton–Crosser model. However, their proposed model was unable to predict the nonlinear behavior of the nanofluid thermal conductivity. Considering the interface effect between the nanoparticles and base fluid, Xue [36] developed a theoretical model for the effective thermal conductivity for nanofluids based on the Maxwell theory and average polarization theory. Xue [36] postulated that the developed model can interpret the anomalous enhancement of the effective thermal conductivity of the nanofluid. Other theoretical studies in the literature indicate additional conflicting results. Based on molecular dynamic simulations and simple liquid–solid interfaces, Xue et al. [38] illustrated that the layering of the liquid atoms at the liquid–solid interface does not have any considerable effect on thermal transport properties.

One can notice from the earlier cited information that neither Brownian motion nor interfacial liquid layering can be a dominant mechanism and that the findings from experimental and analytical investigations on the heat transfer enhancement using nanofluids are in disagreement for both natural and forced convection. Hence, further theoretical and experimental research are essential in order to explain the basis for possible heat transfer enhancement when using nanofluids. Although many possible mechanisms were proposed in the literature, there is no robust description of the anomalous behavior of nanofluids, including higher thermal conductivity and viscosity. The thermal conductivity and viscosity data of nanofluids are still contradictory in various research publications. As such, it is still unclear as to what are the best models to use for the thermal conductivity and viscosity of nanofluids. Therefore, the aim of this chapter is to analyze the variants within the thermophysical characteristics of nanofluids especially with respect to the thermal conductivity and viscosity models and propose possible physical reasons for the deviations between experimental and analytical studies.

12.1.1 Analytical Models for Physical Properties of Nanofluids

12.1.1.1 Density

The density of nanofluid is based on the physical principle of the mixture rule. As such it can be represented as:

$$\rho_{\mathrm{eff}} = \left(\frac{m}{V}\right)_{\mathrm{eff}} = \frac{m_{\mathrm{f}} + m_{\mathrm{p}}}{V_{\mathrm{f}} + V_{\mathrm{p}}} = \frac{\rho_{\mathrm{f}} V_{\mathrm{f}} + \rho_{\mathrm{p}} V_{\mathrm{p}}}{V_{\mathrm{f}} + V_{\mathrm{p}}} = (1 - \phi_{\mathrm{p}})\rho_{\mathrm{f}} + \phi_{\mathrm{p}}\rho_{\mathrm{p}}, \qquad (12.1)$$

where f and p refer to the fluid and nanoparticle, respectively, and $\phi_{\mathrm{p}} = \frac{V_{\mathrm{p}}}{V_{\mathrm{f}} + V_{\mathrm{p}}}$ is the volume fraction of the nanoparticles. To examine the validity of Eq. (12.1), Pak and Cho [16] and Ho et al. [28] conducted experimental studies to measure the density of Al_2O_3–water nanofluids at room temperature as depicted in Fig. 12.1a. The figure shows an excellent agreement between the experimental results and the predictions using Eq. (12.1).

The physical properties at different temperatures are very important for engineering calculations. Ho et al. [28] measured the density of Al_2O_3–water nanofluid at different temperatures and nanoparticle volume fractions. In the present work, a correlation for the density of Al_2O_3–water nanofluid using the experimental data of Ho et al. [28] as a function of temperature and volume fraction of nanoparticles has been developed. The present developed correlation can be expressed as:

$$\rho_{\text{eff}} = 1001.064 + 2738.6191\phi_{\text{p}} - 0.2095T$$
$$0 \le \phi_{\text{p}} \le 0.04,\ 5 \le T(^{\circ}\text{C}) \le 40 \quad (12.2)$$

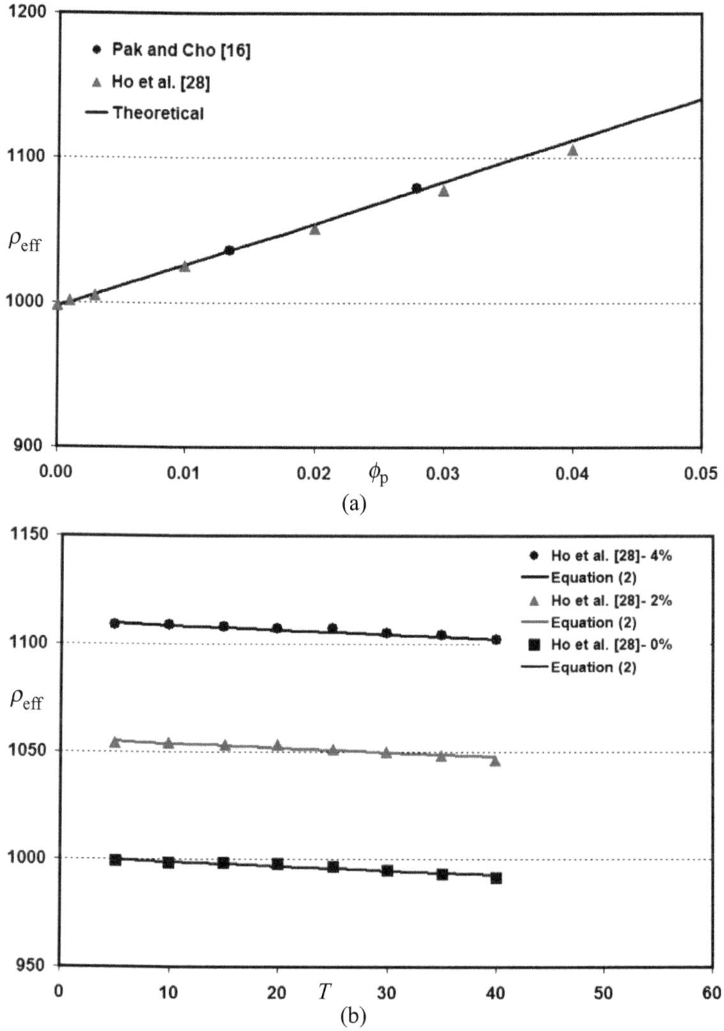

Figure 12.1 Effect of the volume fraction on the density of the Al_2O_3–water nanofluid: (a) room temperature and (b) various temperatures.

The R^2 of the regression is 99.97%, and the maximum relative error is 0.22%. It is clear from Fig. 12.1b that the current regression (Eq. (12.2)) is in excellent agreement with the measurements of Ho et al. [28]. In addition, Fig. 12.1b shows that the rate of decrease of the effective density of Al_2O_3–water nanofluid with increasing temperature is insignificant. This is due to the fact that the density of the Al_2O_3 nanoparticles is even less sensitive to the temperature when compared to the density of water.

12.2 Heat Capacity of Nanofluids

The specific heat of nanofluid can be determined by assuming thermal equilibrium between the nanoparticles and the base fluid phase as follows:

$$\begin{aligned}(\rho c)_{\text{eff}} &= \rho_{\text{eff}}\left(\frac{Q}{m\Delta T}\right)_{\text{eff}} = \rho_{\text{eff}}\frac{Q_{\text{f}}+Q_{\text{p}}}{(m_{\text{f}}+m_{\text{p}})\Delta T}\\ &= \frac{(mc)_{\text{f}}\Delta T+(mc)_{\text{p}}\Delta T}{(m_{\text{f}}+m_{\text{p}})\Delta T} \rightarrow (\rho c)_{\text{eff}}\\ &= \rho_{\text{eff}}\frac{(\rho c)_{\text{f}}V_{\text{f}}+(\rho c)_{\text{p}}V_{\text{p}}}{\rho_{\text{f}}V_{\text{f}}+\rho_{\text{p}}V_{\text{p}}}\\ \Rightarrow c_{\text{eff}} &= \frac{(1-\phi_{\text{p}})\rho_{\text{f}}c_{\text{f}}+\phi_{\text{p}}\rho_{\text{p}}c_{\text{p}}}{\rho_{\text{eff}}}\end{aligned} \tag{12.3}$$

where ρ_p is the density of the nanoparticle, ρ_f is the density of the base fluid, ρ_{eff} is the density of the nanofluid, and c_p and c_f are the heat capacities of the nanoparticle and the base fluid, respectively. However, some authors [16, 22, 39–41] prefer to use a simpler expression given as:

$$c_{\text{eff}} = (1-\phi_{\text{p}})c_{\text{f}} + \phi_{\text{p}}c_{\text{p}}. \tag{12.4}$$

Figure 12.2 shows a comparison of the specific heat of Al_2O_3–water nanofluid at room temperature using both equations with the experimental data of Zhou and Ni [42] for various volume fractions (ϕ_p = 0–21.7%). This figure shows that the specific heat of the nanofluid based on the models given in Eqs. (12.3) and (12.4) decreases with an increase in the volume fraction of nanoparticles. The experimental results were compared with the predictions obtained from the models given in Eqs. (12.3) and (12.4), as shown in Fig. 12.2. Figure 12.2 shows that model I compares very well with the experimental data of Zhou and Ni [42].

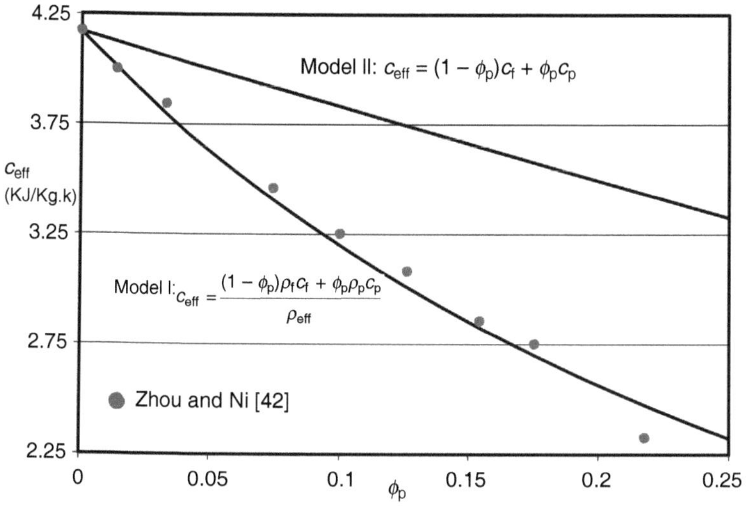

Figure 12.2 Comparison of the heat capacity of Al_2O_3–water nanofluid obtained by models I and II given in Eqs. (12.3) and (12.4) and the experimental data of Zhou and Ni [42].

12.3 Thermal Expansion Coefficient of Nanofluids

The thermal expansion coefficient of nanofluids can be estimated utilizing the volume fraction of the nanoparticles on a weight basis as follows [6]:

$$\beta_{eff} = \frac{(1-\phi_p)(\rho\beta)_f + \phi_p(\rho\beta)_p}{\rho_{eff}} \tag{12.4}$$

where β_f and β_p are the thermal expansion coefficients of the base fluid and the nanoparticle, respectively. A simpler model for the thermal expansion coefficient of the nanofluid is suggested as [43, 44]:

$$\beta_{eff} = (1-\phi_p)\beta_f + \phi_p\beta_p. \tag{12.5}$$

To the best of our knowledge, there is one experimental measurement of the volumetric expansion coefficient of nanoparticles that has been reported in the literature [28]. Ho et al. [28] conducted an experimental study to determine the thermal expansion of Al_2O_3–water nanofluid at various volume fractions of nanoparticles. For water, β_f varies from 1.5×10^{-4} °C^{-1} to 6.2×10^{-4} °C^{-1} over a temperature range of $15°C \leq T \leq 80°C$, which is two orders of magnitude higher than β_p. The values of the thermal expansion of Al_2O_3–water nanofluid predicted by Eqs. (12.4) and (12.5) were compared with the experimental data of Ho et al. [28] at a temperature of 26°C. Figure 12.3a shows that neither Eq. (12.4) nor Eq. (12.5) can be used to properly estimate the thermal expansion of nanofluid as compared to the experimental data of Ho et al. [28].

Ho et al. [28] also studied the effect of temperature and volume fraction of nanoparticles on the thermal expansion coefficient of Al_2O_3–water nanofluid. A correlation for the thermal expansion coefficient of Al_2O_3–water nanofluid as a function of temperature and volume fraction of nanoparticles based on the data presented in Ho et al. [28] has been developed in the current work. This correlation can be presented as:

$$\beta_{\text{eff}} = \left(-0.479\phi_p + 9.3158\times10^{-3}T - \frac{4.7211}{T^2}\right)\times10^{-3}; \tag{12.6}$$

$$0 \leq \phi_p \leq 0.04, \quad 10°C \leq T \leq 40°C$$

The R^2 of the above correlation is 99%. The validity of this correlation (Eq. (12.6)) is depicted in Fig. 12.3b.

12.4 Effective Viscosity of Nanofluids

12.4.1 Analytical Studies

Different models of viscosity have been used by researchers to model the effective viscosity of nanofluid as a function of volume fraction. Einstein [45] determined the effective viscosity of a suspension of spherical solids as a function of volume fraction (volume concentration lower than 5%) using the phenomenological hydrodynamic equations. This equation was expressed by:

$$\mu_{\text{eff}} = (1 + 2.5\phi_p)\mu_f. \tag{12.7}$$

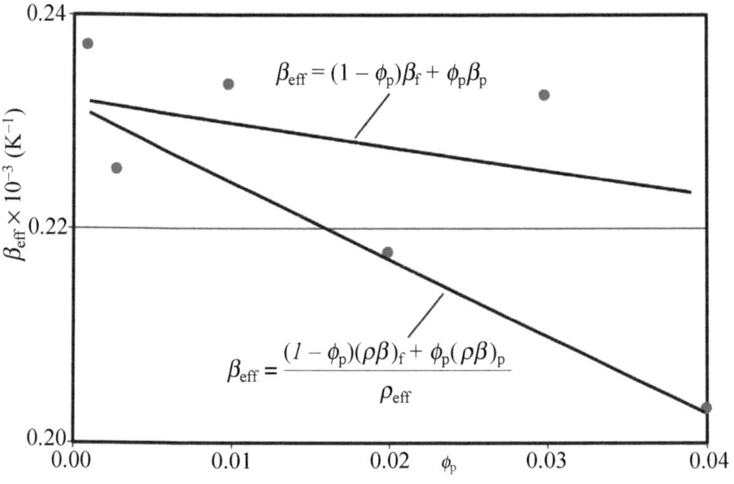

(a)

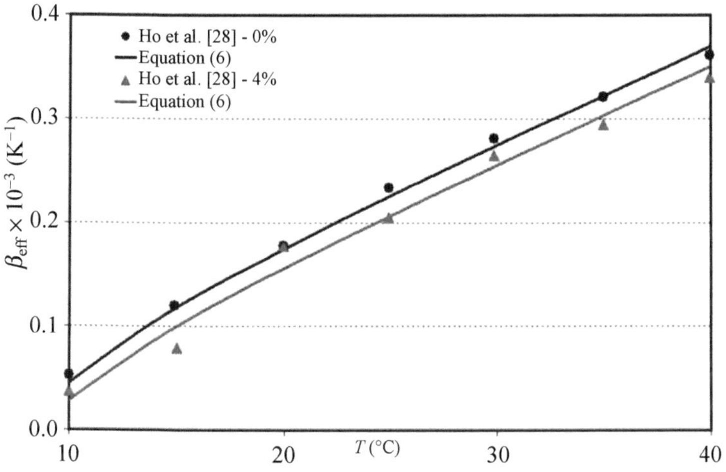

(b)

Figure 12.3 Effect of the volume fraction and temperature on the thermal expansion coefficient of Al_2O_3–water nanofluid: (a) effect of volume fraction as displayed by Eqs. (12.4) and (12.5) at room temperature; (b) temperature effect as displayed by a comparison between Eq. (12.6) and experimental data of Ho et al. [28].

Since Einstein's analysis of the viscosity of a dilute suspension of rigid spheres in a viscous liquid, several equations have been developed in an effort to extend Einstein's formula to suspensions of higher concentrations, including the effect of non-spherical particle concentrations [46–50]. For example, Brinkman [46] presented a viscosity correlation that extended Einstein's equation to concentrated suspensions:

$$\mu_{\text{eff}} = \frac{1}{(1-\phi_{\text{p}})^{2.5}} = (1+2.5\phi_{\text{p}} + 4.375\phi_{\text{p}}^2 + \ldots)\mu_{\text{f}} \tag{12.8}$$

The effect of Brownian motion on the effective viscosity in a suspension of rigid spherical particles was studied by Batchelor [47]. For isotropic structure of suspension, the effective viscosity was given by:

$$\mu_{\text{eff}} = (1+2.5\phi_{\text{p}} + 6.2\phi_{\text{p}}^2)\mu_{\text{f}} \tag{12.9}$$

Lundgren [48] proposed the following equation under the form of a Taylor series in ϕ_{p}:

$$\mu_{\text{eff}} = \frac{1}{1-2.5\phi_{\text{p}}}\mu_{\text{f}} = (1+2.5\phi_{\text{p}} + 6.25\phi_{\text{p}}^2 + O(\phi_{\text{p}}^3))\mu_{\text{f}}, \tag{12.10}$$

It is obvious that if the terms $O(\phi_{\text{p}}^2)$ and higher are neglected, the above correlation reduces to that of Einstein. Table 12.1 summarizes the most common analytical expressions for the viscosity of nanofluids as a function of the volume fraction of the nanoparticles.

12.4.2 Experimental Studies

Compared with the experimental studies on thermal conductivity of nanofluids, there are limited rheological studies reported in the literature [16, 55–63]. Models of the effective viscosity of nanofluids based on the experimental data are limited to certain nanofluids. Masuda et al. [56] were the first to measure the viscosity of several water-based nanofluids for temperatures ranging from room condition to 67°C. Wang et al. [55] obtained, using three different preparation methods, some data for the dynamic viscosity of Al_2O_3–water and Al_2O_3–ethylene glycol mixtures at various temperatures.

Table 12.1 Summary of significant models found in the literature

Models	Effective viscosity	Physical model	Remarks
Einstein [45]	$\mu_{eff} = (1 + 2.5\phi_p)$	–Based on the phenomenological hydrodynamic equations –Considered a suspension containing n solute particles in a total volume V	–Infinitely dilute suspension of spheres (no interaction between the spheres) –Valid for relatively low particle volume fraction ($\phi_p \leq 2\%$)
Brinkman [46]	$\mu_{eff} = \frac{1}{(1-\phi_p)^{2.5}} = (1+2.5\phi_p + 4.375\phi_p^2 + ...)\mu_f$	–Based on Einstein model –Derived by considering the effect of the addition of one solute-molecule to an existing solution	–Spherical particles –Valid for high moderate particle concentrations –Used Einstein's factor ($1 + 2.5\phi_p$)
Batchelor [47]	$\mu_{eff} = (1+\eta\phi_p + k_H\phi_p^2)\mu_f = (1+2.5\phi_p + 6.2\phi_p^2)\mu_f$	–Based on reciprocal theorem in stokes flow problem to obtain an expression for the bulk stress due to the thermodynamic forces –Incorporated both effects: hydrodynamic effects and Brownian motion	–Rigid and spherical particles –Brownian motion –Isotropic structure –Huggins coefficient: $k_H = 6.2$ (5.2 from hydrodynamic effects and 1.0 from Brownian motion)

(*Continued*)

Table 12.1 (*Continued*)

Models	Effective viscosity	Physical model	Remarks
Lundgren [48]	$\mu_{\text{eff}} = \frac{1}{1-2.5\phi_{\text{p}}}\mu_{\text{f}}$ $= (1+2.5\phi_{\text{p}}+6.25\phi_{\text{p}}^2+...)\mu_{\text{f}}$	–Based on a Taylor series expansion in terms of ϕ_{p}	–Dilute concentration of spheres –Random bed of spheres
Graham [49]	$\mu_{\text{eff}} = (1+2.5\phi_p)\mu_{\text{f}}$ $+\left[\frac{4.5}{(h/r_{\text{p}})(2+h/r_{\text{p}})(1+h/r_{\text{p}})^2}\right]\mu_{\text{f}}$	–A cell theory was used to derive the dependence of the zero-shear-rate viscosity on volume concentration for a suspension of uniform, solid, neutrally buoyant spheres	–Agrees well with Einstein's for small ϕ_{p} –r_{p} is the particle radius and h is the interparticle spacing
Simha [50]	$\mu_{\text{eff}} = \left[1+2.5\phi_{\text{p}}+\left(\frac{125}{64\phi_{\text{p max}}}\right)\phi_{\text{p}}^2+...\right]\mu_{\text{f}}$	–Based on Cage model of liquids and solutions	–Spherical particles
Mooney [51]	$\mu_{\text{eff}} = \exp\left(\frac{2.5\phi_{\text{p}}}{1-k\phi_{\text{p}}}\right)\mu_{\text{f}}$ $= \{1+2.5\phi_{\text{p}}+(3.125+2.5k)\phi_{\text{p}}^2+...\}\mu_{\text{f}}$ $1.35 < k < 1.91$	–Einstein's viscosity equation for an infinitely dilute suspension of spheres was extended to apply to a suspension of finite concentration –Based on first-order interaction between particles (crowding effect)	–Rigid spherical spheres –Monodisperse suspension of finite concentration –Not valid at high concentrations –Considered the volume fraction of a suspension to be divided into two portions

Models	Effective viscosity	Physical model	Remarks
Eilers [52]	$\mu_{\text{eff}} = \mu_{\text{f}}\left[1+\frac{1.25\phi_{\text{p}}}{1-\phi_{\text{p}}/0.78}\right]$ $= (1+2.5\phi_{\text{p}}+4.75\phi_{\text{p}}^2+...)\mu_{\text{f}}$	–Based on experimental data	–Suspensions of bitumen spheres –Curve fitting of the experimental data
Saito [53]	$\mu_{\text{eff}} = \left(1+\frac{2.5}{1-\phi_{\text{p}}}\phi_{\text{p}}\right)\mu_{\text{f}}$ $= (1+2.5\phi_{\text{p}}+2.5\phi_{\text{p}}^2+...)\mu_{\text{f}}$	–Developed based on a theory for spherical solute-molecules in which a single solute-molecule is placed in the field of flow, obtained by averaging over all the possible positions of a second solute-molecule	–Spherical rigid particles –Brownian motion –Very small particles
Frankel and Acrivos [54]	$\mu_{\text{eff}} = \left(\frac{9}{8}\cdot\frac{(\phi_{\text{p}}/\phi_{\text{p max}})^{1/3}}{1-(\phi_{\text{p}}/\phi_{\text{p max}})^{1/3}}\right)\mu_{\text{f}}$	–An asymptotic technique was used to derive the functional dependence of effective viscosity on concentration for a suspension of uniform solid spheres, in the limit as concentration approaches its maximum value	–Uniform solid particles

Because the formulas such as the one proposed by Einstein [45] and later improved by Brinkman [46] and Batchelor [47] underestimate the viscosity of the nanofluids when compared to the measured data, Maiga et al. [57, 58] performed a least-square curve fitting of some experimental data of Wang et al. [55], including Al_2O_3 in water and Al_2O_3 in ethylene glycol. Table 12.2 illustrates a summary of the viscosity models at room temperature based on the experimental data available in the literature. Moreover, Fig. 12.4 shows a comparison of the relative dynamic viscosity of Al_2O_3–water nanofluid from various sources at room temperature. This figure shows that the Brinkman model [46], which was derived for two-phase mixture, is to some extent sufficient to estimate the viscosity for relatively low volume fraction of particles (i.e., $\phi_p \leq 2\%$). However, it considerably underestimates the nanofluid viscosity when compared to experimental data at high particle concentrations. The differences in the relative viscosity among the experimental data, as shown in Fig. 12.4, may be due to the difference in the size of the particle clusters, dispersion techniques, and the methods of measurements. This clearly shows the disagreement between the researchers in measuring the viscosity of nanofluids.

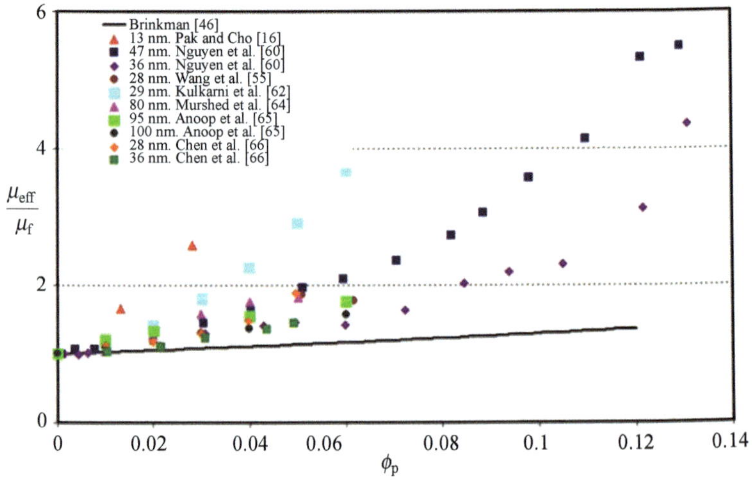

Figure 12.4 Relative viscosity measurement as a function of the volume fraction, ϕ_p, at ambient temperature. (See above-mentioned references for further information.)

Table 12.2 Summary of viscosity models at room temperature based on experimental data

Models	Effective viscosity (regression)	Remarks
Maiga et al. [57]	$\mu_{eff} = (1 + 7.3\phi_p + 123\phi_p^2)\mu_f$	– Least-square curve fitting of Wang et al. [55] data – Al_2O_3–water, d_p = 28 nm
Maiga et al. [57]	$\mu_{eff} = (1 - 0.19\phi_p + 306\phi_p^2)\mu_f; d_p = 28\ \text{nm}$	– Least-square curve fitting of experimental data [55, 56] – Al_2O_3–ethylene glycol
Present work	$\mu_{eff} = (1 + 0.164\phi_p + 302.34\phi_p^2)\mu_f; d_p = 28\ \text{nm}$	– Least-square curve fitting of experimental data [55, 56] – Al_2O_3–ethylene glycol
Buongiorno [67]	$\mu_{eff} = (1 + 39.11\phi_p + 533.9\phi_p^2)\mu_f$	– Curve fitting of Pak and Cho [16] data – Al_2O_3–water, d_p = 13 nm
Buongiorno [67]	$\mu_{eff} = (1 + 5.45\phi_p + 108.2\phi_p^2)\mu_f$	– Curve fitting of Pak and Cho [16] data – TiO_2–water, d_p = 27 nm
Present work	$\mu_{eff} = (1 + 23.09\phi_p + 1525.3\phi_p^2)\mu_f; 0 \le \phi_p \le 0.04$	– Curve fitting of Pak and Cho [16] data – Al_2O_3–water, d_p = 13 nm
Present work	$\mu_{eff} = (1 + 3.544\phi_p + 169.46\phi_p^2)\mu_f; 0 \le \phi_p \le 0.1$	– Curve fitting of Pak and Cho [16] data – TiO_2–water, d_p = 27 nm
Nguyen et al. [60]	$\mu_{eff} = \mu_f \times 0.904e^{0.148\phi_p}; d_p = 47\ \text{nm}$ $\mu_{eff} = (1 + 0.025\phi_p + 0.015\phi_p^2)\mu_f; d_p = 36\ \text{nm}$	– Curve fitting of the experimental data – Al_2O_3–water
Nguyen et al. [60]	$\mu_{eff} = (1.475 - 0.319\phi_p + 0.051\phi_p^2 + 0.009\phi_p^3)\mu_f$	– Curve fitting of the experimental data – CuO–water, d_p = 29 nm
Tseng and Lin [61]	$\mu_{eff} = 13.47\exp(35.98\phi_p)\mu_f; 0.05 \le \phi_p \le 0.12$	– TiO_2–water, shear rate = 100 s^{-1}

12.4.3 Effect of Temperature on the Dynamic Viscosity of Nanofluids

It should be noted that all the aforementioned correlations were developed to relate viscosity as a function of volume fraction only without any temperature dependence considerations. It should be mentioned that there exists a few studies in the literature associated with the effect of temperature on the viscosity of nanofluids. For example, Nguyen et al. [60] investigated experimentally the influence of the temperature on the dynamic viscosities of two particular water-based nanofluids, namely Al_2O_3–water (d_p = 47 nm, 36 nm) and CuO–water (d_p = 29 nm) mixtures. The following formulas were proposed by Nguyen et al. [60] for computing the dynamic viscosity for all three nanofluids tested and particle concentrations of 1% and 4%, respectively:

$$\mu_{\text{eff}} = (1.125 - 0.0007 \times T)\mu_{\text{f}}; \quad \phi_{\text{p}} = 1\%, \tag{12.11}$$

$$\mu_{\text{eff}} = (2.1275 - 0.0215 \times T + 0.0002 \times T^2)\mu_{\text{f}}; \quad \phi_{\text{p}} = 4\% \tag{12.12}$$

As can be seen from Eqs. (12.11) and (12.12), Nguyen et al. [60] did not explicitly express the dynamic viscosity as a function of temperature and volume fraction. Palm et al. [68] proposed equations for the dynamic viscosity (Pa's) by means of the polynomial curve fitting based on the data reported by Putra et al. [13]. The resulting equations as a function of temperature, expressed in Kelvin, are:

For Al_2O_3–water:

$$\mu_{\text{eff}} = 0.034 - 2 \times 10^{-4} T + 2.9 \times 10^{-7} T^2, \quad \phi_{\text{p}} = 1\%, \tag{12.13}$$

$$\mu_{\text{eff}} = 0.039 - 2.3 \times 10^{-4} T + 3.4 \times 10^{-7} T^2, \quad \phi_{\text{p}} = 4\%. \tag{12.14}$$

Tables 12.3 and 12.4 provide a summary of different models of the dynamic viscosity of nanofluids as a function of temperature and volume fraction of nanoparticles. In the present work, a general correlation (Eq. (12.15)) for the effective viscosity of Al_2O_3–water, one of the most commonly studied nanofluids, is developed using various experimental data found in the literature (Fig. 12.5a) as a function of volume fraction, nanoparticles diameter, and temperature as follows:

$$\mu_{\text{eff}} = -0.4491 + \frac{28.837}{T} + 0.574\phi_{\text{p}} - 0.1634\phi_{\text{p}}^2 + 23.053\frac{\phi_{\text{p}}^2}{T^2} + 0.0132\phi_{\text{p}}^3 - 2354.735\frac{\phi_{\text{p}}}{T^3} + 23.498\frac{\phi_{\text{p}}^2}{d_{\text{p}}^2} - 3.0185\frac{\phi_{\text{p}}^3}{d_{\text{p}}^2}; \tag{12.15}$$

$$1\% \leq \phi_p \leq 9\%,\ 20 \leq T(°C) \leq 70,\ 13\ \text{nm} \leq d_p \leq 131\ \text{nm}$$

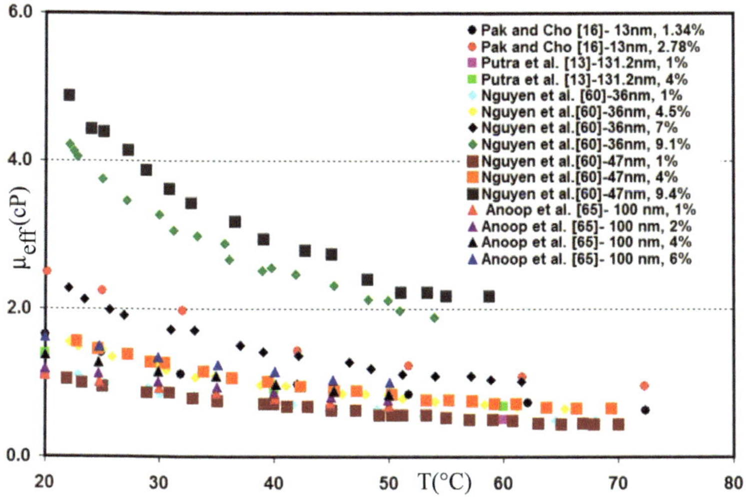

(a)

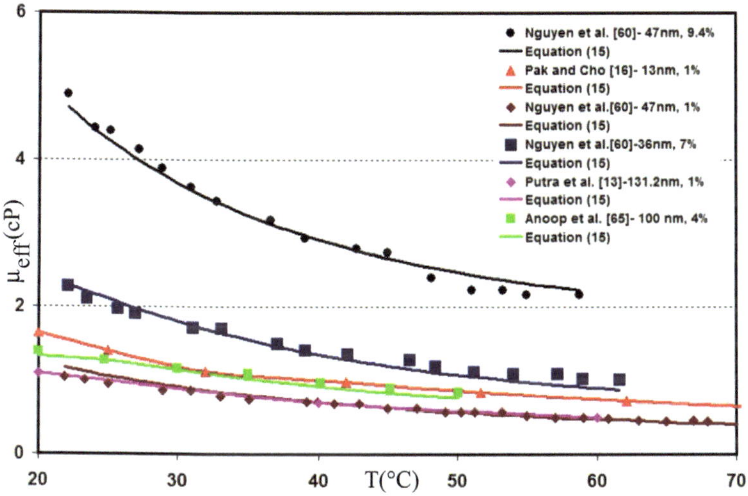

(b)

Figure 12.5 Effect of the volume fraction and temperature on the effective viscosity of Al_2O_3–water nanofluid: (a) experimental measurements; (b) comparison of Eq. (12.15) developed in the current work with the experimental data.

Table 12.3 Effect of temperature and volume fraction on the dynamic viscosity of nanofluids (Al_2O_3–water)

Reference	Model (regression)	Remarks
Present work	$\mu_{\text{eff}} = 0.444 - 0.254\phi_p + 0.0368\phi_p^2 + 26.333\frac{\phi_p}{T} - 59.311\frac{\phi_p^2}{T^2}$ $20°C \le T \le 70°C;\ \phi_p = 1.34\%, 2.78\%$	–Curve fitting of Pak and Cho [16] data; d_p = 13 nm –Units: mPa s
Palm et al. [68]	$\mu_{\text{eff}} = 0.034 - 2\times10^{-4}T(\text{K}) + 2.9\times10^{-7}T^2(\text{K}); \phi_p = 1\%$ $\mu_{\text{eff}} = 0.039 - 2.3\times10^{-4}T(\text{K}) + 3.4\times10^{-7}T^2(\text{K}); \phi_p = 4\%$	–Curve fitting of the experimental data of Putra et al. [13]; d_p = 131.2 nm –Units: Pa s
Nguyen et al. [60]	$\mu_{\text{eff}} = (1.125 - 0.0007\times T(°C))\mu_f; \phi_p = 1\%$ $\mu_{\text{eff}} = (2.1275 - 0.0215\times T(°C) + 0.0002\times T^2(°C))\mu_f;$ $\phi_p = 4\%$	–Units: mPa s
Present work	$\mu_{\text{eff}} = -0.4892 + \frac{26.9036}{T} + 0.6837\phi_p + \frac{24.1141}{T^2} - 0.1785\phi_p^2$ $+0.1818\frac{\phi_p}{T} + 27.015\frac{\phi_p^2}{T^2} + 0.0132\phi_p^3 - 2940.1775\frac{\phi_p}{T^3};$ $1\% \le \phi_p \le 9.4\%,\ 20°C \le T \le 70°C$	–Curve fitting of Nguyen et al. [60] data; d_p = 47 nm –Units: mPa s

Reference	Model (regression)	Remarks
Present work	$\mu_{\text{eff}} = -0.1011 + \frac{18.0162}{T} + 0.3619\phi_{\text{p}} + \frac{164.0837}{T^2}$ $-0.0966\phi_{\text{p}}^2 + 0.1609\frac{\phi_{\text{p}}}{T} + 22.4901\frac{\phi_{\text{p}}^2}{T^2}$ $+0.0078089\phi_{\text{p}}^3 - 2316.3754\frac{\phi_{\text{p}}}{T^3};$ $1\% \le \phi_{\text{p}} \le 9.1\%,\ 20°\text{C} \le T \le 70°\text{C}$	–Curve fitting of Nguyen et al. [60] data; $d_{\text{p}} = 36$ nm –Units: mPa s
Present work	$\mu_{\text{eff}} = -0.4491 + \frac{28.837}{T} + 0.574\phi_{\text{p}} - 0.1634\phi_{\text{p}}^2 + 23.053\frac{\phi_{\text{p}}^2}{T^2} +$ $0.0132\phi_{\text{p}}^3 - 2354.735\frac{\phi_{\text{p}}}{T^3} + 23.498\frac{\phi_{\text{p}}^2}{d_{\text{p}}^2} - 3.0185\frac{\phi_{\text{p}}^3}{d_{\text{p}}^2};$ $1\% \le \phi_{\text{p}} \le 9\%,\ 20°\text{C} \le T \le 70°\text{C},\ 13\text{ nm} \le d_{\text{p}} \le 131\text{ nm}$	–Curve fitting of various experimental data available in the literature [13, 16, 60, 65] –Units: mPa s
Namburu et al. [70, 71]	$\text{Log}(\mu_{\text{eff}}) = Ae^{-BT}$, in mm Pa s $A = -0.29956\phi_{\text{p}}^3 + 6.7388\phi_{\text{p}}^2 - 55.444\phi_{\text{p}} + 263.11$ $B = (-6.4745\phi_{\text{p}}^3 + 140.03\phi_{\text{p}}^2 - 1478.5\phi_{\text{p}} + 20341)/10^6$	–Experimental Al_2O_3–ethylene glycol and water mixture $1\% \le \phi_{\text{p}} \le 10\%$; $d_{\text{p}} = 53$ nm; $238\text{ K} < T < 323\text{ K}$

Table 12.4 Effect of temperature and volume fraction on the dynamic viscosity of nanofluids (TiO_2–water, CuO–water)

Models	Effective viscosity (regression)	Remarks
Duangthongsuk and Wongwises [69]	$\frac{\mu_{\mathrm{eff}}}{\mu_{\mathrm{f}}} = 1.0226 + 0.0477\phi_{\mathrm{p}} - 0.0112\phi_{\mathrm{p}}^2;\ T = 15°\mathrm{C}$ $\frac{\mu_{\mathrm{eff}}}{\mu_{\mathrm{f}}} = 1.013 + 0.092\phi_{\mathrm{p}} - 0.015\phi_{\mathrm{p}}^2;\ T = 25°\mathrm{C}$ $\frac{\mu_{\mathrm{eff}}}{\mu_{\mathrm{f}}} = 1.018 + 0.112\phi_{\mathrm{p}} - 0.0177\phi_{\mathrm{p}}^2;\ T = 35°\mathrm{C}$	– Experimental data – TiO_2–Water, $0.2\% \le \phi_{\mathrm{p}} \le 2\%$ – $d_{\mathrm{p}} = 21$ nm
Present work	$\frac{\mu_{\mathrm{eff}}}{\mu_{\mathrm{f}}} = 1.0538 + 0.1448\phi_{\mathrm{p}} - 3.363\times10^{-3}T - 0.0147\phi_{\mathrm{p}}$ $+6.735\times10^{-5}T^2 - 1.337\frac{\phi_{\mathrm{p}}}{T}$ $15°\mathrm{C} \le T \le 35°\mathrm{C},\ \ 0.2\% \le \phi_{\mathrm{p}} \le 2\%$	– Curve fitting of the experimental data [69] – TiO_2–water – $d_{\mathrm{p}} = 21$ nm
Present work	$\mu_{\mathrm{eff}} = 0.6002 - 0.569\phi_{\mathrm{p}} + 0.0823\phi_{\mathrm{p}}^2$ $+28.8763\frac{\phi_{\mathrm{p}}}{T} - 204.2202\frac{\phi_{\mathrm{p}}^2}{T^2} + 561.3175\frac{\phi_{\mathrm{p}}^3}{T^3}$ $20°\mathrm{C} \le T \le 70°\mathrm{C};$ $\phi_{\mathrm{p}} = 0.99\%, 2.04\%, 3.16\%$	– Curve fitting of Pak and Cho [16] data – TiO_2–Water – $d_{\mathrm{p}} = 27$ nm – Units: mPa s

Models	Effective viscosity (regression)	Remarks
Present work	$\mu_{eff} = -0.4262 + \frac{8.4312}{T} + 0.898\phi_p + \frac{524.7147}{T^2} - 0.2217\phi_p^2$ $-4.7329\frac{\phi_p}{T} + 70.3105\frac{\phi_p^2}{T^2} + 0.0176\phi_p^3 - 5559.4641\frac{\phi_p}{T^3}$; $1\% \le \phi_p \le 9\%,\ 20°C \le T \le 70°C$	– Curve fitting of Nguyen et al. [60] data – CuO–water; d_p = 29 nm – Units: mPa s
Namburu et al. [71]	$\text{Log}(\mu_{eff}) = Ae^{-BT}$, in mm Pa s $A = 1.8375\phi_p^2 - 29.643\phi_p + 165.56$; $B = 4\times10^{-6}\phi_p^2 - 0.001\phi_p + 0.0186$	CuO–ethylene glycol and water Mixture; $1\% \le \phi_p \le 6\%$; d_p = 29 nm; 238 K < T < 323 K
Kulkarni et al. [62, 63]	$\ln\mu_{eff} = A\left(\frac{1}{T}\right) - B$, in mm Pa s $A = 20587\phi_p^2 + 15857\phi_p + 1078.3$; $B = -107.12\phi_p^2 + 53.54\phi_p + 2.8715$	– CuO–water; $0.05 \le \phi_p \le 0.15$; d_p = 29 nm; $278\text{ K} \le T \le 323\text{ K}$ Shear rate = 100 1/s
Koo and Kleinstreuer [72]	$\mu_{eff} = \mu_{static} + \mu_{Brownian}$ $\mu_{Brownian} = 5\times10^4\beta\rho_f\phi_p\sqrt{\frac{kT}{\rho_p d_p}}f(\phi_p,T)$ $f(\phi_p,T) = (-6.04\phi + 0.4705)T + (1722.3\phi_p - 134.63)$ $\beta = \begin{cases} 0.0137(100\phi_p)^{-0.8229}, & \phi_p < 0.01 \\ 0.0011(100\phi_p)^{-0.7272}, & \phi_p > 0.01 \end{cases}$ $1\% < \phi_p < 4\%$; $300\text{ K} < T < 325\text{ K}$	– CuO–water

The R^2 of the regression is 99%. The validity of the above correlation (Eq. (12.15)) is depicted in Fig. 12.5b. As can be seen in Fig. 12.5a, the viscosity of the nanofluid decreases with an increase in the temperature. Moreover, there is no agreement between researchers about the experimentally observed magnitude of the nanofluid's viscosity. Published results indicate a surprising range of variation of the results.

12.5 Thermal Conductivity of Nanofluids

A wide range of experimental and theoretical studies were conducted in the literature to model thermal conductivity of nanofluids. Published results illustrated neither agreement about the mechanisms for heat transfer enhancement nor a unified possible explanation regarding the rather large discrepancies in the results even for the same base fluid and nanoparticles size. Currently, there are no theoretical results available in the literature that predict accurately the thermal conductivity of nanofluids. The existing results were generally based on the definition of the effective thermal conductivity of a two-component mixture as follows [73]:

$$k_{\text{eff}} = \frac{k_{\text{f}}(1-\phi_{\text{p}})(dT/dx)_{\text{f}} + k_{\text{p}}\phi_{\text{p}}(dT/dx)_{\text{p}}}{\phi_{\text{p}}(dT/dx)_{\text{p}} + (1-\phi_{\text{p}})(dT/dx)_{\text{f}}}, \tag{12.16}$$

where $(dT/dx)_{\text{f}}$ is the temperature gradient within the fluid and $(dT/dx)_{\text{p}}$ is the temperature gradient through the particle. The Maxwell model [74] was one of the first models proposed for solid–liquid mixture with relatively large particles. It was based on the solution of heat conduction equation through a stationary random suspension of spheres. The effective thermal conductivity is given by:

$$k_{\text{eff}} = \frac{k_{\text{p}} + 2k_{\text{f}} + 2\phi_{\text{p}}(k_{\text{p}} - k_{\text{f}})}{k_{\text{p}} + 2k_{\text{f}} - \phi_{\text{p}}(k_{\text{p}} - k_{\text{f}})} k_{\text{f}} = k_{\text{f}} + \frac{3\varphi_{\text{p}}(k_{\text{p}} - k_{\text{f}})}{k_{\text{p}} + 2k_{\text{f}} - \phi_{\text{p}}(k_{\text{p}} - k_{\text{f}})}, \tag{12.17}$$

where k_{p} is the thermal conductivity of the particles, k_{f} is the fluid thermal conductivity, and ϕ_{p} is the volume fraction of the suspended particles. The Maxwell model is accurate to order ϕ_{p}^1 and applicable to $\varphi_p \leq 1$ or $\left|\frac{k_p}{k_f} - 1\right| \leq 1$. Bruggeman [75] proposed a model to study the interactions between randomly distributed spherical particles. The Bruggeman model can be expressed as:

$$\frac{k_{eff}}{k_f} = \frac{(3\phi_p - 1)\frac{k_p}{k_f} + \left\{3(1-\phi_p) - 1\right\} + \sqrt{\Delta}}{4};$$
$$\Delta = \left[(3\phi_p - 1)\frac{k_p}{k_f} + \left\{3(1-\phi_p) - 1\right\}\right]^2 + 8\frac{k_p}{k_f}. \tag{12.18}$$

The Bruggeman model [75] is applicable for large volume fraction of spherical particles. For low volume fractions, the Bruggeman model [75] results in approximately the same results as the Maxwell model [74]. For non-spherical particles, Hamilton and Crosser [73] developed a model for the effective thermal conductivity of two-component mixtures. Their model was a function of the thermal conductivity of both the base fluid and the particle, volume fraction of the particles, and the shape of the particles. For the thermal conductivity ratio of two phases larger than 100 ($k_p/k_f > 100$), the thermal conductivity of two-component mixtures can be expressed as follows [73]:

$$k_{eff} = \frac{k_p + (n-1)k_f + (n-1)\phi_p(k_p - k_f)}{k_p + (n-1)k_f - \phi_p(k_p - k_f)} k_f, \tag{12.19}$$

where n is the empirical shape factor given by $n = 3/\psi$, and ψ is the particle sphericity defined by the ratio of the surface area of a sphere with volume equal to that of the particle, to the surface area of the particle. Maxwell's model [74] is a special case of the Hamilton–Crosser model for sphericity equal to one (for spheres, $n = 3$). The shape factor for cylinders is $n = 6$. Apparently, the most notable drawback of the Hamilton–Crosser model [73] is that important physical parameters such as temperature and particle size are not considered. Tables 12.5–12.7 summarize some pertinent models for the effective thermal conductivity of nanofluids, including the effects of Brownian motion and nanolayer.

Table 12.5 Summary of theoretical models for the effective thermal conductivity of nanofluids

Models	Expressions	Physical model	Remarks
Maxwell [74]	$\frac{k_{\text{eff}}}{k_{\text{f}}}=\frac{k_{\text{p}}+2k_{\text{f}}+2\phi_{\text{p}}(k_{\text{p}}-k_{\text{f}})}{k_{\text{p}}+2k_{\text{f}}-\phi_{\text{p}}(k_{\text{p}}-k_{\text{f}})}$	– Based on the conduction solution through a stationary random suspension of spheres	– Spherical particles – Accurate to order ϕ_{p}^1
Bruggeman [75]	$\frac{k_{\text{eff}}}{k_{\text{f}}}=\frac{(3\phi_{\text{p}}-1)\frac{k_{\text{p}}}{k_{\text{f}}}+\{3(1-\phi_{\text{p}})-1\}+\sqrt{\Delta}}{4}$ $\Delta=\left[(3\phi_{\text{p}}-1)\frac{k_{\text{p}}}{k_{\text{f}}}+\{3(1-\phi_{\text{p}})-1\}\right]^2+8\frac{k_{\text{p}}}{k_{\text{f}}}$	– Based on the differential effective medium (DEM) theory to estimate the effective thermal conductivity of composites at high particle concentrations – It consists in building up the composite medium through a process of incremental homogenization	– Applicable to high volume fraction of spherical particles – Suspension with spherical inclusions – No shape factor
Hamilton–Crosser [73]	$\frac{k_{\text{eff}}}{k_{\text{f}}}=\frac{k_{\text{p}}+(n-1)k_{\text{f}}+(n-1)\phi_{\text{p}}(k_{\text{p}}-k_{\text{f}})}{k_{\text{p}}+(n-1)k_{\text{f}}-\phi_{\text{p}}(k_{\text{p}}-k_{\text{f}})}$	– Based on the effective thermal conductivity of a two-component mixture	– Spherical and non-spherical particles – $n = 3$ (spheres), $n = 6$ (cylinders)
Wasp [76]	$\frac{k_{\text{eff}}}{k_{\text{f}}}=\frac{k_{\text{p}}+2k_{\text{f}}+2\phi_{\text{p}}(k_{\text{p}}-k_{\text{f}})}{k_{\text{p}}+2k_{\text{f}}-\phi_{\text{p}}(k_{\text{p}}-k_{\text{f}})}$	– Based on effective thermal conductivity of a two-component mixture	– Special case of Hamilton and Crosser's model with $n = 3$

Models	Expressions	Physical model	Remarks
Jeffery [77]	$\frac{k_{eff}}{k_f} = 1 + 3\eta\phi_p + \phi_p^2\left(3\eta^2 + \frac{3\eta^2}{4} + \frac{9\eta^3}{16}\frac{\kappa+2}{2\kappa+3} + \ldots\right)$ $\kappa = \frac{k_p}{k_f},\ \eta = \frac{\kappa-1}{\kappa+2}$	– Based on the conduction solution through a stationary random suspension of spheres	– High order terms represent pair interactions of randomly dispersed spherical particles – Accurate to order ϕ_p^2
Davis [78]	$\frac{k_{eff}}{k_f} = 1 + \frac{3(\kappa-1)}{(\kappa+2) - \phi_p(\kappa-1)}$ $[\phi_p + f(\kappa)\phi_p^2 + O(\phi_p^3)]$	– Green's theorem was applied to the space occupied by the matrix material (spherical inclusions) – Decaying temperature field was used	– Accurate to order ϕ_p^2 – High order terms represent pair interactions of randomly dispersed particles – $f(\kappa) = 2.5$ for $\kappa = 10$; $f(\kappa) = 0.5$ for $\kappa = \infty$
Lu and Lin [79]	$\frac{k_{eff}}{k_f} = 1 + a\phi_p + b\phi_p^2$	– The effective conductivity of composites containing aligned spheroids of finite conductivity was modeled with the pair interaction – The pair interaction was evaluated by solving a boundary value problem involving two aligned spheroids	– Spherical and non-spherical particles – Spherical particles: $a = 2.25$, $b = 2.27$ for $\kappa = 10$; $a = 3$, $b = 4.51$ for $\kappa = \infty$

Table 12.6 Summary of theoretical models for the effective thermal conductivity of nanofluids (nanolayer effect)

Models	Expressions	Physical model	Remarks
Yu and Choi [33]	$k_{\text{eff}} = \frac{k_{\text{pe}} + 2k_{\text{f}} + 2\phi_{\text{p}}(k_{\text{pe}} - k_{\text{f}})(1+\beta)^3}{k_{\text{pe}} + 2k_{\text{f}} - \phi_{\text{p}}(k_{\text{pe}} - k_{\text{f}})(1+\beta)^3} k_{\text{f}}$ $k_{\text{pe}} = \frac{2(1-\gamma)+(1+\beta)^3(1+2\gamma)\gamma}{-(1-\gamma)+(1+\beta)^3(1+2\gamma)} k_{\text{p}}$ $(\beta = t/r_{\text{p}}$ and $\gamma = k_{\text{layer}}/k_{\text{p}})$	– Modified Maxwell model [74]	– Spherical particles – Nanolayer
Yu and Choi [34]	$k_{\text{eff}} = \left(1 + \frac{nf_{\text{e}}A}{1 - f_{\text{e}}A}\right) k_{\text{f}}$ $A = \frac{1}{3} \sum_{j=a,b,c} \frac{(k_{\text{pj}} - k_{\text{f}})}{k_{\text{pj}} + (n-1)k_{\text{f}}}$ $f_{\text{e}} = \frac{f\sqrt{(a^2+t)(b^2+t)(c^2+t)}}{abc}$	– Modified Hamilton–Crosser model [73]	– Non-spherical particles – Nanolayer
Xue [36]	$9\left(1 - \frac{\phi_{\text{p}}}{\lambda}\right)\frac{k_{\text{eff}} - k_{\text{f}}}{2k_{\text{eff}} + k_{\text{f}}} + \frac{\phi_{\text{p}}}{\lambda} \frac{k_{\text{eff}} - k_{c,x}}{k_{\text{eff}} + B_{2,x}(k_{c,x} - k_{\text{eff}})}$ $+ \frac{\phi_{\text{p}}}{\lambda} 4 \frac{k_{\text{eff}} - k_{c,y}}{(1 - B_{2,x})(k_{c,y} - k_{\text{eff}})} = 0$	– Based on the Maxwell model and the average polarization theory and on the assumption that there is an interfacial shell between the nanoparticles and the base fluid	– Spherical particles – Nanolayer

Models	Expressions	Physical model	Remarks
Xue and Xu [35]	$\left(1-\frac{\phi_p}{\alpha}\right)\frac{k_{eff}-k_f}{2k_{eff}+k_f}+\frac{\phi_p}{\alpha}\frac{(k_{eff}-k_{shell})(2k_{shell}+k_p)-\alpha(k_p-k_{shell})(2k_{shell}+k_{eff})}{(2k_{eff}+k_{shell})(2k_{shell}+k_p)+2\alpha(k_p-k_{shell})(k_{shell}-k_{eff})}=0$	– A modified Bruggeman model [75], including the effect of interfacial shells	– Spherical particles – Nanolayer
Xie et al. [37]	$\frac{k_{eff}-k_f}{k_f}=3\Theta\phi_T+\frac{3\Theta^2\phi_T^2}{1-\Theta\phi_T}$ $\phi_T=\frac{4}{3}\pi(r_p+t)^3N_p=\phi_p(1+\beta)^3,\ \beta=\frac{t}{r_p}$	– Based on Fourier's law of heat conduction	– Low particle loadings – Nanolayer

Table 12.7 Summary of theoretical models for the effective thermal conductivity of nanofluids (Brownian effect)

Models	Expressions	Physical model	Remarks
Wang et al. [19]	$\frac{k_{eff}}{k_f}=\frac{(1-\phi_p)+3\phi_p\int_0^{\infty}\frac{k_{cl}(r)n(r)}{k_{cl}(r)+2k_f}dr}{(1-\phi_p)+3\phi_p\int_0^{\infty}\frac{k_f(r)n(r)}{k_{cl}(r)+2k_f}dr}$	– Based on the effective medium approximation and the fractal theory for predicting the thermal conductivity of nanofluids	– Accounts for the size effect and the surface adsorption of nanoparticles
Xuan et al. [80]	$\frac{k_{eff}}{k_f}=\frac{k_p+2k_f-2\phi_p(k_f-k_p)}{k_p+2k_f+\phi_p(k_f-k_p)}+\frac{\rho_p\phi_p c_p}{2k_f}\sqrt{\frac{k_B T}{3\pi r_c \mu}}$	– Based on the Maxwell model – The theory of Brownian motion and the diffusion-limited aggregation model are applied to simulate random motion and the aggregation process of the nanoparticles	– Includes the effect of random motion, particle size, concentration, and temperature
Jang and Choi [10]	$k_{eff}=k_f(1-\phi_p)+k_p\phi_p+3C\frac{d_f}{d_p}k_f Re_{d_p}^2 Pr\phi_p$	– A theoretical model was developed based on kinetics, Kapitza resistance, and convection – A general expression for the thermal conductivity of nanofluids involving four modes of energy transport in nanofluids was derived	– Considered four modes of energy transport: collision between fluid molecules, thermal diffusion of nanoparticles, collision between nanoparticles due to Brownian motion, and thermal interactions of dynamic nanoparticles with fluid molecules – Collision of nanoparticles due to Brownian motion is neglected

(Continued)

Models	Expressions	Physical model	Remarks
Prasher et al. [81]	$k_{\text{eff}} = (1 + ARe^{m}Pr^{0.333}\phi_{\text{p}}) \times \left[\frac{k_{\text{p}} + 2k_{\text{f}} + 2\phi_{\text{p}}(k_{\text{p}} - k_{\text{f}})}{k_{\text{p}} + 2k_{\text{f}} - \phi_{\text{p}}(k_{\text{p}} - k_{\text{f}})}\right] k_{\text{f}}$	– Based on Maxwell model and heat transfer in fluidized beds	– Accounts for convection caused by the Brownian motion from multiple nanoparticles
Koo and Kleinstreuer [72, 82]	$k_{\text{eff}} = k_{\text{static}} + k_{\text{Brownian}} = \frac{k_{\text{p}} + 2k_{\text{f}} + 2\phi_{\text{p}}(k_{\text{p}} - k_{\text{f}})}{k_{\text{p}} + 2k_{\text{f}} - \phi_{\text{p}}(k_{\text{p}} - k_{\text{f}})} k_{\text{f}} + 5\times10^{4}\beta\phi_{\text{p}}\rho_{\text{p}}c_{\text{p}}\sqrt{\frac{k_{\text{B}}T}{\rho_{\text{p}}D}} f(T, \phi_{\text{p}})$	– Based on the Maxwell model – Curve fitting of the available experimental data to determine the effective conductivity due to Brownian motion	– Considered surrounding liquid traveling with randomly moving nanoparticles
Chon et al. [83]	$\frac{k_{\text{eff}}}{k_{\text{f}}} = 1 + 64.7\phi_{p}^{0.74}\left(\frac{d_{\text{f}}}{d_{\text{p}}}\right)^{0.369}\left(\frac{k_{\text{p}}}{k_{\text{f}}}\right)^{0.747} \times Pr^{0.9955}Re^{1.2321}$ $Pr = \frac{\mu_{\text{f}}}{\rho_{\text{f}}\alpha_{\text{f}}},\ Re = \frac{\rho_{\text{f}}V_{\text{Br}}d_{\text{p}}}{\mu_{\text{f}}} = \frac{\rho_{\text{f}}k_{\text{B}}T}{3\pi\mu_{\text{f}}^{2}l_{\text{f}}}$	– Based on the curve fitting of the experimental data	– Reynolds number is based on the Brownian motion velocity

12.5.1 Experimental Investigations

Several experimental studies were conducted in the literature to measure the thermal conductivity of nanofluids using different techniques such as transient hot wire, steady-state parallel plates, and temperature oscillation. Despite numerous studies motivated by the significant benefits of utilizing nanofluids in various applications, the pertinent mechanisms of the thermal conductivity enhancement of nanofluids have not been well understood. Moreover, published results demonstrate thermal conductivity enhancement varying from anomalously large to small values, as shown in Table 12.8. The table shows a comparison of the experimental thermal conductivity enhancements of Al_2O_3 nanofluids cited in the literature. Al_2O_3 and CuO are the most common nanoparticles used in the literature. Several studies have reported enhancement in the thermal conductivity of nanofluids at room temperature [84]. Figures 12.6a and 12.6b show the effective thermal conductivity measurements at ambient temperature for Al_2O_3–water and CuO–water nanofluids at various volume concentrations and nanoparticle diameters. Figure 12.9 shows that the effective thermal conductivity increases with an increase in the volume fraction. In addition, the size of the particles is found to have a significant effect on the thermal conductivity enhancement. It should be noted that smaller particles exhibit larger surface-area-to-volume ratio than the larger particles. As such, smaller particle diameters can possibly result in a larger augmentation in the effective thermal conductivity. It is interesting to note from Figs. 12.6a and 12.6b that the Hamilton–Crosser model [73] may represent a good approximation for the effective thermal conductivity value for smaller volume fractions ($\phi_p \leq 4\%$).

A general correlation for the effective thermal conductivity of Al_2O_3–water and CuO–water nanofluids at ambient temperature accounting for various volume fractions and nanoparticles diameters is obtained by the present authors using the available experimental data in the literature. This model can be expressed as:

$$\frac{k_{\text{eff}}}{k_{\text{f}}} = 1.0 + 1.0112\phi_{\text{p}} + 2.4375\phi_{\text{p}}\left(\frac{47}{d_{\text{p}}(\text{nm})}\right) - 0.0248\phi_{\text{p}}\left(\frac{k_{\text{p}}}{0.613}\right);$$

$$R^2 = 96.5\% \qquad (12.20)$$

where k_f is the thermal conductivity of water. Figure 12.7 demonstrates that the general correlation, represented by Eq. (12.20), is in good agreement with the experimental measurements of Al_2O_3–water and CuO–water nanofluids.

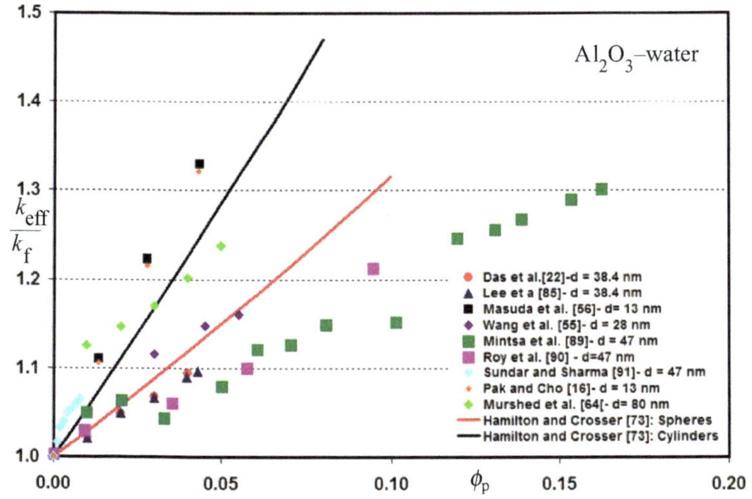

(a)

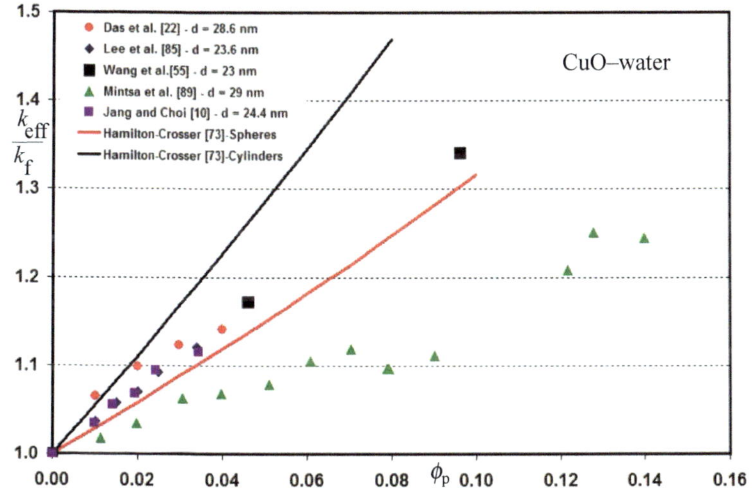

(b)

Figure 12.6 Effect of the volume fraction on the effective thermal conductivity measurements: (a) Al_2O_3–water; (b) CuO–water. (See above-mentioned references for further information.)

Table 12.8 Comparison of the experimental thermal conductivity enhancements of Al_2O_3 nanofluids cited in the literature

Ref.	Base fluid	D_p	ϕ_p (%)	Percent thermal conductivity enhancement	Method
[22]	Water	38.4 nm	4	9.4% (21°C), 24.3% (51°C)	Temperature oscillation
[13]	Water	131 nm	4	24% (51°C)	Steady-state parallel plates
[56]	Water	13 nm	4.4	33%	Transient hot wire
[85]	Water	38.4 nm	4.3	11%	Transient hot wire
[85]	Ethylene glycol	38.4 nm	5	19%	Transient hot wire
[55]	Water	28 nm	5.5	16%	Steady-state parallel plates
[55]	Ethylene glycol	28 nm	5	24.5%	Steady-state parallel plates
[83]	Water	11 nm	1	14.8% (70°C)	Transient hot wire
[83]	Water	47 nm	1	10.2% (70°C)	Transient hot wire
[83]	Water	150 nm	1	4.8% (60°C)	Transient hot wire
[83]	Water	47 nm	4	28.8 % (70°C)	Transient hot wire
[86]	Water	36 nm	6	28.2%	Steady-state parallel plates
[86]	Water	47 nm	6	26.1%	Steady-state parallel plates
[87]	Water	20 nm	5	15%	Transient hot wire
[88]	Water	11 nm	5	8%	Transient hot wire
[88]	Water	20 nm	5	7%	Transient hot wire
[88]	Water	40 nm	5	10%	Transient hot wire

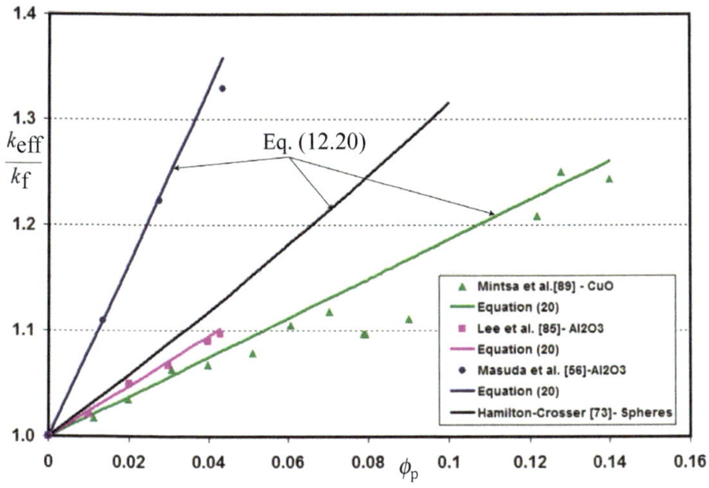

Figure 12.7 Comparison of the general correlation, Eq. (12.20), developed in the current work, with the experimental data (Al_2O_3–water, CuO–water) at room temperature.

Thermal conductivity measurements at different temperatures are essential because the measurements at ambient temperature are not adequate for estimating the heat transfer characteristics. Figure 12.8 shows a comparison of the relative effective thermal conductivity (ratio of the effective thermal conductivity of the nanofluid to the thermal conductivity of the base fluid at the same temperature) results of Al_2O_3-water nanofluid obtained from various experimental results as a function of volume fraction and nanoparticle's diameter. Figure 12.8 shows that temperature has a significant effect on the thermal conductivity enhancement. A general correlation is developed for Al_2O_3-water nanofluid by the present authors using the available experimental data at various temperatures, nanoparticle's diameter, and volume fraction. The developed correlation is expressed in terms of nanoparticle's diameter, volume fraction, dynamic viscosity of water, effective dynamic viscosity of the nanofluid, and temperature as follows:

$$\frac{k_{eff}}{k_f} = 0.9843 + 0.398\phi_p^{0.7383}\left(\frac{1}{d_p(\text{nm})}\right)^{0.2246}\left(\frac{\mu_{eff}(T)}{\mu_f(T)}\right)^{0.0235} - 3.9517\frac{\phi_p}{T} + 34.034\frac{\phi_p^2}{T^3} + 32.509\frac{\phi_p}{T^2} \quad (12.21)$$

$$0 \le \phi_p \le 10\%,\ 11\,\text{nm} \le d \le 150\,\text{nm},\ 20°\text{C} \le T \le 70°\text{C}$$

where the dynamic viscosity (Pa s) of water at different temperatures can be expressed as $\mu_f(T) = 2.414 \times 10^{-5} \times 10^{247.8/(T-140)}$.

The validity of the above correlation (Eq. (12.21)) is depicted in Fig. 12.9. The figure shows a very good agreement between the predicted relative effective thermal conductivity by our model and the experimental data.

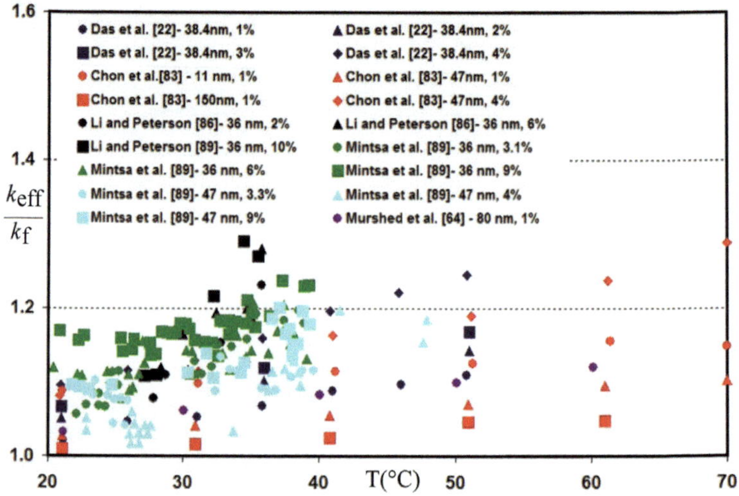

Figure 12.8 Comparison of the experimental data for the thermal conductivity enhancement of Al_2O_3–water nanofluid at different temperatures and volume fractions.

Different models were developed in the past for the effective thermal conductivity of a two-component mixture such as the Hamilton–Crosser model [73]. Although this model gave a good approximation for the effective thermal conductivity of the Al_2O_3–water and CuO–water nanofluids for small volume fractions at room temperature, it does not provide a good approximation of the effective thermal conductivity at various temperatures shown in Fig. 12.9 because this model as well as a number of other models in this area does not properly account for the variations of the effective thermal conductivity with temperature. Therefore, these analytical models cannot be used to determine the effective thermal conductivity of nanofluids at various temperatures. Instead, Eq. (12.21) developed in this work should be used. Table 12.9

summarizes all the correlations that were developed in the present work for nanofluids.

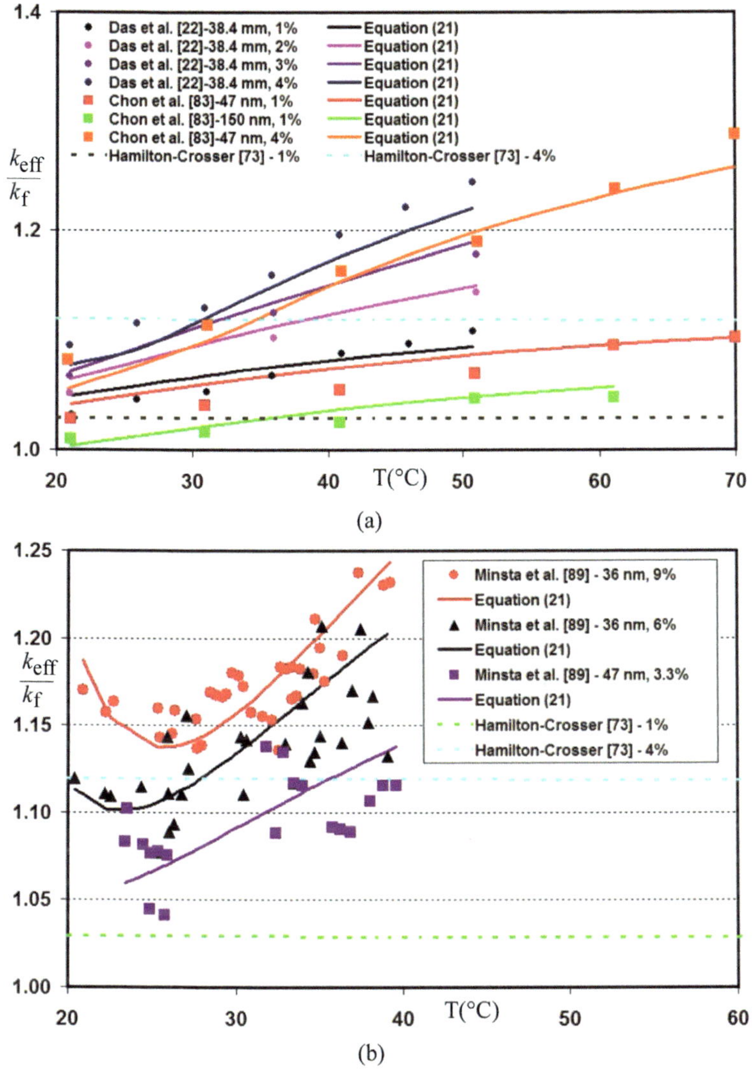

Figure 12.9 Comparison of the general correlation, Eq. (12.21), developed in the current work, with the experimental data (Al_2O_3–water) at various temperatures and volume fractions: (a) Das et al. [22] and Chon et al. [83]; (b) Minsta et al. [89].

Table 12.9 Summary of the correlations developed in the present work

Physical properties	Room temperature	Temperature dependent
Density	$\rho_{eff} = (1-\varphi_p)\rho_f + \varphi_p\rho_p$	$\rho_{eff} = 1001.064 + 2738.6191\phi_p - 0.2095T$ $0 \le \phi_p \le 0.04,\ 5°C \le T \le 40°C$
Specific heat	$c_{eff} = \dfrac{(1-\phi_p)\rho_f c_f + \phi_p\rho_p c_p}{\rho_{eff}}$	N/A
Thermal expansion coefficient	$\beta_{eff} = \dfrac{(1-\phi_p)(\rho\beta)_f + \phi_p(\rho\beta)_p}{\rho_{eff}}$ $\beta_{eff} = (1-\phi_p)\beta_f + \phi_p\beta_p$	$\beta_{eff} = \left(-0.479\phi_p + 9.3158\times10^{-3}T - \dfrac{4.7211}{T^2}\right)\times10^{-3}$ $0 \le \phi_p \le 0.04,\ 10°C \le T \le 40°C$
Viscosity	N/A	Al_2O_3 $\mu_{eff} = -0.4491 + \dfrac{28.837}{T} + 0.574\phi_p - 0.1634\phi_p^2 + 23.053\dfrac{\phi_p^2}{T^2} + 0.0132\phi_p^3 - 2354.735\dfrac{\phi_p}{T^3} + 23.498\dfrac{\phi_p^2}{d_p^2} - 3.0185\dfrac{\phi_p^3}{d_p^2}$ $1\% \le \phi_p \le 9\%,\ 20°C \le T \le 70°C,\ 13\,nm \le d_p \le 131\,nm$
Thermal conductivity	Al_2O_3 and CuO $\dfrac{k_{eff}}{k_f} = 1.0 + 1.0112\phi_p + 2.4375\phi_p\left(\dfrac{47}{d_p(nm)}\right) - 0.0248\phi_p\left(\dfrac{k_p}{0.613}\right)$	Al_2O_3 $\dfrac{k_{eff}}{k_f} = 0.9843 + 0.398\phi_p^{0.7383}\left(\dfrac{1}{d_p(nm)}\right)^{0.2246}\left(\dfrac{\mu_{eff}(T)}{\mu_f(T)}\right)^{0.0235} - 3.9517\dfrac{\phi_p}{T} + 34.034\dfrac{\phi_p^2}{T^3} + 32.509\dfrac{\phi_p}{T^2}$ $0 \le \phi_p \le 10\%,\ 11\,nm \le d \le 150\,nm,\ 20°C \le T \le 70°C$

12.5.2 Natural Convection Heat Transfer Utilizing Nanofluids

Conflicting results were reported in the literature regarding natural convection heat transfer enhancement using nanofluids. The findings of both experimental and analytical investigations are still in disagreement. Analytical studies show an increase in heat transfer with an increase in the volume fraction of nanoparticles, which is not in agreement with the experimental results. Since Rayleigh number, which is the ratio of the buoyancy to the viscous forces, represents a significant parameter in natural convection processes, comparison of nanofluid Rayleigh number to the base fluid Rayleigh number at various volume fractions and temperature is highlighted in this section. Using a scale analysis approach, the viscous and buoyancy forces can be expressed as:

$$\text{Viscous force: } \frac{\nu}{H^3}\alpha \quad (\text{N/kg}), \tag{12.22}$$

$$\text{Buoyancy force: } g\beta\Delta T \quad (\text{N/kg}). \tag{12.23}$$

The ratio of the nanofluid Rayleigh number to that of the base fluid can then be expressed as:

$$\frac{Ra_{\text{nf}}}{Ra_{\text{f}}} = \frac{\beta_{\text{nf}}}{\beta_{\text{f}}}\frac{\nu_{\text{f}}}{\nu_{\text{nf}}}\frac{\alpha_{\text{f}}}{\alpha_{\text{nf}}}. \tag{12.24}$$

Figure 12.10a shows that the ratio of the Rayleigh number of nanofluid to that of the base fluid decreases with an increase in the Al_2O_3 volume fraction. Higher volume fractions of the solid nanoparticles cause an increase in the viscous force of nanofluids and consequently suppress heat transfer. Moreover, Fig. 12.10a shows the effect of varying particle diameter on the Rayleigh number ratio. As the particle diameter increases, the ratio of the Rayleigh numbers decreases because the effective thermal conductivity of nanofluids decreases and the kinematic viscosity increases with an increase in the size of nanoparticles. However, the rate of increase of the kinematic viscosity of the nanofluid with the particle size is larger than the resulting decrease of the effective thermal conductivity. This may provide a physical reason for the reduction in the natural convection heat transfer enhancement with an increase in the volume fraction of nanoparticles at room temperature.

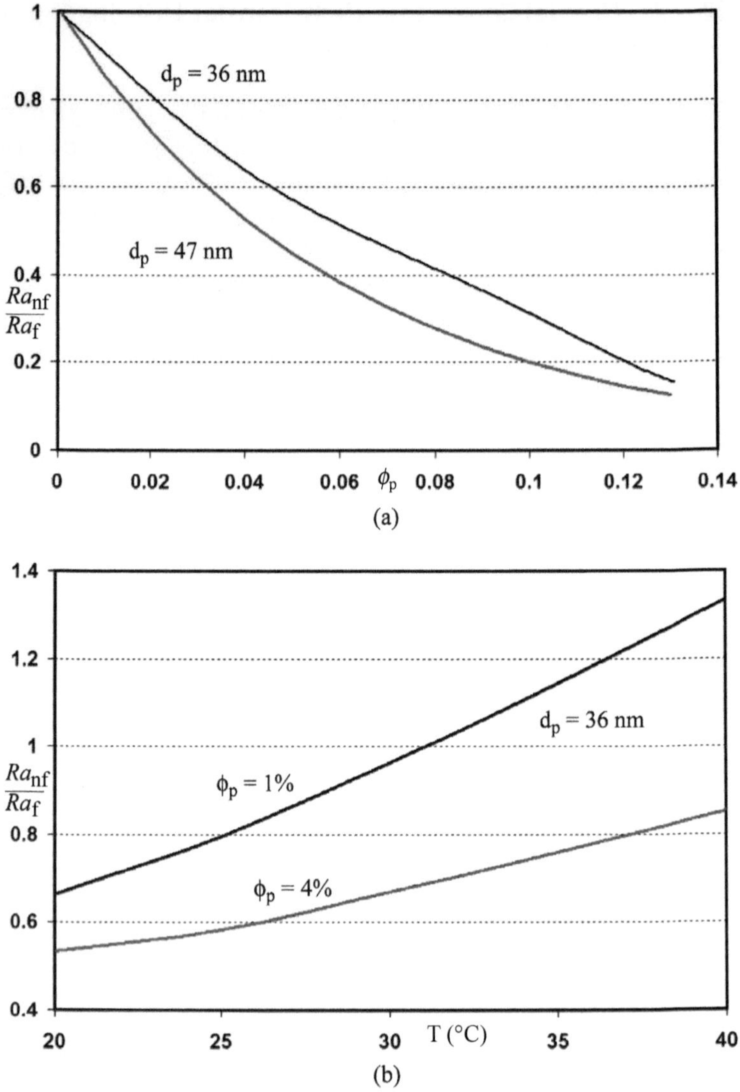

Figure 12.10 Effect of volume fraction and temperature on the ratio of the Rayleigh numbers for different particle diameters (Al_2O_3–water nanofluid): (a) room temperature and (b) various temperatures.

The effect of varying the temperature of nanofluids and volume fraction on the ratio of Rayleigh numbers is illustrated in Fig. 12.10b for nanoparticle diameter of 36 nm. Figure 12.10b shows that the ratio of the nanofluid Rayleigh number to the base fluid Rayleigh number increases with an increase in the temperature. Moreover, this ratio is higher for volume fraction of 1% compared to 4% for various temperatures. This is because the kinematic viscosity and the effective thermal conductivity of nanofluids increase with an increase in the volume fraction of nanoparticles. For a volume fraction of 1%, Fig. 12.10b shows an interesting result associated with the fact that the nanofluid Rayleigh number is smaller than the Rayleigh number of water base below 31°C. For temperatures greater than 31°C, Fig. 12.10b shows that the nanofluid Rayleigh number is higher than that of the base water. Hence, nanofluids may exhibit natural convection heat transfer enhancement at high temperatures. This is associated with the behavior of the kinematic viscosity and the thermal diffusivity for both nanofluid and the base water at various temperatures, as shown in Fig. 12.11. Hence, Figs. 12.10a and 12.10b show that natural convection heat transfer is not exclusively characterized by the effective thermal conductivity of nanofluids but also depends on the viscosity of nanofluids.

12.5.3 Surface Tension

Studies on surface tension of nanofluids are limited in the literature [92–96]. Golubovic et al. [92] showed experimentally that the surface tension of Al_2O_3–water nanofluid does not change for concentrations of nanoparticles considered in their work (0–0.01 g/L) and the surface tension is approximately equal to the surface tension of pure water at $T = 24°C$. Xue et al. [93] showed insignificant effect of carbon nanotube nanofluid on surface tension compared with pure water. Similarly, Kim et al. [94] found that the surface tension of Al_2O_3–water nanofluids (0.1% volume fraction) differed negligibly from that of pure water. The effect of temperature on the surface tension of nanofluids was studied by Murshed et al. [95]. Their experimental results showed that nanofluids having TiO_2 nanoparticles of 15 nm diameter in deionized water exhibit substantially lower surface tension than those of the base fluid (i.e., deionized water). Similarly, Zhu et al. [96] showed that the surface tension of nanofluids was highly dependent on the temperature. One can note from the above that as expected temperature plays a significant role on the surface tension of nanofluids.

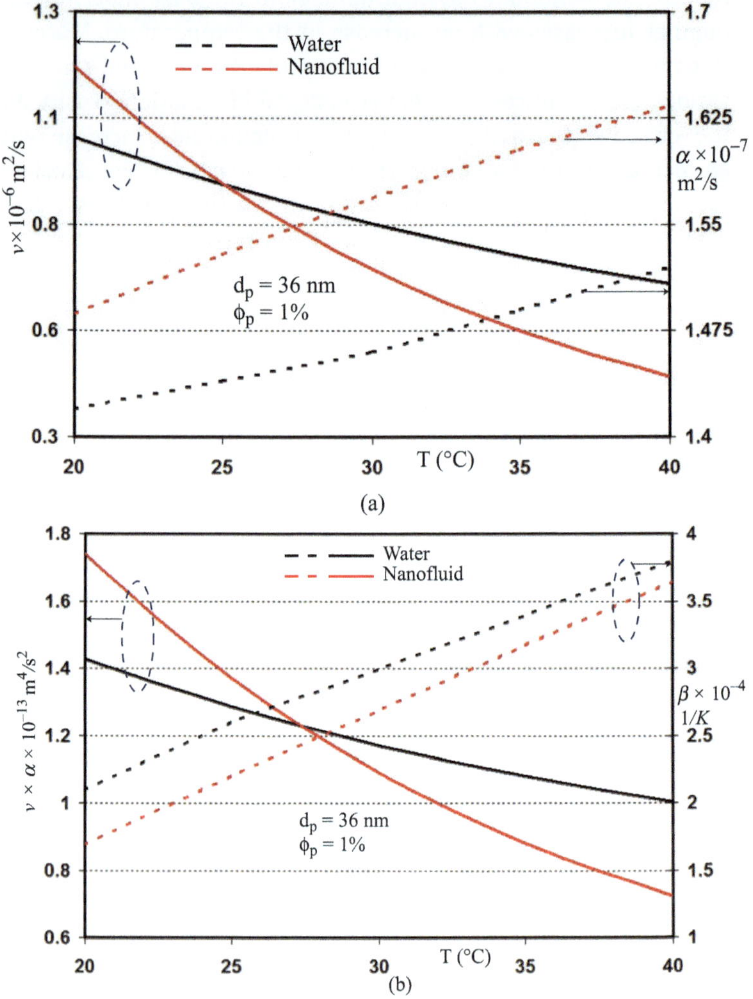

Figure 12.11 Effect of varying temperature on the thermophysical properties: (a) Al_2O_3–water nanofluid and (b) water.

12.5.4 Nucleate Pool Boiling and Critical Heat Flux of Nanofluids

Several experimental studies on the nucleate pool boiling and critical heat flux (CHF) characteristics of nanofluids have been investigated in the literature. Many researchers expected that the addition of nanoparticles would have a potential to enhance the boiling heat transfer characteristics. Conflicting results were reported on the effect of the nanoparticle concentration on the pool boiling characteristics of nanofluids. Some studies have shown that the addition of nanoparticles exhibited a decrease or no change in the nucleate boiling heat transfer rate [20, 21, 97–99], while other studies have illustrated an increase [100–103]. However, most CHF experiments using nanofluids were in agreement in reporting an enhancement in CHF for pool boiling conditions [98, 99, 104, 105]. The mechanism for the CHF enhancement may be attributed to the deposition and sintering of nanoparticles on the boiling surfaces resulting in an increase in the nucleation sites [98]. For example, Kim et al. [94] conducted an experimental study on the CHF characteristics of nanofluids in pool boiling. Their results show that the CHF of nanofluids containing TiO_2 or Al_2O_3 were enhanced up to 100% over that of pure water.

You et al. [98] conducted an experimental study to determine the boiling curve and the CHF at pool boiling conditions from a flat square polished copper heater immersed in Al_2O_3–water nanofluid at various volume fractions ranging between 0.0 g/L and 0.05 g/L. Their results illustrate a drastic enhancement in the CHF for Al_2O_3–water nanofluids (~200% higher than pure water; measured in nanofluids containing 0.005 g/L of Al_2O_3 nanoparticles). However, heat transfer enhancement or degradation was not observed in the nucleate boiling regime for nanofluids. For example, Vassallo et al. [106] experimentally demonstrated a marked increase in the CHF for both nano- and micro-solutions (silica–water) at the same concentration (0.5% volume fraction) compared to the base water. However, from boiling curves data, no heat transfer enhancement of nanofluids was observed in the nucleate boiling regime.

12.5.5 Nucleate Pool Boiling Heat Transfer and CHF Mechanisms of Nanofluids

Several studies have been conducted in the literature to explore the enhancement mechanisms or deterioration of nucleate pool boiling heat transfer using nanofluids. These mechanisms include formation of nanoparticle coatings on the surface during pool boiling of nanofluids [99], decrease in active nucleation sites due to nanoparticle sedimentation on the boiling surface [107], and the wettability change of the surface [20, 21]. The available experimental results on nucleate pool boiling heat transfer coefficient of nanofluids are in disagreement. While the CHF enhancement results by nanofluids are consistent in the literature, the responsible mechanisms are not verified. Golubovic et al. [92] concluded that the major reason behind the increase in CHF in the pool boiling of nanofluids is a decrease in the static surface contact angle.

Many other studies consider the primary reason for CHF enhancement is due to the surface coating effect [99, 100, 106, 108–111]. For example, Bang and Chang [99] conducted an experimental study on boiling heat transfer characteristics of nanofluids with nanoparticles suspended in water using different concentrations of alumina nanoparticles. The CHF performance was enhanced for both horizontal (32%) and vertical (13%) flat surfaces, and the authors related this enhancement to a change in surface characteristics by the deposition of nanoparticles. If this reasoning is accepted, it might be easier to modify the boiling surface in pursuit of a greater number of nucleation sites per area rather than using nanofluids. The CHF enhancement is normally achieved by increasing boiling surface area using a variety of fin shapes and sizes [112, 113]. Anderson and Mudawar [112] illustrated that the surfaces with microgrooves and square microstuds are highly effective in enhancing the nucleate boiling heat transfer coefficient in Fluorinert electronic liquid (FC-72) and increasing CHF values by up to 2.5 times compared to a smooth surface. Honda et al. [114] and Wei et al. [115] showed that CHF values for the nano-roughened surface and micro-pin-finned surfaces were respectively 1.8–2.2 and 2.3 times those for a smooth silicon surface. Ujereh et al. [116] performed experiments to assess the impact of coating silicon and copper substrates with nanotubes on pool boiling performance. Fully coating the substrate surface with carbon nanotubes was found to be highly effective at reducing the incipience superheat and significantly enhancing both the nucleate boiling heat transfer coefficient and CHF.

More robust physical models are essential to explain the effect of nanofluids on nucleate pool boiling and CHF. The thermophysical properties of nanofluids, such as surface tension, density, viscosity, and specific heat, may also have an effect on nucleate pool boiling heat transfer coefficient and CHF. A detailed knowledge of the thermophysical properties of nanofluids, structure of the boiling surface, and coating of nanoparticles can be helpful in resolving the controversies in the pool boiling heat transfer coefficient of nanofluids as well as in illustrating the mechanisms that cause the significant increase in CHF.

12.6 Conclusion

Thermophysical characteristics of nanofluids and their role in heat transfer enhancement are analyzed in this chapter. General correlations for the effective thermal conductivity and viscosity of nanofluids are developed based on the pertinent experimental data in terms of the volume fraction, particle diameter, temperature, and the base fluid physical properties. The effective viscosity of nanofluids is found to increase with an increase in the volume fraction and decrease with an increase in the temperature. The effective thermal conductivity of nanofluids increases with an increase in temperature and volume fraction and decreases with an increase in the particle diameter. At room temperature, classical models can be used to estimate the thermal conductivity and viscosity of nanofluids for low volume fractions. However, these models cannot predict the thermal conductivity at other temperatures. Correlations based on available experimental data accounting for the temperature effect are developed in this chapter. The findings from experimental and analytical investigations are still in disagreement regarding natural convection enhancement utilizing nanofluids. The current chapter illustrates that the viscosity of nanofluids plays a key role in predicting the heat transfer characteristics. Differences in thermal conductivity and viscosity measurements in the literature using different measurement techniques are highlighted. Moreover, for high heat flux applications, the experimental results illustrate conflicting results in pool boiling heat transfer characteristics, while the CHF of nanofluids shows a significant increase with the addition of nanoparticles. While the addition of nanoparticles appears not to have a significant effect on the surface tension of the nanofluid, experiment shows that the surface tension of the nanofluids is significantly dependent on the temperature. Several pertinent correlations for the thermophysical properties of nanofluids were developed in this chapter based on the available experimental data. This chapter clearly illustrates the need for additional investigations in measuring the nanofluid properties.

References

1. S. U. S. Choi, "Nanofluids: From vision to reality through research," *J. Heat Transfer*, 2009; **131**: 1–9.
2. K. V. Wong and O. Leon, "Applications of nanofluids: Current and future," *Adv. Mech. Eng.*, 2010; **2010**: 1–11.
3. V. Bianco, F. Chiacchio, O. Manca, and S. Nardini, "Numerical investigation of nanofluids forced convection in circular tubes," *Appl. Therm. Eng.*, 2009; **29**: 3632–3642.
4. M. Shafahi, V. Bianco, K. Vafai, and O. Manca, "Thermal performance of flat-shaped heat pipes using nanofluids," *Int. J. Heat Mass Transfer*, 2010; **53**: 1438–1445.
5. M. Shafahi, V. Bianco, K. Vafai, and O. Manca, "An investigation of the thermal performance of cylindrical heat pipes using nanofluids," *Int. J. Heat Mass Transfer*, 2010; **53**: 376–383.
6. K. Khanafer, K. Vafai, and M. Lightstone, "Buoyancy-driven heat transfer enhancement in a two-dimensional enclosure utilizing nanofluids," *Int. J. Heat Mass Transfer*, 2003; **46**: 3639–3653.
7. A. R. A. Khaled and K. Vafai, "Heat transfer enhancement through control of thermal dispersion effects," *Int. J. Heat Mass Transfer*, 2005; **48**: 2172.
8. J. A. Eastman, S. U. S. Choi, S. Li, L. J. Thompson, and S. Lee, "Enhanced thermal conductivity through the development of nanofluids," in: *1996 Fall Meeting of the Materials Research Society (MRS)*, Boston, USA.
9. J. A. Eastman, S. U. S. Choi, S. Li, W. Yu, and L. J. Thompson, "Anomalously increased effective thermal conductivities of ethylene glycol-based nanofluids containing copper nanoparticles," *Appl. Phys. Lett.*, 2001; **78**: 718–720.
10. S. P. Jang and S. U. S. Choi, "Role of Brownian motion in the enhanced thermal conductivity of nanofluids," *Appl. Phys. Lett.*, 2004; **84**: 4316–4318.
11. S. Lee and S. U. S. Choi, "Application of metallic nanoparticle suspensions in advanced cooling systems," in: *1996 International Mechanical Engineering Congress and Exhibition*, Atlanta, USA.

12. A. Ali, K. Vafai, and A.-R. A. Khaled, "Comparative study between parallel and counter flow configurations between air and falling film desiccant in the presence of nanoparticle suspensions," *Int. J. Energy Res.*, 2003; **27**: 725–745.

13. N. Putra, W. Roetzel, and S. K. Das, "Natural convection of nanofluids," *Heat Mass Transfer*, 2003; **39**: 775–784.

14. Y. M. Xuan and Q. Li, "Heat transfer enhancement of nanofluids," *Int. J. Heat Fluid Flow*, 2000; **21**: 58–64.

15. Y. M. Xuan and Q. Li, "Investigation on convective heat transfer and flow features of nanofluids," *ASME J. Heat Transfer*, 2003; **125**: 151–155.

16. B. C. Pak and Y. I. Cho, "Hydrodynamic and heat transfer study of dispersed fluids with submicron metallic oxide particles," *Exp. Heat Transfer*, 1999; **11**: 151–170.

17. Y. Yang, Z. Zhang, E. Grulke, W. Anderson, G. Wu, and G. Heat, "Transfer properties of nanoparticle-in-fluid dispersions (nanofluids) in laminar flow," *Int. J. Heat Mass Transfer*, 2005; **48**: 1107–1116.

18. D. S. Wen and Y. L. Ding, "Experimental investigation into the pool boiling heat transfer of aqueous based alumina nanofluids," *J. Nanopart. Res.*, 2005; **7**: 265–274.

19. B. X. Wang, L. P. Zhou, and X. F. Peng, "A fractal model for predicting the effective thermal conductivity of liquid with suspension of nanoparticles," *Int. J. Heat Mass Transfer*, 2003; **46**: 2665–2672.

20. S. K. Das, N. Putra, and W. Roetzel, "Pool boiling characteristics of nanofluids," *Int. J. Heat Mass Transfer*, 2003; **46**: 851–862.

21. S. K. Das, N. Putra, and W. Roetzel, "Pool boiling of nanofluids on horizontal narrow tubes," *Int. J. Multiphase Flow*, 2003; **29**: 1237–1247.

22. S. K. Das, N. Putra, P. Thiesen, and W. Roetzel, "Temperature dependence of thermal conductivity enhancement for nanofluids," *J. Heat Transfer*, 2003; **125**: 567–574.

23. J. Kim, Y. T. Kang, and C. K. Choi, "Analysis of convective instability and heat transfer characteristics of nanofluids," *Phys. Fluids*, 2004; **16**: 2395–23401.

24 A. G. A. Nanna, T. Fistrovich, K. Malinski, and S. U. S. Choi, "Thermal transport phenomena in buoyancy-driven nanofluids," in: *Proceedings of 2005 ASME International Mechanical Engineering Congress and RD&D Exposition*, 15–17 November 2004, Anaheim, California, USA.

25. A. G. A. Nnanna and M. Routhu, "Transport phenomena in buoyancy-driven nanofluids–Part II," in: *Proceedings of 2005 ASME Summer Heat Transfer Conference*, 17–22 July 2005, San Francisco, California, USA.
26. Y. Ding, H. Chen, L. Wang, C.-Y. Yang, Y. He, W. Yang, W. P. Lee, L. Zhang, and R. Huo., "Heat transfer intensification using nanofluids," *J. Particle Powder*, 2007; **25**: 23–36.
27. B. H. Chang, A. F. Mills, and E. Hernandez, "Natural convection of microparticle suspensions in thin enclosures," *Int. J. Heat Mass Transfer*, 2008; **51**: 1332–1341.
28. C. J. Ho, W. K. Liu, Y. S. Chang, and C. C. Lin, "Natural convection heat transfer of alumina–water nanofluid in vertical square enclosures: An experimental study," *Int. J. Therm. Sci.*, 2010; **49**: 1345–1353.
29. X. Wang and A. Mujumdar, "Heat transfer characteristics of nanofluids: A review," *Int. J. Therm. Sci.*, 2007; **46**: 1–19.
30. P. Keblinski, S. R. Phillpot, S. U. S. Choi, and J. A. Eastman, "Mechanisms of heat flow in suspensions of nano-sized particles (nanofluids)," *Int. J. Heat Mass Transfer*, 2002; **45**: 855–863.
31. J. A. Eastman, S. R. Phillpot, S. U. S. Choi, and P. Keblinski, "Thermal transport in nanofluids," *Annu. Rev. Mater. Res.*, 2004; **34**: 219–246.
32. W. Evans, J. Fish, and P. Keblinski, "Role of Brownian motion hydrodynamics on nanofluid thermal conductivity," *Appl. Phys. Lett.*, 2006; **88**(9): 93116.
33. W. Yu and S. U. S. Choi, "The role of interfacial layers in the enhanced thermal of nanofluids: A renovated Maxwell model," *J. Nanopart. Res.*, 2003; **5**(1–2): 167–171.
34. W. Yu and S. U. S. Choi, "The role of interfacial layers in the enhanced thermal conductivity of nanofluids: A renovated Hamilton–Crosser model," *J. Nanopart. Res.*, 2004; **6**(4): 355–361.
35. Q. Xue and W. M. Xu, "A model of thermal conductivity of nanofluids with interfacial shells," *Mater. Chem. Phys.*, 2005; **90**: 298–301.
36. Q. Xue, "Model for effective thermal conductivity of nanofluids," *Phys. Lett. A*, 2003; **307**: 313–317.
37. H. Xie, M. Fujii, and X. Zhang, "Effect of interfacial nanolayer on the effective thermal conductivity of nanoparticle–fluid mixture," *Int. J. Heat Mass Transfer*, 2005; **48**: 2926–2932.
38. L. Xue, P. Keblinski, S. R. Phillpot, S. U. S. Choi, and J. A. Eastman, "Effect of liquid layering at the liquid–solid interface on thermal transport," *Int. J. Heat Mass Transfer*, 2004; **47**(19–20): 4277–4284.

39. S. P. Jang and S. U. S. Choi, "Free convection in a rectangular cavity (Benard convection) with nanofluids," in: *Proceedings of the 2004 ASME International Mechanical Engineering Congress and Exposition*, Anaheim, California, November 13–20, 2004.

40. L. Gosselin and A. K. da Silva, "Combined heat transfer and power dissipation optimization of nanofluid flows," *Appl. Phys. Lett.*, 2004; **85**: 4160.

41. J. Lee and I. Mudawar, "Assessment of the effectiveness of nanofluids for single-phase and two-phase heat transfer in micro-channels," *Int. J. Heat Mass Transfer*, 2007; **50**: 452–463.

42. S. Q. Zhou and R. Ni, "Measurement of the specific heat capacity of water-based Al_2O_3 nanofluid," *Appl. Phys. Lett.*, 2008; **92**: 093123.

43. K. S. Hwang, J. H. Lee, and S. P. Jang, "Buoyancy-driven heat transfer of water-based Al_2O_3 nanofluids in a rectangular cavity," *Int. J. Heat Mass Transfer*, 2007; **50**: 4003–4010.

44. C. J. Ho, M. W. Chen, and Z. W. Li, "Numerical simulation of natural convection of nanofluid in a square enclosure: Effects due to uncertainties of viscosity and thermal conductivity," *Int. J. Heat Mass Transfer*, 2008; **51**: 4506–4516.

45. A. Einstein, "Eine neue bestimmung der molekuldimensionen," *Ann. Phys., Leipzig*, 1906; 19: 289–306.

46. H. C. Brinkman, "The viscosity of concentrated suspensions and solutions," *J. Chem. Phys.*, 1952; **20**: 571.

47. G. Batchelor, "The effect of Brownian motion on the bulk stress in a suspension of spherical particles," *J. Fluid Mech.*, 1977; **83**: 97–117.

48. T. Lundgren, "Slow flow through stationary random beds and suspensions of spheres," *J. Fluid Mech.*, 1972; **51**: 273–299.

49. A. L. Graham, "On the viscosity of suspensions of solid spheres," *Appl. Sci. Res.*, 1981; **37**: 275–286.

50. R. A. Simha, "Treatment of the viscosity of concentrated suspensions," *J. Appl. Phys.*, 1952; **23**: 1020–1024.

51. M. Mooney, "The viscosity of a concentrated suspension of spherical particles," *J. Colloid Sci.*, 1951; **6**: 162–170.

52. V. H. Eilers, "Die viskocitat von emulsionen hochviskoser stoffe als funktion der konzentration," *Kolloid-Zeitschrift*, 1941; **97**: 313–321.

53. N. Saito, "Concentration dependence of the viscosity of high polymer solutions," *J. Phys. Soc. Jpn.*, 1950; **5**: 4–8.

54. N. A. Frankel and A. Acrivos, "On the viscosity of a concentrate suspension of solid spheres," *Chem. Eng. Sci.*, 1967; **22**: 847–853.

55. X. Wang, X. Xu, and S. U. S. Choi, "Thermal conductivity of nanoparticles–fluid mixture," *J. Thermophys. Heat Transfer*, 1999; **13**: 474–480.

56. H. Masuda, A. Ebata, K. Teramae, and N. Hishinuma, "Alteration of thermal conductivity and viscosity of liquid by dispersing ultra-fine particles (dispersion of c-Al_2O_3, SiO_2, and TiO_2 ultra-fine particles)," *Netsu Bussei*, 1993; **4**: 227–233.

57. S. Maiga, S. J. Palm, C. T. Nguyen, G. Roy, and N. Galanis, "Heat transfer enhancement by using nanofluids in forced convection flows," *Int. J. Heat Fluid Flow*, 2005; **26**: 530–546.

58. S. Maiga, C. T. Nguyen, N. Galanis, G. Roy, T. Mar'e, and M. Coqueux, "Heat transfer enhancement in turbulent tube flow using Al_2O_3 nanoparticle suspension," *Int. J. Num. Meth. Heat Fluid Flow*, 2006; **16**: 275–292.

59. P. K. Namburua, D. P. Kulkarni, D. Misra, and D. K. Das, "Viscosity of copper oxide nanoparticles dispersed in ethylene glycol and water mixture," *Exp. Therm. Fluid Sci.*, 2007; **32**: 397–402.

60. C. T. Nguyen, F. Desgranges, G. Roy, N. Galanis, T. Mar′e, S. Boucher, H. A. Mintsa, "Temperature and particle-size dependent viscosity data for water-based nanofluids–hysteresis phenomenon," *Int. J. Heat Fluid Flow*, 2007; **28**: 1492–1506.

61. W. J. Tseng and K. C. Lin, "Rheology and colloidal structure of aqueous TiO_2 nanoparticle suspensions," *Mater. Sci. Eng.*, 2003; **A355**: 186–192.

62. D. P. Kulkarni, D. K. Das, and S. L. Patil, "Effect of temperature on rheological properties of copper oxide nanoparticles dispersed in propylene glycol and water mixture," *J. Nanosci. Nanotechnol.*, 2007; **7**: 2318–2322.

63. D. P. Kulkarni, D. K. Das, and G. Chukwa, "Temperature dependent rheological of copper oxide nanoparticles suspension (nanofluid)," *J. Nanosci. Nanotechnol.*, 2006; **6**: 1150–1154.

64. S. M. S. Murshed, K. C. Leong, and C. Yang, "Investigations of thermal conductivity and viscosity of nanofluids," *Int. J. Therm. Sci.*, 2008; **47**: 560–568.

65. K. B. Anoop, S. Kabelac, T. Sundararajan, and S. K. Das, "Rheological and flow characteristics of nanofluids: Influence of electroviscous effects and particle agglomeration," *J. Appl. Phys.*, 2009; **106**: 034909.

66. H. Chen, Y. Ding, Y. He, and Ch. Tan, "Rheological behavior of ethylene glycol based titania nanofluids, *Chem. Phys. Lett.*, 2007; **444**: 333–337.

67. J. Buongiorno, "Convective transport in nanofluids," *ASME J. Heat Transfer*, 2006; **128**: 240–250.

68. S. J. Palm, G. Roy, and C. T. Nguyen, "Heat transfer enhancement with the use of nanofluids in radial flow cooling systems considering temperature-dependent properties," *Appl. Therm. Eng.*, 2006; **26**: 2209–2218.

69. W. Duangthongsuk and S. Wongwises, "Measurement of temperature-dependent thermal conductivity and viscosity of TiO_2–water nanofluids," *Exp. Therm. Fluid Sci.*, 2009; **33**: 706–714.

70. P. K. Namburu, D. K. Das, K. M. Tanguturi, and R. S. Vajjha, "Numerical study of turbulent flow and heat transfer characteristics of nanofluids considering variable properties," *Int. J. Therm. Sci.*, 2009; **48**: 290–302.

71. P. K. Namburu, D. P. Kulkarni, D. Misra, and D. K. Das, "Viscosity of copper oxide nanoparticles dispersed in ethyleneglycol and water mixture," *Exp. Therm. Fluid Sci.*, 2007; **32**: 397–402.

72. J. Koo and C. Kleinstreuer, "Laminar nanofluid flow in microheat-sinks," *Int. J. Heat Mass Transfer*, 2005; **48**: 2652–2661.

73. R. L. Hamilton and O. K. Crosser, "Thermal conductivity of heterogeneous two-component systems," *I&EC Fundam.*, 1962; **1**: 182–191.

74. J. C. A. Maxwell, *Treatise on Electricity and Magnetism*, 2nd ed., Clarendon Press, Oxford, UK, 1881.

75. D. A. G. Bruggeman, "Berechnung verschiedener physikalischer konstanten von heterogenen substanzen, I. Dielektrizitatskonstanten und leitfahigkeiten der mischkorper aus isotropen substanzen," *Ann. Phys., Leipzig*, 1935; **24**: 636–679.

76. E. J. Wasp, J. P. Kenny, and R. L. Gandhi., *Solid–Liquid Slurry Pipeline Transportation*, Trans Tech Publications, Berlin, 1977.

77. D. J. Jeffrey, "Conduction through a random suspension of spheres," *Proc. Roy. Soc. (Lond.)*, 1973; **A335**: 355–367.

78. R. H. Davis, "The effective thermal conductivity of a composite material with spherical inclusions," *Int. J. Thermophys.*, 1986; **7**: 609–620.

79. S. Lu and H. Lin, "Effective conductivity of composites containing aligned spherical inclusions of finite conductivity," *J. Appl. Phys.*, 1996; **79**: 6761–6769.

80. Y. Xuan, Q. Li, and W. Hu, "Aggregation structure and thermal conductivity of nanofluids," *AIChE J.*, 2003; **49**(4): 1038–1043.

81. R. Prasher, P. Bhattacharya, and P. E. Phelan, "Thermal conductivity of nanoscale colloidal solutions (nanofluids)," *Phys. Rev. Lett.*, 2005; **94**(2): 025901.

82. J. Koo and C. Kleinstreuer, "A new thermal conductivity model for nanofluids," *J. Nanopart. Res.*, 2004; **6**(6): 577–588.

83. C. H. Chon, K. D. Kihm, S. P. Lee, and S. U. S. Choi, "Empirical correlation finding the role of temperature and particle size for nanofluid (Al_2O_3) thermal conductivity enhancement," *Appl. Phys. Lett.*, 2005; **87**: 153107.

84. J. Lee, P. E. Gharagozloo, B. Kolade, J. K. Eaton, and K. E. Goodson, "Nanofluid convection in microtubes," *ASME J. Heat Transfer*, 2010; **132**: 354–360.

85. S. Lee, S. U. S. Choi, S. Li, and J. A. Eastman, "Measuring thermal conductivity of fluids containing oxide nanoparticles," *ASME J. Heat Transfer*, 1999; **121**: 280–289.

86. C. H. Li and G. P. Peterson, "The effect of particle size on the effective thermal conductivity of Al_2O_3–water nanofluids," *J. Appl. Phys.*, 2007; **101**: 044312.

87. X. Zhang, H. Gu, and M. Fujii, "Effective thermal conductivity and thermal diffusivity of nanofluids containing spherical and cylindrical nanoparticles," *J. Appl. Phys.*, 2006; **100**: 1–5.

88. E. V. Timofeeva, A. N. Gavrilov, J. M. McCloskey, and Y. V. Tolmachev, "Thermal conductivity and particle agglomeration in alumina nanofluids: Experiment and theory," *Phys. Rev. E*, 2007; **76**: 061203.

89. H. A. Minsta, G. Roy, C. T. Nguyen, and D. Doucet, "New temperature-dependent thermal conductivity data for water-based nanofluids," *Int. J. Therm. Sci.*, 2009; **48**: 363–371.

90. G. Roy, C. T. Nguyen, and P.-R. Lajoie, "Numerical investigation of laminar flow and heat transfer in a radial flow cooling system with the use of nanofluids," *Superlatt. Microstruct.*, 2004; **35**: 497–511.

91. L. S. Sundar and K. V. Sharma, "Experimental determination of thermal conductivity of fluid containing oxide nanoparticles," *Int. J. Dynam. Fluids*, 2008; **4**: 57–69.

92. M. Golubovic, H. D. M. Hettiarachchi, W. M. Worek, W. J. Minkowycz, "Nanofluids and critical heat flux, experimental and analytical study," *Appl. Therm. Eng.*, 2009; **29**: 1281–1288.

93. H. S. Xue, J. R. Fan, R. H. Hong, and Y. C. Hu, "Characteristic boiling curve of carbon nanotube nanofluid as determined by the transient calorimeter technique," *Appl. Phys. Lett.*, 2007; **90**: 184107.

94. H. D. Kim, J. Kim, M. H. Kim., "Experimental studies on CHF characteristics of nanofluids at pool boiling," *Int. J. Multiphase Flow*, 2007; **33**: 691–706.

95. S. M. Murshed, S. H. Tan, and N. T. Nguyen, "Temperature dependence of interfacial properties and viscosity of nanofluids for droplet-based microfluidics," *J. Phys. D: Appl. Phys.*, 2008; **41**: 085502.

96. D. S. Zhu, S. Y. Wu, and N. Wang, "Surface tension and viscosity of aluminum oxide nanofluids," in: *The 6th International Symposium on Multiphase Flow, Heat Mass Transfer and Energy Conversion, AIP Conference Proceedings*, March 1, 2010, **1207**: 460–464.

97. G. P. Narayan, K. B. Anoop, and S. K. Das, "Mechanism of enhancement/deterioration of boiling heat transfer using stable nanoparticle suspensions over vertical tubes," *J. Appl. Phys.*, 2007; **102**: 074317.

98. S. M. You, J. H. Kim, and K. H. Kim, "Effect of nanoparticles on critical heat flux of water in pool boiling heat transfer," *Appl. Phys. Lett.*, 2003; **83**: 3374–3376.

99. I. C. Bang and S. H. Chang, "Boiling heat transfer performance and phenomena of Al_2O_3–water nanofluids from a plain surface in a pool," *Int. J. Heat Mass Transfer*, 2005; **48**: 2407–2419.

100. J. P. Tu, N. Dinh, and T. Theofanous, "An experimental study of nanofluid boiling heat transfer," in: *Proceedings of 6th International Symposium on Heat Transfer*, Beijing, China, 2004.

101. D. S. Wen and Y. L. Ding, "Experimental investigation into the pool boiling heat transfer of aqueous-based γ-alumina nanofluids," *J. Nanopart. Res.*, 2005; **7**: 265–274.

102. D. S. Wen, Y. L. Ding, and R. A. Williams, "Pool boiling heat transfer of aqueous based TiO_2 nanofluids," *J. Enhanced Heat Transfer*, 2006; **13**: 231–244.

103. Z. H. Liu, J. G. Xiong, and R. Bao, "Boiling heat transfer characteristics of nanofluids in a flat heat pipe evaporator with micro-grooved heating surface," *Int. J. Multiphase Flow*, 2007; **33**: 1284–1295.

104. M. H. Kim, J. B. Kim, and H. D. Kim, "Experimental studies on CHF characteristics of nano-fluids at pool boiling," *Int. J. Multiphase Flow*, 2007; **33**: 691–706.

105. S. J. Kim, T. McKrell, J. Buongiorno, and L. W. Hu, "Experimental study of flow critical heat flux in alumina–water, zinc-oxide–water and diamond–water nanofluids," *J. Heat Transfer*, 2009; **131** (4): 043204.

106. P. Vassallo, R. Kumar, and S. D. Amico, "Pool boiling heat transfer experiments in silica–water nanofluids," *Int. J. Heat Mass Transfer*, 2004; **47**: 407–411.

107. C. H. Li, B. X. Wang, and X. F. Peng, "Experimental investigations on boiling of nanoparticle suspensions," in: *Proc. 5th Int. Conf. Boiling Heat Transfer*, 2003, Montego Bay, Jamaica.

108. S. J. Kim, I. C. Bang, J. Buongiorno, and L. W. Hu, "Effects of nanoparticle deposition on surface wettability influencing boiling heat transfer in nanofluids," *Appl. Phys. Lett.*, 2006; **89**: 153107.

109. S. J. Kim, I. C. Bang, J. Buongiorno, and L. W. Hu, "Surface wettability change during pool boiling of nanofluids and its effect on critical heat flux," *Int. J. Heat Mass Transfer*, 2007; **50**: 4105–4116.

110. K. Sefiane, "On the role of structural disjoining pressure and contact line pinning in critical heat flux enhancement during boiling of nanofluids," *Appl. Phys. Lett.*, 2006; **89**: 044106.

111. R. Hegde, S. S. Rao, and R. P. Reddy, "Critical heat flux enhancement in pool boiling using alumina nanofluids," *Heat Transfer Asian Res.*, 2010; **39**: 323–331.

112. T. M. Anderson and I. Mudawar, "Microelectronic cooling by enhanced pool boiling of a dielectric fluorocarbon liquid," *ASME J. Heat Transfer*, 1989; **111**: 752–759.

113. I. Mudawar and T. M. Anderson, "Optimization of enhanced surfaces for high flux chip cooling by pool boiling," *ASME J. Electron. Packag.*, 1993; **115**: 89–99.

114. H. Honda, H. Takamastu, and J. J. Wei, "Enhanced boiling of FC-72 on silicon chips with micro-pin-fins and submicron-scale roughness," *ASME J. Heat Transfer*, 2002; **124**: 383–389.

115. J. J. Wei, H. Honda, and L. J. Guo, "Experimental study of boiling phenomena and heat transfer performances of FC-72 over micro-pin-finned silicon chips," *Heat Mass Transfer*, 2005; **41**: 744–755.

116. S. Ujereh, T. Fisher, and I. Mudawar, "Effects of carbon nanotube arrays on nucleate pool boiling," *Int. J. Heat Mass Transfer*, 2007; **50**: 4023–4038.

Chapter 13

Thermal Performance of Flat-Shaped Heat Pipes using Nanofluids

Maryam Shafahi,[a] Vincenzo Bianco,[b] Kambiz Vafai,[a] and Oronzio Manca[b]

[a]*Department of Mechanical Engineering, University of California Riverside, Riverside, CA 92521, USA*

[b]*Dipartimento di Ingegneria Aerospaziale e Meccanica, Seconda Università degli Studi di Napoli, Via Roma 29, 81031 Aversa (CE), Italy*

vafai@engr.ucr.edu

Analytical models are utilized to investigate the thermal performance of rectangular- and disk-shaped heat pipes using nanofluids. The liquid pressure, liquid velocity profile, temperature distribution of the heat pipe wall, temperature gradient along the heat pipe, thermal resistance, and maximum heat load are obtained for the flat-shaped heat pipes utilizing a nanofluid as the working fluid. The flat-shaped heat pipe's thermal performance using a nanofluid is substantially enhanced compared with the one using a

---

This chapter has been reprinted from the *International Journal of Heat and Mass Transfer*, Vol. 53, Maryam Shafahi, Vincenzo Bianco, Kambiz Vafai, and Oronzio Manca, Thermal performance of flat-shaped heat pipes using nanofluids, 1438–1445, Copyright 2009, with permission from Elsevier.

Nanotechnology and Energy
Edited by Kaufui V. Wong

ISBN 978-981-4774-24-6 (Hardcover), 978-1-315-16357-4 (eBook)
www.panstanford.com

regular fluid. The nanoparticles presence within the working fluid results in a decrease in the thermal resistance and an increase in the maximum heat load capacity of the flat-shaped heat pipe. The existence of an optimum nanoparticle concentration level and wick thickness in maximizing the heat removal capability of the flat-shaped heat pipe was established.

13.1 Introduction

Heat pipes have been utilized in heat transfer related applications for many years. Depending on their application area, they can operate over a wide range of temperatures with a high heat removal capability. Heat pipes have been found to be useful in a number of technologies such as electronic cooling, spacecraft thermal control, transportation systems, automotive industry, permafrost stabilization, biorelated applications, solar systems, and manufacturing [1–6].

Mathematical and experimental models of the transport mechanisms for flat-shaped heat pipes have been developed by Vafai et al. [2–6], and a comprehensive analytical solution for a cylindrical heat pipe incorporating the effect of liquid–vapor coupling and non-Darcian transport was presented by Zhu and Vafai [7].

The idea of utilizing nanoparticles within the working fluid of a heat pipe has become a subject of interest in recent years [8–12]. There are a number of studies on thermal conductivity of the nanofluids [13–15]. The nanoparticles within the fluid change its thermal conductivity, viscosity, and density. It has been shown experimentally that for a given concentration level, the thermal conductivity of the nanofluids increases with a decrease in particle diameter [15–17]. However, there are some investigations that do not confirm this enhancement in the thermal conductivity [18, 19].

Most of the investigations on the effect of using nanofluids within a heat pipe have been carried out experimentally. The thermal performance of a nanofluid-based cylindrical heat pipe was studied analytically by Shafahi et al. [8]. In the current work, analytical models established by Vafai et al. [3, 4] are utilized to study the effect of using a nanofluid within the flat-shaped heat pipes. Disk- and rectangular-shaped heat pipes have substantial advantages over conventional cylindrical heat pipes. A more flexible geometry, substantially higher heat removal capability, ability to incorporate asymmetrical heating and a built-in secondary transport to assimilate overload and prevent or delay a dry out are some of the advantages of the flat-shaped heat pipes analyzed by Vafai et al. [2–6]. A wide range of nanoparticles such as silver [20–23], CuO [10, 24], diamond [11, 25], titanium [9, 12], nickel oxide [26], and gold [27] have been used to study the influence of nanofluid on the heat pipe performance. It has been observed experimentally that the heat pipe thermal resistance [11, 12, 20, 22–24, 27] and temperature gradient [20, 23, 25] decrease when using a nanofluid as the working fluid.

In this work, a comprehensive analytical solution for disk- and rectangular-shaped heat pipes [2, 4] has been modified to incorporate the presence of nanoparticles within the working fluid. The liquid pressure and velocity profile, temperature distribution, temperature gradient along the heat pipe, thermal resistance, and maximum heat removal capability are obtained for different concentration levels of nanoparticles within flat-shaped heat pipes. Three of the most common types of nanoparticles, namely Al_2O_3, CuO, and TiO_2, are considered in this chapter. There is substantial enhancement in thermal performance of a flat-shaped heat pipe utilizing a nanofluid. The effect of nanoparticle size on the performance of the flat-shaped heat pipe is investigated, and it is shown that smaller nanoparticles produce the most enhancement in the thermal performance of a heat pipe. The maximum heat capability is established for various wick thicknesses and nanoparticle concentrations, and the potential for reducing the size of a heat pipe for a given heat load by utilizing a nanofluid is studied. Our results demonstrate the existence of an optimum concentration level and wick thickness for maximizing the heat removal capability of a flat-shaped heat pipe.

Nomenclature

A_c	condenser area [m^2]
A_e	evaporator area [m^2]
A_{tot}	total external surface area of the heat pipe [m^2]
$f(x)$	position of the maximum value of vapor velocity in y direction [m]
h	convective heat transfer coefficient [W/m^2 K]
h_b^+	dimensionless half width of any of the vapor channels
h_v	height of vapor space for the heat pipe [m]
h_{tot}	thickness of the heat pipe [m]
h_w	thickness of the wick [m]

h_w^+	dimensionless thickness of the wick $\left(\frac{h_w}{h_v}\right)$
K	permeability of the wick [m^2]
K^+	dimensionless permeability $\left(K^+ = \frac{K}{h_w^2}\right)$
k	thermal conductivity [W/m K]
k_{eff}	effective thermal conductivity of the liquid saturated wick [W/m K]
k_{layer}	nanolayer thermal conductivity [W/m K]
k_{wall}	thermal conductivity of the heat pipe wall [W/m K]
l	length of the rectangular-shaped heat pipe [m]
l^+	dimensionless length of the rectangular-shaped heat pipe
p	pressure [Pa]
p_l	liquid pressure [Pa]
p_l^+	dimensionless liquid pressure
p_v	vapor pressure [Pa]
p_v^+	dimensionless vapor pressure
Q_{max}	maximum heat input [W]
r	dimensional radial coordinate [m]
r^+	dimensionless radial coordinate
r_p	nanoparticles radius [nm]
r_c	effective pore radius of the wick [m]
R	radius of the disk-shaped heat pipe [m]
R^+	dimensionless radius of the disk-shaped heat pipe
R_t	heat pipe's thermal resistance [K/W]
Re_h	injection Reynolds number $(v_1 h_v/\nu_v)$
T	temperature [K]
T_b	bulk temperature [K]
u_l	liquid velocity [m/s]
v_1	vapor injection velocity [m/s]
v_1^+	dimensionless vapor injection velocity $(v_1 h_v/\nu_v)$
v_2	vapor suction velocity [m/s]
v_2^+	dimensionless vapor suction velocity $(v_1 h_v/\nu_v)$
w	nanolayer thickness [nm]
x	dimensional coordinate for the rectangular-shaped heat pipe [m]
x^+	dimensionless coordinate for the rectangular-shaped heat pipe

Greek symbols

ϕ	particle concentration
ε	porosity of the wick
μ	dynamic viscosity [Ns/m^2]
ν	kinematic viscosity [m^2/s]
μ^+	dimensionless dynamic viscosity
ρ	density [kg/m^3]
ρ^+	dimensionless density $\left(\frac{\rho_l}{\rho_v}\right)$
σ_l	surface tension of the working fluid [N/m]
Subscripts	
0	heat pipe with pure fluid
bf	base fluid
l	liquid phase
max	maximum
nf	nanofluid
p	particle
v	vapor
w	wick
Superscript	
+	dimensionless quantity

13.2 Analysis and Formulation

A simplified schematic of a rectangular- or disk-shaped heat pipe's cross section is shown in Fig. 13.1. It is assumed that vapor and liquid flow are steady, and laminar and transport properties of the vapor and liquid are considered to be constant. The vapor injection and suction rates are considered to be uniform in the evaporator [2].

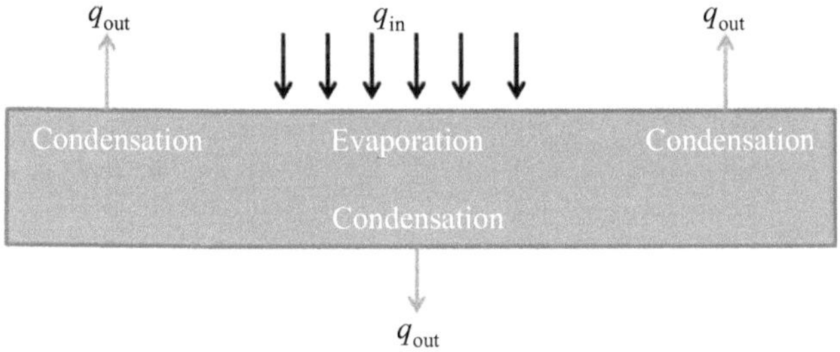

Figure 13.1 Simple schematic of a rectangular- or disk-shaped heat pipe's cross section.

13.2.1 Pressure Distribution

Pressure distribution is obtained by integrating the momentum equation while applying the appropriate boundary conditions [2, 4]. This results in the following expressions for the vapor and liquid pressure distribution in disk- and rectangular-shaped heat pipes, respectively.

13.2.1.1 Disk-shaped heat pipe's vapor pressure distribution

$$p_v^+(r^+)=\begin{cases} p_v^+(0)-\dfrac{24}{25}\left(\dfrac{1-\lambda^2}{2-\lambda^2}Re_h R^+\right)^2\left(\dfrac{r^+}{R^+}\right)^2 & 0\le r^+\le \lambda R^+ \\ p_v^+(0)-\dfrac{8}{25}\left(\dfrac{\lambda^2}{2-\lambda^2}Re_h R^+\right)^2\left[3\left(\dfrac{r^+}{R^+}\right)^2+\left(\dfrac{R^+}{r^+}\right)^2 \right. & \\ \left. -4\ln\dfrac{r^+}{R^+}-2\left(3-2\ln\lambda-\dfrac{1}{\lambda^2}\right)\right] & \lambda R^+\le r^+\le R^+ \end{cases} \tag{13.1}$$

where λ is the square root of the ratio of the evaporator area to the total area: $r^+=\dfrac{rv_v}{h_v^2 v_1}$ and $p_v^+=\dfrac{p_v}{\rho_v v_1^2}$ for the disk-shaped heat pipe.

13.2.1.2 Rectangular-shaped heat pipe's vapor pressure distribution

The vapor flow in the rectangular-shaped heat pipe does not have a symmetrical velocity profile. The maximum vapor velocity position moves toward the lower plate due to vapor injection from the upper wall. An approximate solution for f^+ is obtained using Bernoulli's equation [2]:

$$\frac{df^+(x^+)}{dx^+}=\begin{cases} \left[-\dfrac{9}{2}(1-\lambda)f^+(x^+)+5\dfrac{(2-\lambda)}{Re_h}\dfrac{1}{f^+(x^+)}-\dfrac{5}{2}\lambda\right]\dfrac{1}{(1-\lambda)x^+} \\ \qquad\qquad 0< x^+\le \lambda l^+ \\ \left[-\lambda f^+(x^+)+10\dfrac{(2-\lambda)}{Re_h}\dfrac{1}{f^+(x^+)}\right]\dfrac{1}{7\lambda(l^+-x^+)} \\ \qquad\qquad l^+\le x^+< l^+ \end{cases} \tag{13.2}$$

where the dimensionless coordinate for rectangular-shaped heat pipe's vapor phase is $x^+ = \dfrac{v_v x}{h_v^2 U_{ov}}$ and U_{ov} is the maximum vapor velocity.

As such, the vapor pressure distribution can be presented as:

$$\Delta p_v^+(x^+) = \begin{cases} -\dfrac{4(1-\lambda)}{(2-\lambda)} Re_h \left(\left[\dfrac{16(1-\lambda)}{25(2-\lambda)} Re_h + \dfrac{1}{2(h_b^+)^2} \right] (x^+)^2 + \displaystyle\int_0^{x^+} \dfrac{x^+}{f^+(x^+)(1-f^+(x^+))} dx^+ \right) & 0 \le x^+ \le \lambda l^+ \\ \Delta p_v^+(\lambda l^+) - \dfrac{4\lambda}{(2-\lambda)} Re_h \left(\left[\dfrac{16\lambda}{25(2-\lambda)} Re_h - \dfrac{1}{2(h_b^+)^2} \right] \left[(x^+ - l^+)^2 - (\lambda l^+ - l^+)^2 \right] - \displaystyle\int_0^{x^+} \dfrac{x^+ - l^+}{f^+(x^+)(1-f^+(x^+))} dx^+ \right) & \lambda l^+ \le x^+ < l^+ \end{cases} \tag{13.3}$$

where the dimensionless vapor pressure is $p_v^+ = \dfrac{p_v}{\rho_v U_{ov}^2}$ and λ is the ratio of the evaporator length to the heat pipe length for the rectangular-shaped heat pipe. The liquid pressure distribution for the rectangular- and disk-shaped heat pipes can be presented as [2, 4]:

13.2.1.3 Disk-shaped heat pipe's liquid pressure distribution

$$p_l^+(r^+) = \begin{cases} p_v^+(0) + \dfrac{v^+ Re_h (R^+)^2}{4K^+(h_w^+)^3} \dfrac{1-\lambda^2}{2-\lambda^2} \left[\left(\dfrac{r^+}{R^+} \right)^2 + \dfrac{2\lambda^2}{1-\lambda^2} \ln\lambda \right] - \left(\dfrac{4}{5} \dfrac{\lambda^2}{2-\lambda^2} Re_h R^+ \right)^2 \left(2\ln\lambda + \dfrac{1}{\lambda^2} - 1 \right) & 0 \le r^+ \le \lambda R^+ \\ p_v^+(0) + \dfrac{v^+ Re_h (R^+)^2}{4K^+(h_w^+)^3} \dfrac{\lambda^2}{2-\lambda^2} \left(1 - \left(\dfrac{r^+}{R^+} \right)^2 - 2\ln\left(\dfrac{R^+}{r^+} \right) \right) - \left(\dfrac{4}{5} \dfrac{\lambda^2}{2-\lambda^2} Re_h R^+ \right)^2 \left(2\ln\lambda + \dfrac{1}{\lambda^2} - 1 \right) & \lambda R^+ \le r^+ \le R^+ \end{cases} \tag{13.4}$$

where the dimensionless liquid pressure is $p_l^+ = \frac{p_l}{\rho_l v_l^2}$ and $v^+ = \frac{v_l}{v_v}$ for the disk-shaped heat pipe.

13.2.1.4 Flat-shaped heat pipe's liquid pressure distribution

$$\Delta p_l^+(x^+) = \begin{cases} \Delta p_v^+(l^+) - \frac{h_w^+ \mu^+ (1-\lambda) Re_h}{2(2-\lambda)K^+} \{\lambda(1-\lambda)(l^+)^2 \\ \quad + [(\lambda l^+)^2 - (x^+)^2]\} & 0 \le x^+ \le \lambda l^+ \\ \\ \Delta p_v^+(l^+) - \frac{h_w^+ \mu^+ \lambda Re_h}{2(2-\lambda)K^+} (l^+ - x^+)^2 & \lambda l^+ \le x^+ \le l^+ \end{cases} \tag{13.5}$$

where $\mu^+ = \frac{\mu_v}{\mu_l}$; the dimensionless coordinate for the rectangular-shaped heat pipe's liquid phase is $x^+ = \frac{v_l x}{U_{ol} h_w^2}$ and U_{ol} is maximum liquid phase velocity. It can be seen that liquid pressure is related to the liquid viscosity, which is affected when using nanoparticles within the working fluid. The classical Brinkman model [28] is used to obtain the nanofluid viscosity:

$$\mu_{nf} = \frac{\mu_{bf}}{(1-\phi)^{2.5}} \tag{13.6}$$

where μ_{bf} is the viscosity of the base fluid, which is water for our investigation. Nanofluid's density is calculated based on a simple representation of the particle and fluid phases proposed by Pak and Cho [29].

$$\rho_{nf} = \rho_p \phi + (1-\phi)\rho_{bf} \tag{13.7}$$

where ρ_{nf}, ρ_p, and ρ_{bf} are the nanofluid, particle, and base fluid densities, respectively.

13.2.2 Velocity Distribution

Using Darcy's and applying the mass balance for the lower and upper wicks, the liquid velocity distribution for the rectangular- and disk-shaped heat pipes can be presented by [2, 4]:

Liquid velocity distribution for the disk-shaped heat pipe

$$\mu_l^{+}(r^{+})=\begin{cases}\dfrac{v_1^{+}-v_2^{+}}{4\rho^{+}h_w^{+}}r^{+} & 0\le r^{+}\le \lambda R^{+}\\[2ex] -\dfrac{v_2^{+}}{2\rho^{+}h_w^{+}}\dfrac{(r^{+})^2-(R^{+})^2}{r^{+}} & \lambda R^{+}\le r^{+}\le R^{+}\end{cases} \tag{13.8}$$

where $\rho^{+} = \rho_l/\rho_v$.

Liquid velocity distribution for the rectangular-shaped heat pipe

$$\mu_l^{+}(r^{+})=\begin{cases}\dfrac{h_w^{+}\mu^{+}(v_1^{+}-v_2^{+})}{2}x^{+} & 0\le^{+}\le \lambda l^{+}\\[2ex] h_w^{+}\mu^{+}v_2^{+}(l^{+}-x^{+}) & \lambda l^{+}\le x^{+}\le l^{+}\end{cases} \tag{13.9}$$

13.2.3 Temperature Distribution

The wall temperature profile is obtained based on the assumption of uniform temperature along the condenser and evaporator sections. Based on simple energy balances, the condenser and evaporator wall temperatures as well as the vapor temperature can be, respectively, presented as:

Condenser wall temperature

$$T_{\text{wall,e}} = T_b + \frac{Q}{A_c h} \tag{13.10}$$

Evaporator wall temperature

$$T_{\text{wall,e}} = T_v + \frac{Q}{A_e}\left(\frac{h_{\text{wall}}}{k_{\text{wall}}} + \frac{h_w}{k_{\text{eff}}}\right) \tag{13.11}$$

where h_{wall} and h_w are the wall and wick thicknesses, respectively.

Vapor temperature

$$T_v = T_b + \frac{Q}{A_c}\left(\frac{1}{h} + \frac{h_{\text{wall}}}{k_{\text{wall}}} + \frac{h_w}{k_{\text{eff}}}\right) \tag{13.12}$$

The heat pipe's temperature distribution is dependent on the wick thermal conductivity, which can be presented as [30]:

$$k_{\text{eff}} = \frac{k_{\text{l}}[(k_{\text{l}} + k_{\text{s}}) - (1-\varepsilon)(k_{\text{l}} - k_{\text{s}})]}{[(k_{\text{l}} + k_{\text{s}}) + (1-\varepsilon)(k_{\text{l}} - k_{\text{s}})]} \tag{13.13}$$

where k_{s} and k_{l} are solid and liquid conductivities. For this work, k_{l} is replaced with k_{nf}. It should be noted that k_{eff} increases with an increase in the thermal conductivity of nanofluid. The nanofluid conductivity is obtained from Yu and Choi [31] model that can be presented as:

$$k_{\text{nf}} = \frac{k_{\text{pe}} + 2k_{\text{l}} + 2(k_{\text{pe}} - k_{\text{l}})(1+\beta)^3\phi}{k_{\text{pe}} + 2k_{\text{l}} + (k_{\text{pe}} - k_{\text{l}})(1+\beta)^3\phi} k_{\text{l}} \tag{13.14}$$

To account for the effect of nanolayer, solid particle conductivity is modified as:

$$k_{\text{pe}} = \frac{[2(1-\alpha) + (1+\beta)^3(1+2\alpha)]\alpha}{-(1-\alpha) + (1+\beta)^3(1+2\alpha)} k_{\text{p}} \tag{13.15}$$

$$\alpha = \frac{k_{\text{layer}}}{k_{\text{p}}}; \quad \beta = \frac{w}{r_{\text{p}}} \tag{13.16}$$

where k_{p} and k_{layer} are the thermal conductivities of the nanoparticle and nanolayer, respectively.

13.2.4 Maximum Heat Transfer Capability

For a heat pipe under steady-state operation, capillary pressure of the wick establishes a stable circulation for the working fluid. The maximum heat transport capillary limit of the heat pipe is achieved when the total pressure loss reaches the maximum capillary pressure of the porous wick, which can be presented as:

$$\Delta p_{\text{cap}} = \Delta p_{\text{v,total}} + \Delta p_{\text{l,total}} + \Delta p_{\text{g,total}} \tag{13.17}$$

$$(\Delta p_{\text{cap}})_{\text{max}} = \frac{2\sigma_{\text{l}}}{r_{\text{c}}} \tag{13.18}$$

where σ_{l} is the liquid surface tension, and r_{c} is the effective pore radius of the wick. The total vapor, liquid, and gravitational pressure drops should be less than the maximum capillary pressure head. Neglecting the gravitational head, Eq. (13.17) can be represented as:

$$(\Delta p_{\text{cap}})_{\text{max}} \geq \Delta p_{\text{v,total}} + \Delta p_{\text{l,total}} \tag{13.19}$$

13.3 Results and Discussion

The analytical models for disk- and rectangular-shaped heat pipes [2, 4] were utilized to study the influence of using a nanofluid on the thermal performance of these heat pipes. Results are obtained for disk- and rectangular-shaped heat pipes, which include nanoparticles within the working fluid. The diameter of the disk-shaped heat pipe is taken equal to the length of the rectangular-shaped heat pipe, which is 0.46 m for this work. Other nominal dimensions are taken from Vafai and Wang [2] and Vafai et al. [4] works. Three of the most common nanoparticles, namely, Al_2O_3, CuO, and TiO_2, with a range of diameters (10, 20, and 40 nm) are utilized to form the nanofluid. The influence of the nanofluid on the liquid pressure, temperature distribution, liquid velocity, temperature gradient, thermal resistance, heat pipe size, and maximum heat removal capability of the flat-shaped heat pipes is investigated.

Figure 13.2 displays the liquid pressure distribution for disk- and rectangular-shaped heat pipes for a range of CuO-based nanofluid concentration levels. Initially, the liquid pressure drop decreases by utilizing the nanoparticles within the water due to an increase in density. After reaching a critical concentration level, a reversal effect occurs, resulting in an increase in the pressure drop as the increase in viscosity of the nanofluid becomes more prominent. The competing effects of the density and viscosity of the nanofluid create an optimum level of nanoparticle concentration in minimizing the pressure drop for the flat heat pipes.

The influence of nanofluid concentration level on the liquid velocity profile is shown in Fig. 13.3. An increase in the concentration level decreases the velocity in both disk- and rectangular-shaped heat pipes. It can be seen through Eqs. (13.8) and (13.9) that liquid velocity is inversely proportional to either the density or the viscosity of the liquid.

Figure 13.4 displays the wall temperature distribution for flat- and disk-shaped heat pipes utilizing the CuO-based nanofluid as the working fluid. The input heat and the nanoparticle diameter are 9 kW and 20 nm, respectively. As seen in Fig. 13.4, the evaporator temperature drops in the presence of nanoparticles. An increase in the concentration level of nanoparticles increases the thermal conductivity of the nanofluid, thus further decreasing the evaporator temperature.

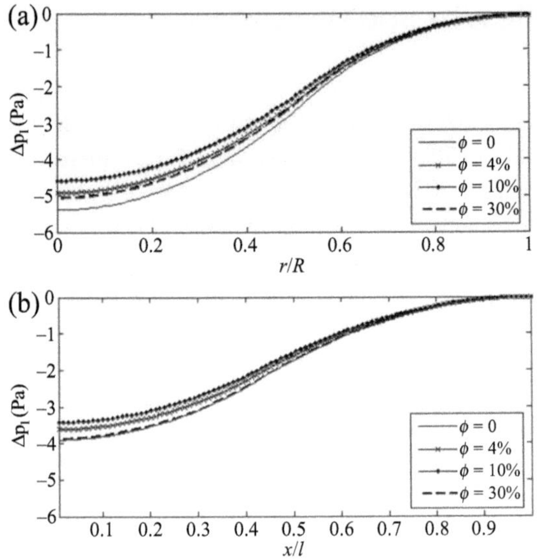

Figure 13.2 Liquid pressure distribution for different concentration levels of CuO nanoparticles within the heat pipe; $Q = 9$ kW: (a) disk-shaped heat pipe and (b) rectangular-shaped heat pipe.

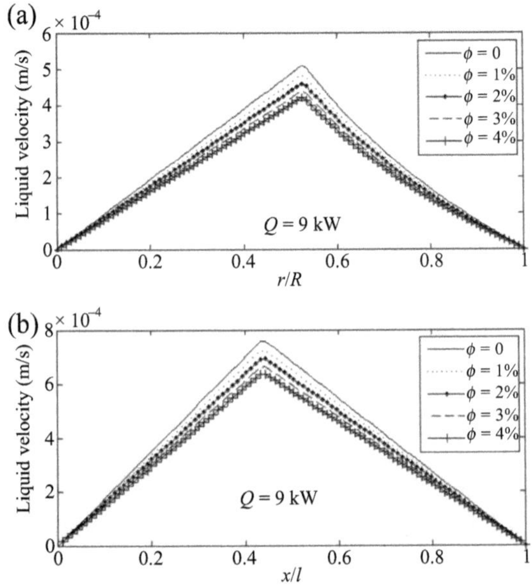

Figure 13.3 Liquid velocity profiles for various concentration levels of CuO nanoparticles within the flat heat pipes: (a) disk-shaped heat pipe and (b) rectangular-shaped heat pipe, $Q = 9$ kW.

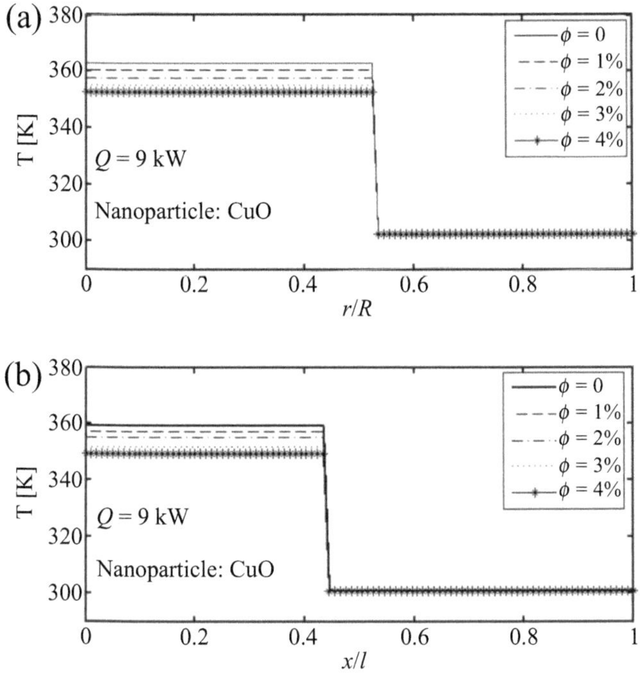

Figure 13.4 Heat pipe wall temperature distribution for different nanoparticle concentration levels; Q = 9 kW; d_p = 20 nm: (a) disk-shaped heat pipe and (b) rectangular-shaped heat pipe.

Figures 13.5 and 13.6 show the reduction in the evaporator and condenser temperature differences for different nanoparticle concentration levels for various heat loads. The effect of variations in nanoparticle diameter is also displayed in Figs. 13.5 and 13.6. As can be seen, the smaller the nanoparticle or the larger the concentration level, the more pronounced the reduction in the temperature difference between the condenser and the evaporator. Thermal resistance of the flat-shaped heat pipes, defined as $T_e - T_c/Q$, becomes noticeably smaller when a nanofluid is utilized, as seen in Figs. 13.7 and 13.8. As seen in these figures, the thermal resistance of a flat-shaped heat pipe can be reduced up to 83% of its initial resistance.

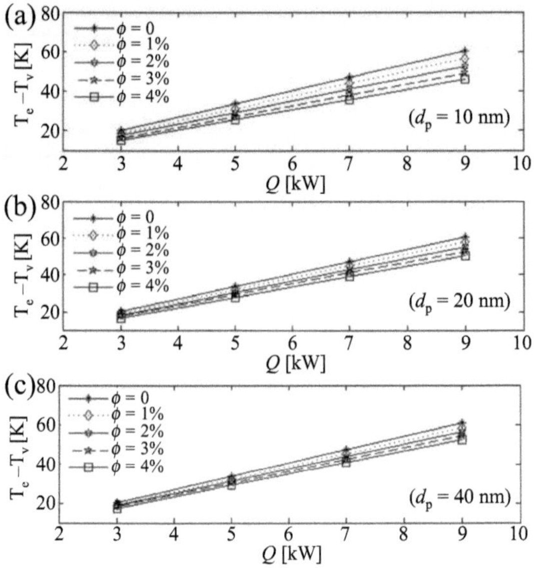

Figure 13.5 Temperature difference between evaporator and condenser in disk-shaped heat pipe for different CuO nanoparticle concentration levels: (a) d_p = 10 nm, (b) d_p = 20 nm, and (c) d_p = 40 nm.

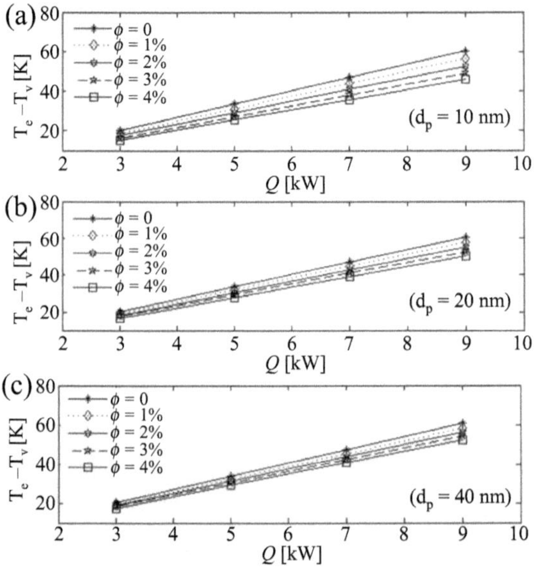

Figure 13.6 Temperature difference between evaporator and condenser for the rectangular-shaped heat pipe for different CuO concentration levels: (a) d_p = 10 nm, (b) d_p = 20 nm, and (c) d_p = 40 nm.

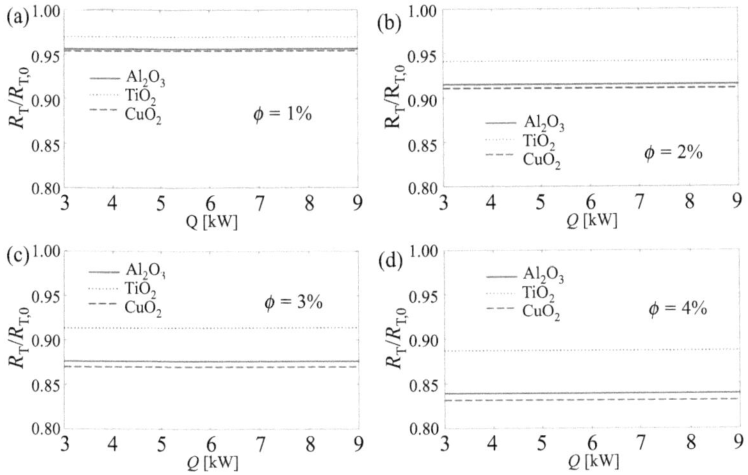

Figure 13.7 Effect of nanofluids (d_p = 20 nm) on disk-shaped heat pipe thermal resistance for various heat inputs: (a) ϕ = 1%, (b) ϕ = 2%, (c) ϕ = 3%, (d) ϕ = 4%.

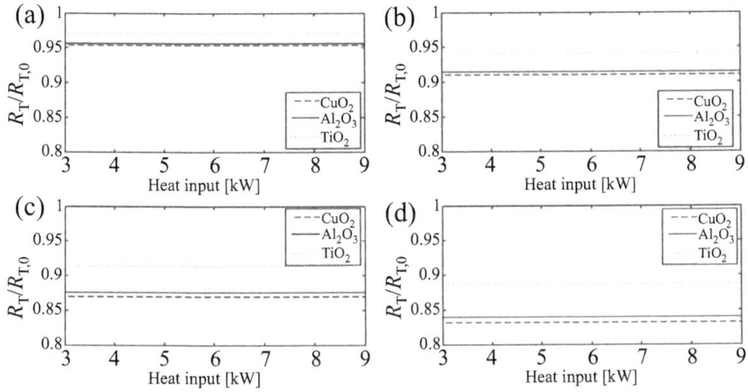

Figure 13.8 Effect of nanofluids (d_p = 20 nm) on rectangular-shaped heat pipe thermal resistance for various heat inputs: (a) ϕ = 1%, (b) ϕ = 2%, (c) ϕ = 3%, (d) ϕ = 4%.

Figures 13.9 and 13.10 demonstrate that using the nanoparticles provides the possibility of a reduction in the size of flat-shaped heat pipe while maintaining the same heat load removal capability. The results are shown for different concentration levels of Al_2O_3, CuO, and TiO_2 with different diameters. For these figures, the applied heat input was kept at 9 kW and the evaporator temperature was kept at

90°C. As can be seen, the smallest particle size results in the largest reduction in the size of the heat pipe. For CuO-based nanofluid, a 30% and 20% reduction in size can be realized for the disk- and rectangular-shaped heat pipes, respectively.

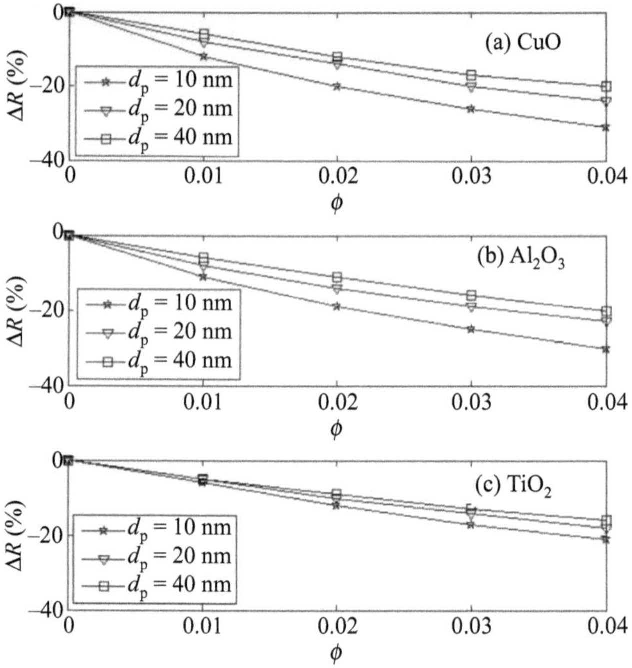

Figure 13.9 Percentage reduction level of a disk-shaped heat pipe when using different size nanoparticles and different concentration levels, $T_{\text{evaporator wall}}$ = 90°C and Q = 9 kW.

The effect of changing the wick thickness and nanoparticle concentration level on the maximum heat removal capability of a flat-shaped heat pipe is shown in Fig. 13.11. The existence of an optimum point for both wick thickness and nanoparticle concentration level can be seen in this figure. The existence of an optimum thickness for the wick was established earlier in the work of Vafai et al. [4]. The appearance of an optimum nanoparticle concentration level based on the competing effects of density and viscosity of the nanofluid level was also established in the work of Shafahi et al. [8] for cylindrical heat pipes.

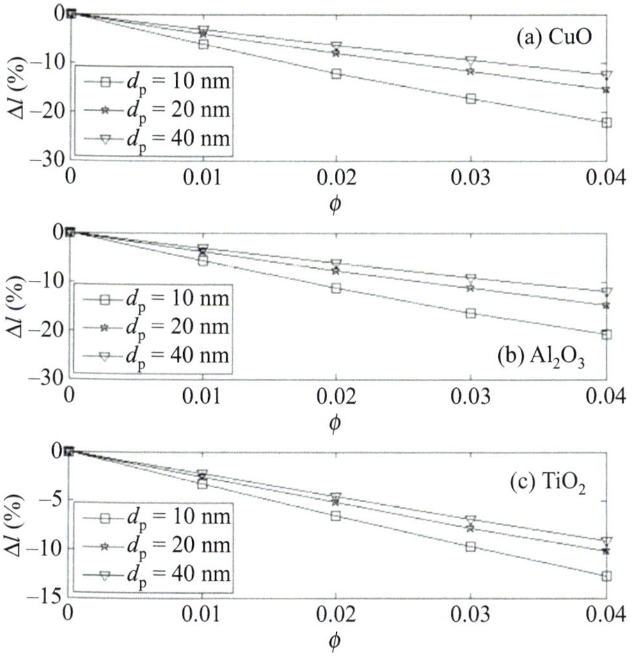

Figure 13.10 Percentage reduction level of a rectangular-shaped heat pipe when using different size nanoparticles and different concentration levels, $T_{\text{evaporator wall}} = 90°C$ and $Q = 9$ kW.

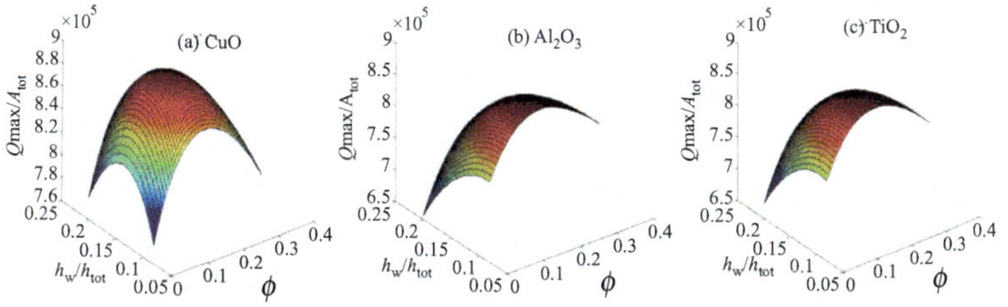

Figure 13.11 Three-dimensional representation of disk-shaped heat pipe maximum heat removal capability in terms of nanoparticle concentration levels and dimensionless thickness of the wick: (a) CuO, (b) Al_2O_3, and (c) TiO_2.

13.4 Conclusion

The performance of flat-shaped heat pipes utilizing nanoparticles within the working fluid was investigated in this work. Some of the more widely used nanoparticles, such as Al_2O_3, CuO, and TiO_2, with a range of nanoparticle diameters were considered. The flat-shaped heat pipe's liquid pressure, liquid velocity profile, temperature distribution, temperature gradient, and maximum heat removal capability were investigated. Furthermore, the possibility of reducing the flat-shaped heat pipe's dimension using a nanofluid was explored. Results show that the presence of nanoparticles in the working fluid leads to a reduction in the speed of the liquid, smaller temperature difference along the heat pipe, and the possibility of reduction in size under the same operational conditions. Using a nanofluid reduces the thermal resistance of the flat-shaped heat pipe similar to what has been observed experimentally. The maximum heat removal capability of the flat-shaped heat pipe was displayed for a range of wick thicknesses and nanoparticle concentration levels. The existence of an optimum nanoparticle concentration level and wick thickness in maximizing the heat removal capability of the flat-shaped heat pipe was established.

References

1. A. Faghri, *Heat Pipe Science and Technology*, Taylor & Francis, Washington, DC, 1995.
2. K. Vafai and W. Wang, "Analysis of flow and heat transfer characteristics of an asymmetrical flat-plate heat pipe," *Int. J. Heat Mass Transfer*, 1991; **35**: 2087–2099.
3. N. Zhu and K. Vafai, "Analytical modeling of the startup characteristics of asymmetrical flat-plate and disk-shaped heat pipes," *Int. J. Heat Mass Transfer*, 1998; **41**: 2619–2637.
4. K. Vafai, N. Zhu, and W. Wang, "Analysis of asymmetric disk-shaped and flat-plate heat pipes," *J. Heat Transfer*, 1995; **117**: 209–218.
5. Y. Wang and K. Vafai, "An experimental investigation of the transient characteristics on a flat-plate heat pipe during startup and shutdown operations," *ASME J. Heat Transfer*, 2000; **122**: 525–535.
6. Y. Wang and K. Vafai, "An experimental investigation of the thermal performance of an asymmetrical flat plate heat pipe," *Int. J. Heat Mass Transfer*, 2000; **43**: 2657–2668.
7. N. Zhu and K. Vafai, "Analysis of cylindrical heat pipes incorporating the effects of liquid–vapor coupling and non-Darcian transport: A closed form solution," *Int. J. Heat Mass Transfer*, 1999; **42**: 3405–3418.

8. M. Shafahi, V. Bianco, K. Vafai, and O. Manca, "An investigation of the thermal performance of cylindrical heat pipes using nanofluids," *Int. J. Heat Mass Transfer*, 2010; 53: 376–383.
9. P. Naphon, D. Thongkum, and P. Assadamongkol, "Heat pipe efficiency enhancement with refrigerant–nanoparticles mixtures," *Energy Convers. Manage.*, 2009; **50**: 772–776.
10. X. F. Yang, Z. H. Liu, and J. Zhao, "Heat transfer performance of a horizontal micro-grooved heat pipe using CuO nanofluid," *J. Micromechan. Microeng.*, 2008; **18**: 035038.
11. H. B. Ma, C. Wilson, Q. Yu, K. Park, U. S. Choi, and M. Tirumala, "An experimental investigation of heat transport capability in a nanofluid oscillating heat pipe," *J. Heat Transfer*, 2006; **128**: 1213–1216.
12. P. Naphon, P. Assadamongkol, and T. Borirak, "Experimental investigation of titanium nanofluids on the heat pipe thermal efficiency," *Int. Commun. Heat Mass Transfer*, 2008; **35**: 1316–1319.
13. P. D. Shima, J. Philip, and B. Raj, "Role of microconvection induced by Brownian motion of nanoparticles in the enhanced thermal conductivity of stable nanofluids," *Appl. Phys. Lett.*, 2009; **94**: 223101.
14. Y. Xuan, Y. Huang, and L. Qiang, "Experimental investigation on thermal conductivity and specific heat capacity of magnetic microencapsulated phase change material suspension," *Chem. Phys. Lett.*, 2009; **479**: 264–269.
15. W. Evans, J. Fish, and P. Keblinski, "Role of Brownian motion hydrodynamics on nanofluid thermal conductivity," *Appl. Phys. Lett.*, 2006; **88**: 093116.
16. K. B. Anoop, T. Sundararajan, and S. K. Das, "Effect of particle size on the convective heat transfer in nanofluid in the developing region," *Int. J. Heat Mass Transfer*, 2009; **52**: 2189–2195.
17. J. Jung and J. Y. Yoo, "Thermal conductivity enhancement of nanofluids in conjunction with electrical double layer (EDL)," *Int. J. Heat Mass Transfer*, 2009; **52**: 525–528.
18. M. Beck, Y. Yuan, P. Warrier, and A. S. Teja, "The effect of particle size on the thermal conductivity of alumina nanofluids," *J. Nanoparticle Res.*, 2009; **11**: 1129–1136.
19. K. Fang, C. Weng, and S. Ju, "An investigation into the structural features and thermal conductivity of silicon nanoparticles using molecular dynamics simulations," *Nanotechnology*, 2006; **17**: 3909–3914.

20. S. W. Kang, W. C. Wei, S. H. Tsai, and S. Y. Yang, "Experimental investigation of silver nano-fluid on heat pipe thermal performance," *Appl. Therm. Eng.*, 2006; **26**: 2377–2382.

21. S. W. Kang, W. C. Wei, S. H. Tsai, and C. C. Huang, "Experimental investigation of nanofluids on sintered heat pipe thermal performance," *Appl. Therm. Eng.*, 2009; **29**: 973–979.

22. Y. Chen, W. Wei, S. Kang, and C. Yu, "Effect of nanofluid on flat heat pipe thermal performance," in: *24th IEEE SEMI-THERM Symposium*, 16–19, March 16–20, 2008.

23. Y. Lin, S. Kang, and H. Chen, "Effect of silver nano-fluid on pulsating heat pipe thermal performance," *Appl. Therm. Eng.*, 2008; **28**; 1312–1317.

24. Z. Liu, J. Xiong, and R. Bao, "Boiling heat transfer characteristics of nanofluids in a flat heat pipe evaporator with micro-grooved heating surface," *Int. J. Multiphase Flow*, 2007; **33**: 1284–1295.

25. H. B. Ma, C. Wilson, B. Borgmeyer, K. Park, Q. Yu, S. U. S. Choi, and M. Tirumala, "Effect of nanofluid on the heat transport capability in an oscillating heat pipe," *Appl. Phys. Lett.*, 2006; **88**: 143116.

26. R. R. Riehl, "Analysis of loop heat pipe behavior using nanofluid," in: *Heat Powered Cycles International Conference (HPC)*, New Castle, UK, paper 06102, September 11–14, 2006.

27. C. Y. Tsaia, H. T. Chiena, P. P. Dingb, B. Chanc, T. Y. Luhd, and P. H. Chena, "Effect of structural character of gold nanoparticles in nanofluid on heat pipe thermal performance," *Mater. Lett.*, 2004; **58**: 1461–1465.

28. H. C. Brinkman, "The viscosity of concentrated suspensions and solution," *J. Chem. Phys.*, 1952; **20**: 571–581.

29. B. C. Pak and Y. I. Cho, "Hydrodynamic and heat transfer study of dispersed fluids with submicron metallic oxide particles," *Exp. Heat Transfer*, 1998; **11**: 151–170.

30. S. W. Chi, *Heat Pipe Theory and Practice*, Hemisphere, Washington, DC, 1976.

31. W. Yu and S. U. S. Choi, "The role of interfacial layers in the enhanced thermal conductivity of nanofluids: A renovated Maxwell model," *J. Nanoparticle Res.*, 2003; **5**: 167–171.

Chapter 14

Heat Transfer Characteristics and CHF Prediction in Nanofluid Boiling

Jingliang Bi,[a,b] Kambiz Vafai,[b] and David M. Christopher[a]
[a]*Key Laboratory for Thermal Science and Power Engineering of Ministry of Education, Department of Thermal Engineering, Tsinghua University, Beijing 100084, China*
[b]*Department of Mechanical Engineering, University of California Riverside, Riverside, CA 92521, USA*
vafai@engr.ucr.edu

A composite model for the average heat flux using both pure fluids and nanofluids is established incorporating microlayer evaporation, transient conduction due to bubble departure, and microconvection due to bubble growth and movement. Our model also takes into account bubble influence area interference. The average heat flux and the critical heat flux (CHF) when the surface dryout fraction exceeds a certain value are predicted. Nanofluid properties were calculated and the contact angles obtained from experiments were utilized in the present model. The analytical heat flux and CHF values using both regular fluids and nanofluids were compared with

This chapter has been reprinted from the *International Journal of Heat and Mass Transfer*, Vol. 80, Jingliang Bi, Kambiz Vafai, and David M. Christopher, Heat transfer characteristics and CHF prediction in nanofluid boiling, 256–265, Copyright 2014, with permission from Elsevier.

Nanotechnology and Energy
Edited by Kaufui V. Wong

ISBN 978-981-4774-24-6 (Hardcover), 978-1-315-16357-4 (eBook)
www.panstanford.com

the experimental results and existing theories and were found to be in close agreement. It was found that CHF enhancement value increases to 3.54 as the contact angle decreases from 80° to 20°, which is in agreement with recent experimental results.

14.1 Introduction

The heat transfer characteristics and thermal properties of nanofluids have been investigated more intensely in the past couple of decades for different heat transfer applications by various researchers [1–3]. Water is the most popular base fluid, while glycol–ethylene, oil, bio-fluids, and polymer solutions have also been used. A variety of materials are commonly used as nanoparticles, such as alumina, titania, silica, zirconia, graphite, carbon nanotubes, diamond, copper, and gold. [4]. In addition to single-phase heat transfer applications, nanofluid is also used in two-phase heat transfer applications.

CHF is found to be substantially enhanced with nanofluid boiling, while contradictory results with respect to the average wall heat transfer rates have been reported [5, 8–15]. Das et al. [5] experimentally studied pool boiling of alumina–water nanofluid and observed deterioration of the average heat flux. The nanoparticle size range varied from 50 to 150 nm, and the mean diameter was 38 nm. The concentration of the liquid varied from 1% to 4%. They argued that the nanoparticles, which were much smaller than the surface roughness (0.2–1.2 μm), deposit on the relatively uneven surface during boiling and hence created a smoother surface. This causes degradation of the boiling characteristics, and the surface smoothening overshadows the thermal conductivity enhancement of the nanofluids. Coursey and Kim [6] employed water and ethanol-based nanofluids with aluminum nanoparticle concentrations from 0.001 to 10 g/L. Nanofluids were found to degrade or had little influence on the boiling performance. Larger concentrations (≥0.5 g/L) resulted in a moderate (up to 37%) increase in the CHF.

The main reason for the deterioration of heat flux is assumed to be the deposition of nanoparticles on the surface, which changes the surface wettability. Kim et al. [7] experimentally studied alumina-, silica-, and zirconia-based nanofluids and found that when the concentration was smaller than 0.1% by volume, CHF could be greatly enhanced due to the buildup of a porous layer of nanoparticles on the surface.

The nanofluid boiling experimental results [5, 8–20] are summarized in Table 14.1. In this table, nanoparticle size and concentration are specified under the column "nanofluid properties." The heater properties, the contact angle, and roughness characteristics are listed under the column "surface property." The average boiling heat flux (BHF), CHF, the boiling heat transfer coefficient (BHTC), as well as the maximum deterioration and enhancement ratios values at specified nanofluid concentrations are also illustrated in Table 14.1.

Although pool boiling with nanofluids has been investigated extensively experimentally, there are very few reports on an analytical study of nanofluid boiling. Models on nanofluid boiling are needed. Compared to boiling with pure liquid, nanofluid boiling involves the change of liquid thermal properties and the movement of particles. Experiments show that at low concentrations, changes in liquid thermal properties is not significant [12, 14, 21, 22]. Nanoparticles deposited on the surface alter the surface conditions [23] and the surface contact angle [8, 11–14]. Surface contact angle is an important characteristic, and it can substantially influence the active nucleation sites [25]. In our analysis, we will use the contact angle as a parameter to define surface characteristics.

Nomenclature

A	Hamaker constant, area
$\overline{A}$	area for each bubble
c	capacity
C_1	empirical constant in Eq. (14.47)
C_{sf}	coefficient in Eq. (14.58)
D	bubble diameter
f	bubble departure frequency
F	dryout fraction
Gr	Grashof number
h	heat transfer coefficient
h_{ev}	evaporation coefficient
h_{fg}	latent heat
Ja	Jacob number
k	thermal conductivity
$K_{overlap}$	enhancement ratio in overlapping area
N_n	active nucleation site density

$\overline{N_n}$	average cavity density
Nu	Nusselt number
N_{sd}	nucleation site density
P	pressure
Pr	Prandtl number
q	heat transfer rate
q''	heat flux
Q	total wall heat transfer
r	radial coordinate
R_0	minimum radius of the microlayer
R_{max}	maximum radius of the microlayer
R_{b}	bubble radius
R_{c}	critical cavity radius
Re_{b}	bubble Reynolds number
R_{g}	gas constant based on molecular weight
t	time
T	temperature
u	radial velocity
U_{b}	bubble interface velocity
v	vertical velocity
y	vertical coordinate
Greek symbols	
ΔT	temperature difference
β	thermal expansion coefficient
δ	microlayer thickness
θ	surface contact angle
κ	surface curvature
λ'	statistical parameter
μ	dynamic viscosity of the base fluid
ρ	density
ρ^{+}	non-dimensional density difference
σ	surface tension
φ	heater surface inclination angle
ϕ	nanofluid concentration
Subscripts	
0	equilibrium boundary layer
b	bubble
c	contact line

CHF	critical heat flux
d	departure
f	nanofluid
g	growth
int	interface
l	base liquid
max	maximum
mc	microconvection
me	microlayer evaporation
nc	natural convection
overlap	overlapping
re	rewetting
R	radius
s	solid nanoparticle
tc	transient conduction
v	vapor
w	wall

A number of researchers have proposed several theories on the boiling process and the associated heat transfer [26–31]. However, the literature lacks a dominant model representation of boiling heat flux. The earliest researchers simplified the boiling process to a single-phase convection process, including Rohsenow [26], Tien [27], and Zuber [28]. Forster and Greif [29] proposed that departing bubble worked as a micropump to push hot liquid away and cold liquid in. Mikic and Rohsenow [31] postulated that when bubbles depart, they take away the superheated liquid layer adjacent to the heating surface and cold liquid rushes in and transient conduction occurs.

Different mechanisms are active during different bubble periods. Judd and Huang [32] first combined microlayer evaporation, transient conduction, and natural convection to set up a comprehensive model for boiling heat transfer. Benjamin and Balakrishnan [33] proposed a similar model and accounted for the activation period for transient conduction and microlayer evaporation. These two models considered the situations at low to moderate heat flux regions where bubbles did not interfere with each other. Since the boiling heat transfer attributes has not been fully understood, more work needs to be done to better reveal the heat transfer mechanism in nucleate boiling.

Table 14.1 Synthesis of the pertinent nanofluid boiling experiments in the last decade

References	Nanofluid properties	Surface properties	CHF	Average boiling heat flux
Shahmoradi et al. [8]	Al_2O_3 (40 nm)–water (0.001–0.1 vol.%)	Flat plate heater (θ from 45° to 20°, *Ra* from 5.1 nm to 376 nm; 77 nm to 197 nm)	CHF ↑ 47% for 0.1 vol.%	BHF ↓ 40% for 0.1 vol.%
Jung et al. [9]	Al_2O_3 (46 nm) + H_2O/LiBr (0.01–0.1 vol.%)	Plate copper heater (θ from 78° ↓ to 36°)	CHF ↑ by 48.5% at 0.1 vol.%	BHF ↓ 10%
Vazquez and Kumar [10]	Silica (10 nm)–water 0.1 to 2 vol.%	Nichrome wires and ribbons	CHF ↑ 270% at 0.4 vol.%	BHF ↑ 190%
Jung et al. [11]	Al_2O_3 (45 nm)–water (10^{-5}–0.1 vol.%).	Plate copper heater	CHF ↑ 116%	BHF ↓ 60%
Hegde et al. [12]	CuO (50 nm)–water, (0.01–0.5%)	NiCr wire (from 0.34 μm to 0.09 (0.05 vol.%) and 0.23 μm (0.3%))	CHF ↑ 130% (1.3–3 MW/m^2)	BHF ↓ 50%
Kole and Dey [13]	ZnO (30–40 nm)–ethylene glycol (EG)	Copper cylindrical block (90 nm)	CHF ↑ by 117% at 2.6%	BHF ↑ by 22% at 1.6% and ↓ further loading
Bolukbasi and Ciloglu [14]	SiO_2 (34 nm)–water (0.001–0.1 vol.%)	Cylindrical brass (typical 90 nm, *Ra* ↑ to 620 nm, θ 70° to 42.07° and 9.97°)	CHF ↑ 28%	BHF ↓ 40%

References	Nanofluid properties	Surface properties	CHF	Average boiling heat flux
Gerardi et al. [15]	Diamond (34 nm) and silica–water (173 nm) (0.1 vol.% for silica and 0.01 vol.% for diamond)	Upward indium-tin-oxide surface (*h* from 80° to 16°, *Ra* from 30 to 900–2100 nm)	CHF ↑ 100%	BHTC ↓ 50%
Wen et al. [16]	Al_2O_3 (50–900 nm)–water (0.001–0.1%)	Rectangular brass plates (rough: 420 nm and smooth: 25 nm)	—	Twofold ↑ in BHTC at 0.001% nanofluids
Suriyawong and Wongwises [17]	TiO_2–water (21 nm), 0.00005–0.01 vol.%	Horizontal circular plates, copper (*Ra* 0.2 μm) and aluminum (*Ra* 4 μm)	—	BHTC 15% ↑ for copper and a 4% ↑ for aluminum at 0.0001 vol.%; ↓ at higher conc levels
Kwark et al. [18]	(Al_2O_3, CuO, and diamond)–water (2.7×10^{-5}–2.7×10^{-2}% vol.).	1 cm × 1 cm × 0.3 cm copper block	CHF ↑ 31%	BHF no change <500 KW/m^2 and ↓ at high heat flux
Liu et al. [19]	CuO (30 nm)–water	Copper block	CHF ↑ when conc <1%; constant since	BHF ↑ when conc <1%; ↓ >1%
You et al. [20]	Al_2O_3–water (0–0.05 g/l)	Polished copper surfaces	CHF ↑ 200%	Same
Das et al. [5]	Al_2O_3 (50–150 nm)–water (1–4%)	Rectangular stainless steel (0.4 and 1.15 μm)	—	BHF↓

CHF is also an important characteristic in boiling. CHF is the end of nucleate boiling and refers to the limit of safe operation of a heat transfer system. In nuclear reactors or electrical components, loss of liquid contact with surface will dramatically decrease the heat transfer coefficient and causes damage to the surfaces. To investigate CHF and its mechanism, a number of theories have been proposed to predict the CHF.

Zuber [34] proposed that Taylor and Helmholtz instabilities were responsible for the occurrence of CHF. As the heat flux increases, the velocity of vapor jets above the nucleating bubbles increase dramatically, occupying the space of liquid between the jets toward the surface, thus initiating an instability process. Zuber's correlation is the most widely used to predict pool boiling CHF as:

$$q_{\mathrm{CHF}} = 0.131\rho_{\mathrm{v}}^{0.5} h_{\mathrm{fg}} [\sigma g(\rho_{\mathrm{f}} - \rho_{\mathrm{v}})]^{0.25} \tag{14.1}$$

Haramura and Katto [35] proposed a hydrodynamic model, that is, as the heat flux exceeds a certain value, a vapor blanket forms on the liquid film, preventing bulk liquid from feeding the film resulting in the dryout of the liquid film, thus substantially enhancing occurrence of CHF.

Kandlikar [36] took the effect of dynamic receding contact angle into consideration and developed a theoretical model to predict the CHF in saturated pool boiling. As the surface heat flux increases, the force resulting from the evaporation at the bubble interface exceeds the gravitational and surface tension forces that are holding the bubble, resulting in the spread of vapor bubble over the surface, causing CHF to occur. The following correlation by Kandlikar [36] predicts the effect of surface orientation angle and system pressure on CHF:

$$\begin{aligned} q''_{\mathrm{CHF}} &= \rho_{\mathrm{v}}^{1/2} h_{\mathrm{fg}} \left(\frac{1+\cos\theta}{16} \right) \\ &\times \left[\frac{2}{\pi} + \frac{\pi}{4}(1+\cos\theta)\cos\varphi \right]^{\frac{1}{4}} \left[g\sigma(\rho_{\mathrm{f}} - \rho_{\mathrm{v}}) \right]^{\frac{1}{4}} \end{aligned} \tag{14.2}$$

where θ is the receding contact angle and φ is the heater surface angle. The above equation predicts a continuous increase in CHF with decreasing surface contact angle.

The present chapter incorporates the main mechanisms in the boiling process and extends the analysis to high heat flux region where bubble-influenced areas interfere with each other. The CHF value can also be predicted by the present model. The obtained analytical results were compared with recent experimental results and classical equations and were found to be in good agreement.

14.2 Nanofluid Properties

The density of a nanofluid can be expressed based on the mixture rule as:

$$\rho_{\mathrm{f}} = (1-\phi)\rho_{\mathrm{l}} + \phi\rho_{\mathrm{s}} \tag{14.3}$$

Similarly, specific heat and thermal expansion coefficient of nanofluid can be expressed [37] as:

$$C_{\mathrm{f}} = \frac{(1-\phi)\rho_{\mathrm{l}}C_{\mathrm{l}} + \phi\rho_{\mathrm{s}}C_{\mathrm{s}}}{\rho_{\mathrm{f}}} \tag{14.4}$$

$$\beta_{\mathrm{f}} = \frac{(1-\phi)(\rho_{\mathrm{l}}\beta_{\mathrm{l}}) + \phi\rho_{\mathrm{s}}\beta_{\mathrm{s}}}{\rho_{\mathrm{f}}} \tag{14.5}$$

The dynamic viscosity can be expressed as [38]:

$$\mu_{\mathrm{f}} = (1 + 2.5\phi + 6.2\phi^2)\mu_{\mathrm{l}} \tag{14.6}$$

and the thermal conductivity can be expressed based on the correlation proposed by Khanafer and Vafai [37]:

$$k_{\mathrm{f}} = k_{\mathrm{l}}\left[1.0 + 1.0112\phi + 2.4375\phi\left(\frac{47}{d_{\mathrm{s}}(\mathrm{nm})}\right) - 0.0248\phi\left(\frac{k_{\mathrm{s}}}{0.613}\right)\right] \tag{14.7}$$

14.3 Model Development

Boiling is a complex phenomenon as it involves both phase change and forced convection due to bubble movement. A number of researchers have divided the analysis of pool boiling heat transfer into different parts [32, 33]. In prior works, microlayer evaporation and transient conduction during bubble waiting time were analyzed in detail. However microconvection has not received much attention. In the present work, the total heat flux is divided into four parts: microlayer evaporation, transient conduction under bubble, transient conduction after bubble departs, and enhanced natural convection. The interference of bubble influence areas was also analyzed in the present chapter.

14.3.1 Total Bubble Heat Transfer in One Bubble Ebullition Cycle

14.3.1.1 Microlayer evaporation

Microlayer evaporation is considered to be an important mechanism during the bubble growth phase. The evaporation of this thin liquid layer under the bubble generates the vapor needed for the bubble growth, as shown in Fig. 14.1a. However, due to the very small size of the microlayer, it is difficult to directly observe and measure it. An analytical model is presented here to predict the microlayer shape and the heat transfer rate. The microlayer is considered within a cylindrical coordinate system. The continuity equation for the microlayer can be expressed as:

$$\frac{1}{r}\frac{\partial}{\partial r}(ru_{\mathrm{f}})+\frac{\partial v_{\mathrm{f}}}{\partial_{f}}=0 \qquad (14.8)$$

The radial momentum equation in the microlayer assumes the advection term is negligible and that the velocity gradient in the radial direction is much smaller than the velocity gradient in the vertical direction, so the lubrication approximation can be invoked to arrive at:

$$\frac{\partial P_{\mathrm{f}}}{\partial r}=\mu_{\mathrm{f}}\frac{\partial^{2}u_{\mathrm{f}}}{\partial y^{2}} \qquad (14.9)$$

The velocity at $y = 0$ and the shear stress at the interface are zero, so the boundary conditions can be expressed as:

$$u_{\mathrm{f}}(y=0)=0,\ \frac{\partial u_{\mathrm{f}}}{\partial y}(\delta)=0 \qquad (14.10)$$

Integrating Eq. (14.9) subject to the boundary conditions in Eq. (14.10), results in the following expression for the velocity along the microlayer:

$$u_{\mathrm{f}} = -\frac{1}{\mu_{\mathrm{f}}}\frac{\partial P_{\mathrm{f}}}{\partial r}\left(\delta y - \frac{y^2}{2}\right) \tag{14.11}$$

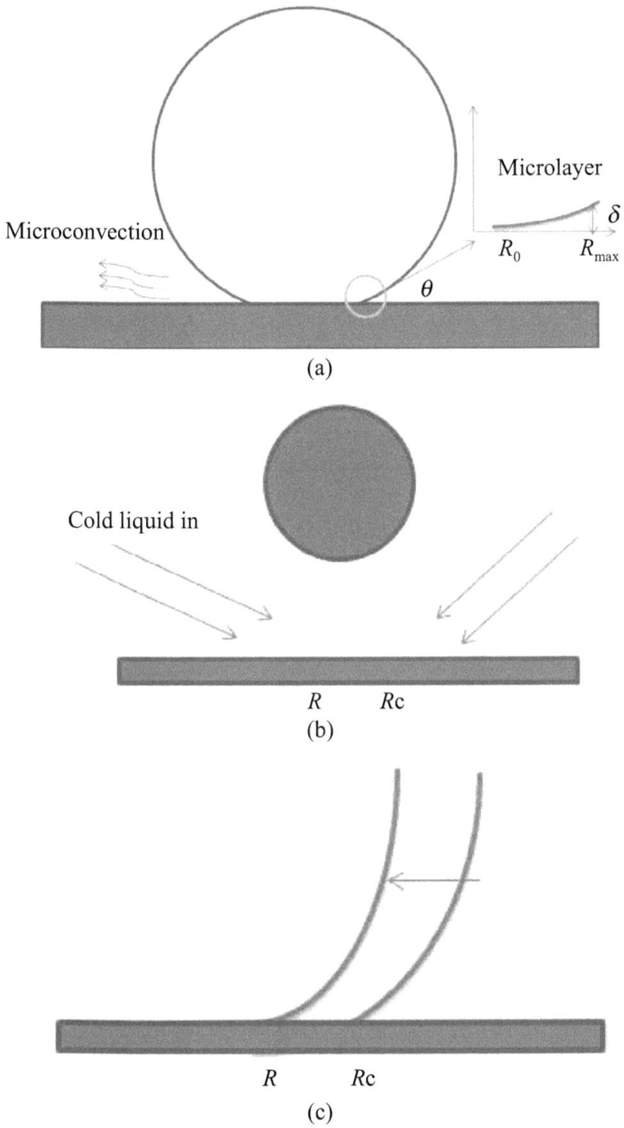

Figure 14.1 Illustration of (a) microlayer and microconvection, (b) liquid rewetting surface after bubble departure, and (c) transient conduction before bubble departure.

Integrating Eq. (14.8) and solving for v_f gives:

$$v_f(\delta)-0=-\frac{1}{r}\frac{\partial}{\partial r}r\int_0^{\delta}u_f dy \tag{14.12}$$

Substituting Eq. (14.11) in Eq. (14.12), the vertical velocity can be expressed as:

$$v_f(\delta)=\frac{\delta^3}{3r\mu_f}\frac{\partial}{\partial r}\left(r\frac{\partial P_f}{\partial r}\right) \tag{14.13}$$

Since the microlayer is quite thin, the velocity due to bubble evaporation perpendicular to the bubble interface is essentially vertical:

$$v_f(\delta)=\frac{q''}{\rho_f h_{fg}} \tag{14.14}$$

Equations (14.13) and (14.14) can be combined and rearranged as:

$$\frac{3r\mu_f q''}{\rho_f h_{fg}\delta^3}=\frac{d}{dr}\left(r\frac{dP_f}{dr}\right)=\frac{dP_f}{dr}+r\frac{d^2P_f}{dr^2} \tag{14.15}$$

The heat flux through the microlayer is based on Fourier's law,

$$q''=k_f\frac{(T_w-T_{int})}{\delta} \tag{14.16}$$

While the modified Clausius–Clapeyron equation [39] for the heat transfer across the interface can be written as:

$$q''=h_{ev}\left[T_{int}-T_v+\frac{(P_f-P_v)T_v}{\rho_v h_{fg}}\right] \tag{14.17}$$

As such, the interface temperature T_{int} can be obtained by combining Eqs. (14.16) and (14.17) as:

$$\left(\frac{k_f}{\delta}+h_{ev}\right)T_{int}=\left[\frac{k_f}{\delta}T_w+h_{ev}T_v-h_{ev}\frac{(P_f-P_v)T_v}{\rho_v h_{fg}}\right] \tag{14.18}$$

The pressure drop across the interface is given by the modified Young–Laplace's equation obtained by a force balance, which includes the surface tension, disjoining pressure, and the vapor recoil force:

$$P_{\text{f}} = P_{\text{v}} - \sigma\kappa - \frac{A}{\delta^3} + \frac{q''^2}{\rho_{\text{v}} h_{\text{fg}}^2} \tag{14.19}$$

where A is the Hamaker constant and is chosen as 10^{-21} based on neglecting the recoil pressure term in Eq. (14.19). This gives:

$$P_{\text{f}} = P_{\text{v}} - \sigma\kappa - \frac{A}{\delta^3} \tag{14.20}$$

In the dryout region, $q'' = 0$ and $T_{\text{int}} = T_{\text{w}}$. Substituting for a zero curvature and heat flux in the above equation, the thickness of the dryout area is found to be:

$$\delta_0^3 = \frac{A}{\left(\dfrac{T_{\text{int}}}{T_{\text{v}}} - 1\right) h_{\text{fg}} \rho_{\text{v}}} \tag{14.21}$$

The interfacial curvature is calculated as:

$$\kappa = \frac{1}{r}\frac{\partial}{\partial r}\left[\frac{r\dfrac{\partial \delta}{\partial r}}{\sqrt{1 + \left(\dfrac{\partial \delta}{\partial r}\right)^2}}\right] \tag{14.22}$$

Calculating the derivative of P_{f} in Eq. (14.19) gives:

$$\frac{dP_{\text{f}}}{dr} = -\sigma\kappa' + 3\frac{A}{\delta^4}\delta' \tag{14.23}$$

and the second derivative is also calculated to be:

$$\frac{d^2 P_{\text{f}}}{dr^2} = -\sigma\kappa'' + 3\frac{A}{\delta^4}\delta'' - 12\frac{A}{\delta^5}\delta'^2 \tag{14.24}$$

Combining Eqs. (14.15), (14.23), and (14.24) to replace the first and second derivatives of P_{f} results in an expression for the second derivative of the curvature:

$$F_1(\delta) = \kappa''$$

$$= \left(-\frac{3\mu_1 q''}{\rho_{\text{f}} h_{\text{fg}} \delta^3} - \frac{\sigma\kappa'}{r} + 3\frac{A}{r\delta^4}\delta' + 3\frac{A}{\delta^4}\delta'' - 12\frac{A}{\delta^5}\delta'^2 \right) \Big/ \sigma \tag{14.25}$$

Calculating the second derivative of κ from Eq. (14.22) gives:

$$\frac{d^2\kappa}{dr^2}=\frac{2\delta'}{r^3(1+\delta'^2)^{\frac{1}{2}}}+\frac{r\delta'''-2\delta''}{r^2(1+\delta'^2)^{\frac{3}{2}}}-\frac{3\delta'\delta''^2+3r\delta''^3+9r\delta'\delta''\delta'''}{r(1+\delta'^2)^{\frac{5}{2}}}$$
$$+\frac{15\delta'^2\delta''^3}{(1+\delta'^2)^{\frac{7}{2}}}+\frac{\delta''''}{(1+\delta'^2)^{\frac{3}{2}}}$$
$$=F_2(\delta)+\frac{\delta''''}{(1+\delta'^2)^{\frac{3}{2}}} \tag{14.26}$$

Combining Eqs. (14.25) and (14.26), the fourth derivative at δ can be obtained as:

$$\delta''''=[F_1(\delta)-F_2(\delta)](1+\delta'^2)^{\frac{3}{2}} \tag{14.27}$$

The slope of the microlayer at $R_{\max}$ is tan (θ), and the radius of the outer edge of the microlayer is R sin(θ); hence, the boundary conditions at the microlayer can be presented as:

$$\delta(R_0)=\delta_0,\delta'(R_0)=0$$

$$\delta'(R_{\max})=\tan(\theta),R_{\max}=\frac{D}{2}\sin(\theta) \tag{14.28}$$

14.3.1.2 Transient conduction after bubble departs

Experiments have shown that the transient conduction is the dominant heat transfer mode in nucleate pool boiling [40, 41]. Mikic and Rohsenow [31] were the first to propose a model and provide an expression representing transient conduction during the bubble waiting time. Their model assumes that when a bubble departs from the surface, it carries away the liquid with a radius of a bubble departure diameter, while the cold liquid rushes toward the surface. This process is illustrated in Fig. 14.1b. The heat transfer from the heating surface to the bulk liquid is considered transient conduction from a hot surface to a semi-infinite body, which is the bulk liquid in this case. The instantaneous heat flux at any time is expressed as:

$$q''_{\text{tc}}(t)=\frac{k_{\text{f}}\Delta T}{\sqrt{\pi\alpha t}} \tag{14.29}$$

In this work, the transient conduction t_{tc} is taken to be three times that of the growth time t_g [33]. Integrating Eq. (14.29) over the time period t_{tc} and an influence area of four times the bubble projecting area as suggested by various researchers [31–33], the total heat transfer during the waiting period at low heat flux region can be obtained as:

$$Q_{tc} = \int_0^{\frac{3}{4f}} q''_{tc} dt \cdot \pi D^2 = \int_0^{\frac{3}{4f}} \frac{k_f \Delta T}{\sqrt{\pi \alpha t}} dt \cdot \pi D^2 = \frac{\sqrt{3}\pi k_f \Delta T D^2}{\sqrt{\pi \alpha f}} \tag{14.30}$$

Equation (14.30) is valid only when the influence areas do not interfere with each other. To take into account the bubble influence area interference, a dryout fraction parameter F is defined as the ratio of bubble projecting area to the entire heating surface as:

$$F = \frac{\pi}{4} D_d^2 N_{sd} \tag{14.31}$$

When $\pi D_d^2 \leq 1/N_{sd}$, the bubble influence area is smaller than the average area for each bubble on the heating surface and the bubble interfaces are considered not to overlap. For this condition, the dryout fraction F is less than 25%, according to Eq. (14.31). When the dryout fraction exceeds 25%, the influence areas will be overlapping. The overlapping area is a function of the nucleation site density and the bubble influence area. The total overlapping area over the heating surface within a unit area (1 m^2) is:

$$\bar{A}_{\text{total}} = \pi D_d^2 \cdot N_{sd} - 1 \tag{14.32}$$

For each bubble, the average overlapping area can be approximated as the total overlapping area divided by the number of active nucleation sites in one unit area. Since each overlapping area is shared by two interfering bubbles, the average overlapping area for each bubble can be represented as:

$$\bar{A}_{\text{overlap}} = \frac{1}{2}(\pi D_d^2 - 1/N_{sd}) \tag{14.33}$$

Therefore, the non-overlapping area for each bubble is:

$$\bar{A}_{\text{non-overlap}} = 1/N_{sd} - \frac{1}{2}(\pi D_d^2 - 1/N_{sd}) = 3/2N_{sd} - \frac{1}{2}\pi D_d^2 \tag{14.34}$$

The effect of overlapping of more than three bubbles was considered to be balanced by a heat transfer augmentation of the overlapping area due to repeated quenching.

In the overlapping area, due to repeated quenching, the transient conduction heat flux is larger than the heat flux at non-overlapping areas. This enhancement ratio was determined by considering the two interfering bubble influence areas, with each bubble departing and entering during the waiting period. The enhancement ratio is defined as:

$$K_{\text{overlap}} = \frac{\int_0^{\frac{3}{4f}} q''_{\text{tc}} dt + \int_0^{\frac{1}{4f}} q''_{\text{tc}} dt}{\int_0^{\frac{3}{4f}} q''_{\text{tc}} dt} = \frac{\sqrt{3}+1}{\sqrt{3}} = 1.58 \tag{14.35}$$

So in the high heat flux region, the total heat transfer after bubble departs from the surface is:

$$\begin{aligned} Q_{\text{tc}} &= \frac{\sqrt{3} k_{\text{f}} \Delta T}{\sqrt{\pi \alpha f}} (K_{\text{overlap}} \cdot \bar{A}_{\text{overlap}} + \bar{A}_{\text{non-overlap}}) \\ &= \frac{\sqrt{3} k_{\text{f}} \Delta T}{\sqrt{\pi \alpha f}} \left(\frac{K_{\text{overlap}}}{2} \left(\pi D^2 - \frac{1}{N_{\text{sd}}} \right) + \left(\frac{3}{2N_{\text{sd}}} - \frac{\pi D^2}{2} \right) \right) \end{aligned} \tag{14.36}$$

14.3.1.3 Transient conduction before the bubble departs

This mode of heat transfer occurs when the bubble contact line starts to shrink and the surface under the bubble is rewetted. The enhanced heat flux has been observed by various researchers [42–45]. The bubble starts to recede from R_c, which is the radius of the contact line. The rewetting process ends when the radius diminishes to zero as the bubble departs. The rewetting process before bubble departure is illustrated in Fig. 14.1c.

Considering when the liquid front approaches R from R_c ($0 \le R \le R_c$), the heat transfer rate at time t_R in this rewetting process can be obtained by integrating the heat flux from the contact line to where the liquid front is encountered, resulting in

$$q(t_R) = \int_{R_c}^{R} \frac{k_f \Delta T}{\sqrt{\pi \alpha}} \frac{2\pi r dr}{\sqrt{t_R - t_{re}}} \tag{14.37}$$

Assuming the bubble contact line receding velocity u_c is constant, the total rewetting time is $t_{re} = R_c/u_c$, and the time when the contact line approaches R is $t_R = (R_c - R)/u_c$. Substituting t_R and t_{re} into Eq. (14.37), the heat transfer rate at the moment t_R is obtained as:

$$q(t_R) = \frac{4\pi k_f \Delta T u_c}{\sqrt{\pi\alpha}}\left(R_c t_R^{1/2} - \frac{2}{3}u_c t_R^{3/2}\right) \quad (14.38)$$

Integrating the heat transfer rate over the time period t_{re}, the total heat transfer during the bubble rewetting period can be represented as:

$$Q_{re} = \int_0^{t_{re}} q_{re}(t) = \frac{8\pi k_f \Delta T u_c}{3\sqrt{\pi\alpha}}\left(R_c t_{re}^{3/2} - \frac{2}{5}u_c t_{re}^{5/2}\right) \quad (14.39)$$

The radius of the contact can be written as $R_c = D/2 \sin(\theta)$. The rewetting time t_{re} is about half of the growth time t_g [43], and $t_w = 3t_g$ is often considered in the nucleate boiling case [33], so $t_{re} = 1/8f$. Substituting t_{re} into Eq. (14.39), the heat transfer during the rewetting period can be expressed as:

$$Q_{re} = \frac{8\pi k_f \Delta T R_c^2}{5\sqrt{\alpha}} t_{re}\sqrt{\pi f} = 0.05\frac{k_f \Delta T D_d^2 \sin^2\theta}{\sqrt{\alpha f}}\sqrt{\pi} \quad (14.40)$$

14.3.1.4 Microconvection and natural convection

The natural convection heat transfer coefficient over a horizontal flat plate can be expressed as:

$$Nu = 0.31(Gr \cdot Pr)^{1/3} \quad (14.41)$$

$$q''_{nc} = h \cdot \Delta T = \frac{Nu \cdot k_f}{d}\Delta T \quad (14.42)$$

$$Q_{nc} = q''_{nc}(1/N_{sd} - \pi D_d^2)/f = \frac{0.31(Gr \cdot Pr)^{1/3} \cdot k_f(1/N_{sd} - \pi D_d^2)}{df}\Delta T \quad (14.43)$$

In nucleate boiling, the rapid bubble growth perturbs the liquid and enhances the natural convection. In addition to the bubble growth perturbing the liquid slightly, bubbles can also oscillate slightly during the bubble growth period. The natural convection is enhanced due to bubble growth, oscillation, and departure. This part of enhanced convection around the bubble is referred to as microconvection, and to evaluate its effect, we utilize the local Nusselt number expression for forced convection over a flat plate.

The Reynolds number was calculated using Forster and Zuber's definition [46] as:

$$Re_{\mathrm{b}} = \frac{\rho_{\mathrm{v}} U_{\mathrm{b}} D}{\mu_{\mathrm{f}}} \tag{14.44}$$

The bubble growth velocity can be written as:

$$R_{\mathrm{b}} = Ja\left(\frac{\alpha t}{\pi}\right)^{1/2}, \; U_{\mathrm{b}} = dR_{\mathrm{b}}/dt \tag{14.45}$$

where $Ja = \rho_{\mathrm{f}} c_{\mathrm{f}} (T_{\mathrm{w}} - T_{\mathrm{sat}})/(\rho_{\mathrm{v}} h_{\mathrm{fg}})$. Substituting Eq. (14.45) in Eq. (14.44), the Reynolds number can be represented as:

$$Re_{\mathrm{b}} = \frac{\rho_{\mathrm{f}}}{\mu_{\mathrm{f}}}\left(\frac{\Delta T C_{\mathrm{f}} \rho_{\mathrm{f}} \sqrt{\alpha}}{h_{\mathrm{lv}} \rho_{\mathrm{v}} \sqrt{\pi}}\right)^2 \tag{14.46}$$

The Nusselt number due to microconvection is expressed based on laminar flow over a flat plate by:

$$Nu_{\mathrm{mc}} = C_1 Re_{\mathrm{b}}^{m} (Pr)^{1/3} \tag{14.47}$$

where C_1 is 0.332 as is the case for the forced convection over a flat plate, and m is 0.67 as in Rohsenow's model [26].

For the low heat flux region, where bubble influence areas do not interfere with each other, the total area of the microconvection is considered to be within the influence area and outside the bubble contact area and the total time t_{g} is one-fourth of the total ebullition cycle. As such, the microconvection can be expressed as:

$$Q_{\mathrm{mc}} = \frac{0.332 Re_{\mathrm{b}}^{0.67} (Pr)^{1/3} \cdot k_{\mathrm{f}}}{4f \cdot D_{\mathrm{d}}} \Delta T \left(\pi D_{\mathrm{d}}^2 - \frac{\pi D_{\mathrm{d}}^2 \sin^2 \theta}{4}\right) \tag{14.48}$$

If the nucleation site density is too large, the bubble influence areas will interfere with each other. Considering the overlapping area will be influenced by two independent bubbles, microconvection over this area will be repeated; hence, the enhancement ratio can be considered to be doubled. As such the total microconvection can be expressed as:

$$q''_{\mathrm{mc}} = \frac{Nu_{\mathrm{mc}} \cdot k_{\mathrm{f}}}{D_{\mathrm{d}}} \Delta T \tag{14.49}$$

$$Q_{\mathrm{mc}} = \frac{1}{4f}\left(2q''_{\mathrm{mc}}\overline{A}_{\mathrm{overlap}} + q''_{\mathrm{mc}}\left(\overline{A}_{\mathrm{non\text{-}overlap}} - \frac{\pi D_{\mathrm{d}}^2 \sin^2\theta}{4}\right)\right)$$

$$= \frac{0.332 Re_{\mathrm{b}}^{0.67}(Pr)^{1/3}\cdot k_{\mathrm{f}}}{8fD_{\mathrm{d}}}\Delta T\left(\pi D_{\mathrm{d}}^2 + \frac{1}{N_{\mathrm{sd}}} - \frac{\pi D_{\mathrm{d}}^2 \sin^2\theta}{2}\right) \quad (14.50)$$

14.3.2 Nucleation Site Density

The active nucleation sites N_{sd} can be predicted using Hibiki and Ishii's equation [47].

$$N_{\mathrm{n}} = \overline{N_{\mathrm{n}}}\left\{1 - \exp\left(-\frac{\theta^2}{8\mu^2}\right)\right\}\left[\exp\left\{f(\rho^+)\frac{\lambda'}{R_{\mathrm{c}}}\right\} - 1\right] \quad (14.51)$$

where

$\overline{N_{\mathrm{n}}}$ = 4.72 × 10^5 site/m^2, μ = 0.722 rad, and λ' = 2.50 × 10^{-6} m

$$R_{\mathrm{c}} = \frac{2\sigma\{1 + (\rho_{\mathrm{g}}/\rho_{\mathrm{f}})\}/P_{\mathrm{f}}}{\exp\{h_{\mathrm{fg}}(T_{\mathrm{g}} - T_{\mathrm{sat}})/(R_{\mathrm{g}}T_{\mathrm{g}}T_{\mathrm{sat}})\} - 1} \quad (14.52)$$

$$f(\rho^+) = -0.01064 + 0.48246\rho^+ - 0.22712\rho^{+2} + 0.05468\,\rho^{+3} \quad (14.53)$$

and $\rho^* = \Delta\rho/\rho_{\mathrm{g}}$, $\rho^+ = \log(\rho^*)$ and R_{g} is the gas constant based on the molecular weight. This correlation predicts a decrease in active nucleation sites with a decrease in the contact angle. In nanofluid boiling, the nanoparticles deposit on the surface and tend to decrease the surface contact angle by filling in the cavities and smoothing the surface. The nucleation site density variations with contact angle change agree with the nanofluid boiling experiments.

14.3.3 Bubble Departure Frequency

Bubble departure diameter is obtained using a relationship provided by Carey [48]:

$$D_{\mathrm{d}} = \frac{1.65 d^*\sigma}{g(\rho_{\mathrm{l}} - \rho_{\mathrm{v}})} + \left[\frac{15.6\rho_{\mathrm{l}}}{g(\rho_{\mathrm{l}} - \rho_{\mathrm{v}})}\right]^{1/3}\left[\frac{\beta_{\mathrm{d}} k_{\mathrm{l}}(T_{\mathrm{w}} - T_{\mathrm{sat}})}{h_{\mathrm{fg}}\rho_{\mathrm{v}}}\right] \quad (14.54)$$

where d^* = 6.0 × 10^{-3} mm and β_{d} = 6.0. The bubble departure frequency can be obtained using the bubble growth equation and can be expressed as:

$$f = \frac{1}{\pi} Ja^2 \frac{\alpha}{D_d^2} \tag{14.55}$$

14.3.4 Average Heat Flux and CHF Calculation

The total heat transfer in one bubble ebullition cycle can be obtained by adding the five heat transfer contributions. The average heat flux at the surface can be obtained as

$$q'' = (Q_{me} + Q_{re} + Q_{tc} + Q_{mc} + Q_{nc}) \cdot f \cdot N_{sd} \tag{14.56}$$

Over the low heat flux region, Q_{tc} and Q_{mc} are expressed using Eqs. (14.30) and (14.48), and for the high heat flux region, Q_{tc} and Q_{mc} are expressed using Eqs. (14.36) and (14.50). Q_{nc} is expressed in Eq. (14.43) for low to moderate heat flux, while it becomes negligible over high heat flux region. Q_{me} is expressed by the radial integration of q''_{me} given by Eq. (14.16) and Q_{re} is expressed by Eq. (14.40).

Based on our model results, transient conduction is the dominant mode of heat transfer at higher superheats, accounting for nearly 70% of the total heat transfer at higher superheats, while microconvection is about 10% of the total heat transfer. Microlayer evaporation accounts for less than 10% of the total heat transfer, while natural convection accounts for about 60% of the total heat transfer at very low superheats. However, this percentage decreases dramatically when the superheat increases.

The critical heat flux can also be obtained by our analytical model. When the dryout fraction F exceeds a certain value, most of the surface is occupied by vapor, thus retarding the liquid from the bulk pool region to flow down to wet the surface. Also, when the bubbles are too close together, they will begin to touch each other and coalesce resulting in a large vapor blanket that covers the surface. CHF is considered to occur when all the bubbles merge and the liquid cannot flow down to wet the liquid. According to Zuber [28], when

$$N_{sd} D_b^2 = 1 \tag{14.57}$$

static bubble interactions will occur. Under this condition, the dryout fraction F is $\pi/4$ according to Eq. (14.31). As such, the heat flux at a dryout fraction of $\pi/4$ is considered to be CHF.

14.4 Results and Discussion

14.4.1 Average Heat Transfer Characteristics

Rohsenow's correlation [26] is widely used to represent pool boiling, as shown in Eq. (14.58).

$$q'' = \mu_f h_{fg} \left(\frac{g(\rho_f - \rho_v)}{\sigma} \right)^{1/2} \left[\frac{C_f (T_w - T_{sat})}{Pr_f^s C_{sf} h_{fg}} \right]^3 \tag{14.58}$$

For water, the exponent of *Pr*, s is chosen as 1.0. Dimensionless factor C_{sf} accounts for the particular combination of surface and liquid characteristics. Usually C_{sf} can range from 0.0065 to 0.0215. The value 0.013 is often chosen for water–copper and water–platinum combination. Rohsenow's correlation is widely used in engineering applications since it was proposed in 1952. However, Rohsenow's model was based on analogy to forced convection and did not consider some main mechanisms, such as microlayer evaporation and transient conduction, in nucleate pool boiling. Also, Rohsenow's correlation does not explain the mechanism of how C_{sf} is chosen, while the value of C_{sf} can greatly affect the predicted heat flux.

Surface contact angle is an important parameter to represent surface conditions as suggested by a number of researchers [7–9, 25]. In the present work, water is chosen as the base fluid. The present model calculates different heat flux characteristics at different contact angles. The predicted results of the present analytical model were compared with the results calculated by Rohsenow's correlation in Fig. 14.2. Researchers have reported different contact angles [7–9, 14, 15] in the range of 60–80° when the surface is clean, while in the presence of nanofluids, the contact angle is in the range of 16–45°. As can be seen in Fig. 14.2, the average heat flux when the contact angle equals to 60° is very close to the average heat flux calculated by Rohsenow's correlation when C_{sf} equals to 0.0136. The heat flux at a contact angle of 70° corresponds to the heat flux when C_{sf} equals to 0.013, and the results calculated with a C_{sf} of 0.0126 are close to the predicted results based on the present model when the contact angle is 80°. The results show that the average heat flux decreases with a decrease in the contact angle at the same surface temperature. When the superheat is 15 K, the average heat flux decreases from 521 kW/m^2 when the contact angle is 80° to 408 kW/m^2 when the contact angle is 60°. The deterioration is up to 21.6%. However, the maximum heat flux increases with a decrease in the contact angle. The maximum heat flux reaches 690 kW/m^2 when the contact angle is 80°. When the contact angle decreases to 60°, the maximum heat flux increases to 917 kW/m^2 at a superheat of 18.84 K.

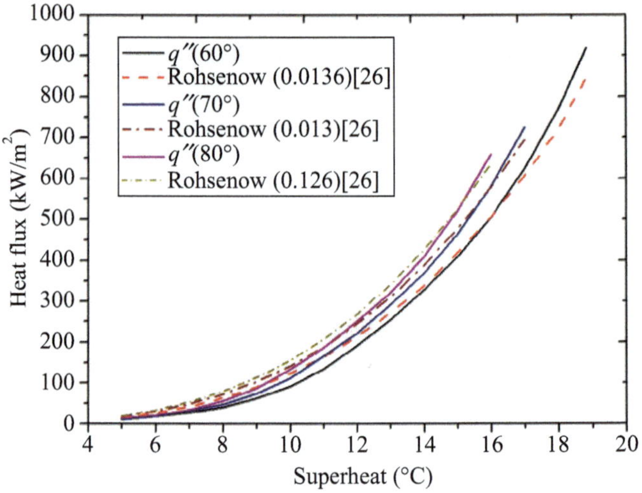

Figure 14.2 Comparison of calculated average heat flux using the present analytical model in the range of contact angles from 60° to 80° with Rohsenow's correlations at different values of C_{sf}.

Figure 14.2 shows some differences between our model and Rohsenow's equation. The differences between our model and Rohsenow's correlation are as follows:

- Even though Rohsenow's correlation is widely used in engineering applications, there is a need for a more robust result for the average heat flux. Furthermore, Rohsenow's correlation applies for clean surfaces. When it is used to estimate the heat flux, it will not be accurate. Also, Rohsenow's correlation is an empirical correlation, while our model is based on the pertinent physical mechanisms.
- Our model establishes a relationship between the average heat flux and the surface contact angle, while Rohsenow's

correlation essentially shows the average heat flux changes for different surfaces with different C_{sf}. Furthermore, it does not explain how C_{sf} is chosen and the correlation does not directly relate the average heat flux and the surface contact angle.

- Our model was based on four heat transfer mechanisms that occur during different stages of nucleate boiling. Rohsenow's correlation is mainly based on an analogy of nucleate boiling to forced convection, which may not accurately reflect the many complicated heat transfer mechanisms during the boiling process.
- Rohsenow's correlation is mainly based on $q'' \propto T^3$, while our model incorporates four different heat transfer mechanisms.

Based on a number of prior research works [7–9, 14, 15], the surface contact angle decreases after the introduction of a nanofluid instead of a regular fluid due to the deposition of nanoparticles in the boiling process as shown in Table 14.1. The present model is validated by comparing our results against a number of recent experimental results. The experimental data were selected from the works of Gerardi et al. [15] and of Bang and Chang [49]. Gerardi et al. [15] reported a contact angle of 80° for boiling with pure water and 16° for boiling with nanofluids. So 80° and 16° were used in our analytical model, respectively. Bang and Chang [49] did not report contact angles, so the contact angle for the base fluid water in the present model is assumed to be 60° and the contact angle for nanofluid is 30° based on previous publications [7–9, 14]. The water properties are used for the cases where the contact angle is 60° and 80°, and the properties of the nanofluid are used for the cases where the contact angle is 16° and 30°. The comparisons between our analytical model and the experimental results are shown in Fig. 14.3. As can be seen in Fig. 14.3, the analytical results are very close to the experimental results.

Figure 14.3 shows that the average heat flux when the contact angle is 60° is very close to the experimental results using pure water in Bang and Chang [49]. The predicted heat flux at a contact angle of 30° is very close to the experimental results with nanofluid from Bang and Chang. The pure water heat fluxes from Gerardi et al. [15] are also close to the predicted results when the contact angle is 80°, and the experimental results using nanofluid are close to analytical results when the contact angle is 16°. The change in nanofluid properties had little effect on the heat transfer. That is even though the thermal conductivity increases due to the addition of the nanoparticles, the decrease in the active nucleation sites is more dominant than the property change of the boiling liquid, so the average heat flux still decreases even as the thermal properties of the liquid enhances.

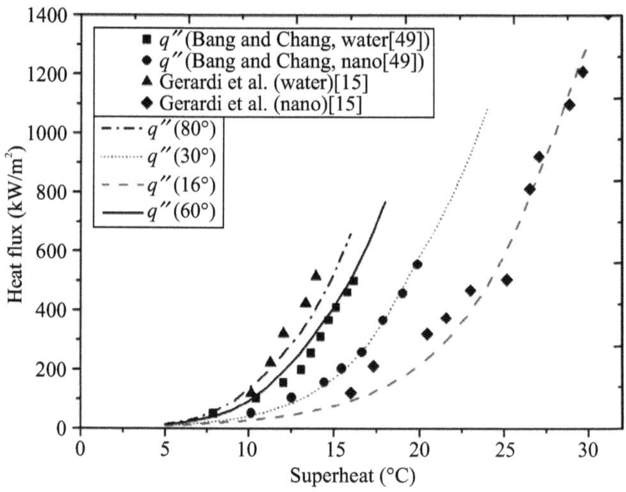

Figure 14.3 Comparison of calculated average heat flux using the present analytical model with the experimental results for pure water and nanofluid boiling (experimental results from Refs. [8, 49]).

The heat transfer coefficient and the Nusselt number at different wall heat fluxes are plotted and compared with the experimental results from Suriyawong et al. [17] in Fig. 14.4. Again, since the cited reference did not provide contact angle values, it is assumed that 80° is the contact angle for a clean surface and 45° is the contact angle for the nanofluid boiling surface. The pure water experimental results agree well with the analytical results when the contact angle is 80°, and the experimental results utilizing 0.1% nanofluid agree well with the predicted analytical results when the contact angle is 45°.

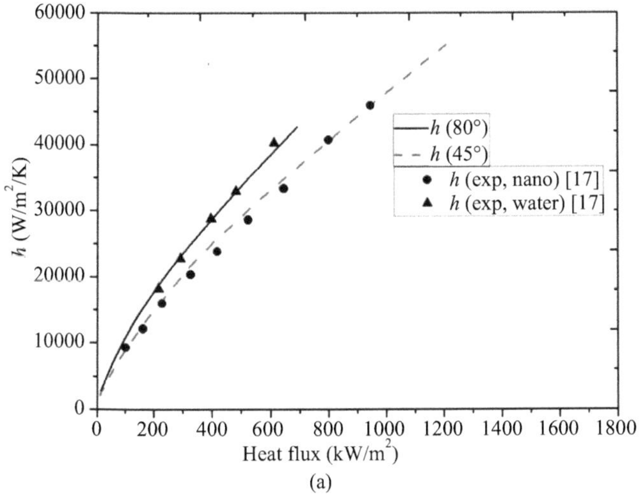

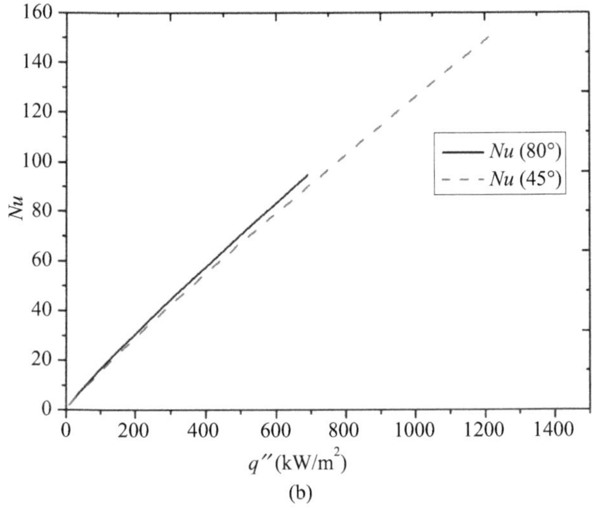

Figure 14.4 Variations of (a) heat transfer coefficient and (b) Nusselt number with average wall heat flux predicted by the present analytical model for both pure water and nanofluid boiling (experiment data from Suriyawong et al. [17]).

The heat transfer coefficients and Nusselt number decrease with a decrease in contact angle. When the wall heat flux is 320 kW/m^2, the heat transfer coefficient decreases from 2.47×10^4 W/m^2 K at a contact angle of 80° to 2.14×10^4 W/m^2 K at a contact angle of 45". The heat transfer coefficient deterioration is about 13.3% at a wall heat flux of 320 kW/m^2. The Nusselt number decreases from 47.2 to 45.1 at this wall heat flux. Although heat transfer deteriorates at the same wall heat flux, the maximum heat transfer coefficient and *Nu* at a smaller contact angle are higher than the heat transfer coefficient at a larger contact angle. Therefore, it appears that in nanofluid boiling, a larger maximum heat transfer coefficient and *Nu* can be obtained. The maximum heat transfer coefficient at a contact angle of 80° is 4.3×10^4 W/m^2 K, while the maximum heat transfer coefficient at a contact of 45° is 5.5×10^4 W/m^2 K. The Nusselt numbers are 94.7 and 151, respectively, for these two contact angles. Our analytical results confirm these experimental trends.

14.4.2 CHF Predictions

The critical heat flux can be enhanced in nanofluid boiling as shown by various researchers [8–15, 18–20]. Many researchers suggest that when the contact angle decreases, the critical heat flux increases [7–9, 14, 15]. Experimental results for CHF from several researchers were compared with the CHF predicted by the present model. Different critical heat fluxes at different contact angles were plotted in Fig. 14.5 and compared with the experimental results from Kim et al. [7] and Kim and Kim [50]. Kandlikar's model [36] and Zuber's correlation [34] were also plotted in Fig. 14.5.

Figure 14.5 shows that the critical heat flux increases from 691 kW/m^2 at a contact angle of 80°, which is measured for pure water on the heating surface, to 2450 kW/m^2 at a contact angle of 20°, which is the measured value for a contact angle after the nanofluid deposition [15]. The present model predicts a higher CHF enhancement than the Kandlikar's model. The enhancement ratio is up to 3.54 if the CHF at a contact angle of 80° is chosen as the base value. Kandlikar's model predicts an enhancement ratio of 1.95 when the contact angle decreases from 80° to 20°. Zuber's model did not consider the surface conditions. The CHF predicted by Zuber's correlation is almost the same with the CHF calculated with the present model at a 53° contact angle. The enhancement ratio predicted by the present analytical model is closer to the CHF enhancement reported in the literature [12, 13, 20, 24]. The enhancement ratio variations of the present model with a change of contact angle were plotted in Fig.

14.6, which shows a continuous increase in CHF with a decrease in the contact angle.

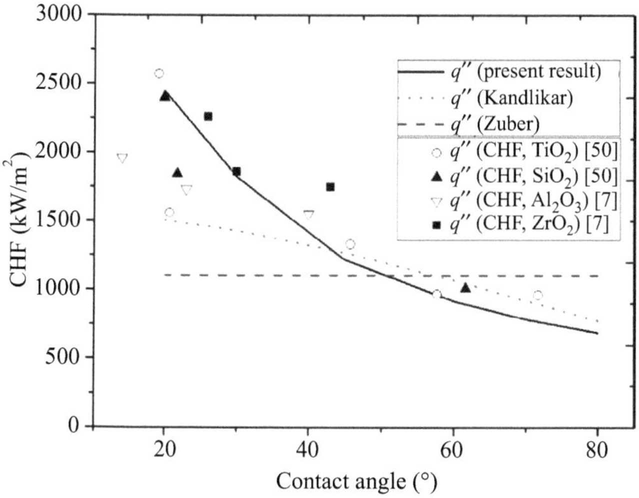

Figure 14.5 Comparison of critical heat flux calculated from the present analytical model against the experimental results and classical correlations [34, 36] in pure water and nanofluid boiling (experimental data from Refs. [7, 50]).

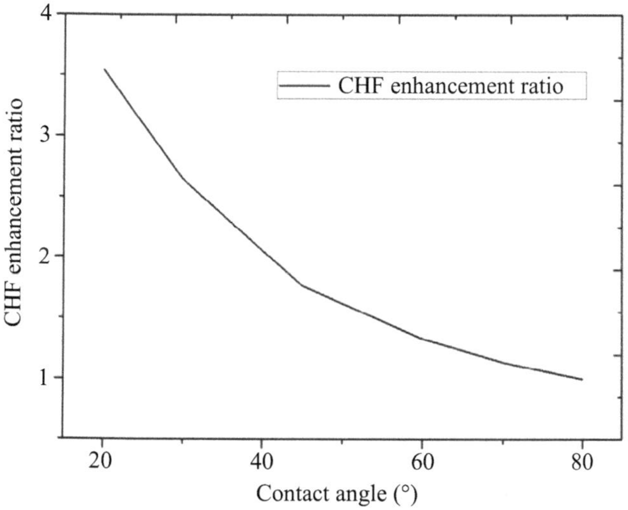

Figure 14.6 Enhancement ratio variations of the critical heat flux for the present model versus the contact angle.

The variations of CHF with departure diameters are plotted in Fig. 14.7. When the contact angle is lower, the surface superheat needed for the CHF occurrence is larger, and larger bubble diameters are needed. Larger bubble diameters and bubble departure frequencies at higher superheats result in higher heat fluxes.

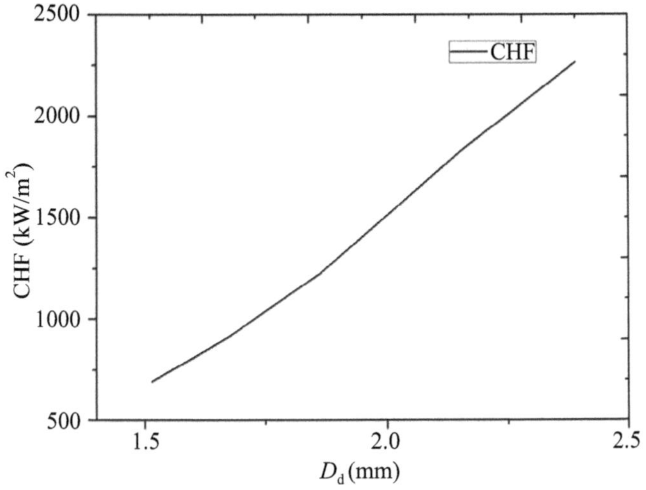

Figure 14.7 Critical heat flux variations versus the departure diameter D_d.

14.5 Conclusion

In this work, a composite model is proposed to compute the average surface heat flux and the average heat flux variations with increasing superheats. The CHF was also predicted for different surface contact angles. Prior theoretical models on nucleate boiling are based on a single mechanism. However, four pertinent mechanisms affect the boiling process. Few researchers have systematically proposed a composite model analyzing these four main mechanisms. The main mechanisms in pool boiling were considered in the presented analytical model. Microlayer evaporation, transient conduction before and after bubble departs from the surface, microconvection induced by bubble growth and movement were all analyzed in the present model.

Our model has established a relationship between the average heat flux and surface contact angle. Our model has also extended the single bubble region to bubble interference region and enables quantification of the average heat flux at very high superheats. Transient conduction was found to be the main mechanism at larger superheats and natural convection accounts for a large percentage of heat transfer at very low superheats.

It was found that even though the thermal conductivity and other thermal properties were enhanced for nanofluids, the decreasing number of active nucleation sites at lower contact angles was more predominant and resulted in a decrease in the average surface heat fluxes for the same superheat. However, a larger heat transfer coefficient and *Nu* can be obtained for nanofluid boiling at a higher superheat. The results from our model were in good agreement with the available experimental results. Our model establishes a relationship between the average heat flux and surface contact angle leading to the determination of CHF. It was also established that CHF enhancement ratio can be up to 3.54 when the surface contact angle decreases from 80° to 20°. This enhancement ratio agrees with a number of experimental results.

Conflict of interest

None declared.

Acknowledgments

This work was supported by the Science Fund for Creative Research Groups of National Natural Science Foundation of China, Grant No. 51321002 and National Basic Research Program of China ("973" Project) Grant No. 2011CB706904. The authors acknowledge the financial support from China Scholarship Council.

References

1. A. V. Kuznetsov and D. A. Nield, "Natural convective boundary-layer flow of a nanofluid past a vertical plate," *Int. J. Therm. Sci.*, 2010; **49**: 243–247.
2. M. Shafahi, V. Bianco, K. Vafai, and O. Manca, "An investigation of the thermal performance of cylindrical heat pipes using nanofluids," *Int. J. Heat Mass Transfer*, 2010; **53**: 376–383.

3. D. A. Nield and A. V. Kuznetsov, "The Cheng–Minkowycz problem for natural convective boundary-layer flow in a porous medium saturated by a nanofluid," *Int. J. Heat Mass Transfer*, 2009; **52**: 5792–5795.
4. H. Kim, "Enhancement of critical heat flux in nucleate boiling of nanofluids: A state-of-art review," *Nanoscale Res. Lett.*, 2011; **6**: 415.
5. S. K. Das, N. Putra, and W. Roetze, "Pool boiling characteristics of nanofluids," *Int. J. Heat Mass Transfer*, 2003; **46**: 851–862.
6. J. S. Coursey and J. Kim, "Nanofluid boiling: The effect of surface wettability," *Int. J. Heat Fluid Flow*, 2008; **29**: 1577–1585.
7. S. J. Kim, I. C. Bang, J. Buongiorno, and L. W. Hu, "Surface wettability change during pool boiling of nanofluids and its effect on critical heat flux," *Int. J. Heat Mass Transfer*, 2007; **50**: 4105–4116.
8. Z. Shahmoradi, N. Etesami, and M. N. N. Esfahany, "Pool boiling characteristics of nanofluid on flat plate based on heater surface analysis," *Int. Commun. Heat Mass Transfer*, 2013; **47**: 113–120.
9. J. Y. Jung, E. S. Kim, Y. Nam, and Y. T. Kang, "The study on the critical heat flux and pool boiling heat transfer coefficient of binary nanofluids (H_2O/LiBr + Al_2O_3)," *Int. J. Refrig.*, 2013; **36**: 1056–1061.
10. D. M. Vazquez and R. Kumar, "Surface effects of ribbon heaters on critical heat flux in nanofluid pool boiling," *Int. Commun. Heat Mass Transfer*, 2013; **41**: 1–9.
11. J. Y. Jung, E. S. Kim, and Y. T. Kang, "Stabilizer effect on CHF and boiling heat transfer coefficient of alumina/water nanofluids," *Int. J. Heat Mass Transfer*, 2012; **55**: 1941–1946.
12. R. N. Hegde, S. S. Rao, and R. P. Reddy, "Experimental studies on CHF enhancement in pool boiling with CuO–water nanofluid," *Int. J. Heat Mass Transfer*, 2012; **48**: 1031–1041.
13. M. Kole and T. K. Dey, "Investigations on the pool boiling heat transfer and critical heat flux of ZnO–ethylene glycol nanofluids," *Appl. Therm. Eng.*, 2012; **37**: 112–119.
14. A. Bolukbasi and D. Ciloglu, "Pool boiling heat transfer characteristics of vertical cylinder quenched by SiO_2–water nanofluids," *Int. J. Therm. Sci.*, 2011; **50**(6): 1013–1021.
15. C. Gerardi, J. Buongiorno, L. W. Hu, and T. McKrell, "Infrared thermometry study of nanofluid pool boiling phenomena," *Nanoscale Res. Lett.*, 2011; **6**: 232–248.
16. D. Wen, M. Corr, X. Hu, and G. Lin, "Boiling heat transfer of nanofluids: The effect of heating surface modification," *Int. J. Therm. Sci.*, 2011; **50**(4): 480–485.

17. A. Suriyawong and S. Wongwises, "Nucleate pool boiling heat transfer characteristics of TiO_2–water nanofluids at very low concentrations," *Exp. Therm. Fluid Sci.*, 2010; **34**: 992–999.

18. S. M. Kwark, R. Kumar, G. Moreno, J. Yoo, and S. M. You, "Pool boiling characteristics of low concentration nanofluids," *Int. J. Heat Mass Transfer*, 2010; **53**: 972–981.

19. Z. Liu, J. Xiong, and R. Bao, "Boiling heat transfer characteristics of nanofluids in a flat heat pipe evaporator with micro-grooved heating surface," *Int. J. Multiphase Flow*, 2007; **33**: 1284–1295.

20. S. M. You, J. H. Kim, and K. H. Kim, "Effect of nanoparticles on critical heat flux of water in pool boiling heat transfer," *Appl. Phys. Lett.*, 2003; **83**: 3374–3376.

21. D. Wen and Y. Ding, "Experimental investigation into the pool boiling heat transfer of aqueous based c-alumina nanofluids," *J. Nanoparticle Res.*, 2005; **7**: 265–274.

22. H. Kim, J. Kim, and M. Kim, "Experimental study on CHF characteristics of water–TiO_2 nano-fluids," *Nucl. Eng. Technol.*, 2006; **38**(1): 61–68.

23. O. Ahmed and M. S. Hamed, "Experimental investigation of the effect of particle deposition on pool boiling of nanofluids," *Int. J. Heat Mass Transfer*, 2012; **55**(13–14): 3423–3436.

24. T. Okawa, M. Takamura, and T. Kamiya, "Boiling time effect on CHF enhancement in pool boiling of nanofluids," *Int. J. Heat Mass Transfer*, 2012; **55**(9–10): 2719–2725.

25. C. H. Wang and V. K. Dhir, "Effect of surface wettability on active nucleation site density during pool boiling of water on a vertical surface," *ASME J. Heat Transfer*, 1993; **115**: 659–669.

26. W. M. Rosenhow, "A method of correlating heat transfer data for surface boiling of liquids," *Trans. ASME J. Heat Transfer*, 1952; **4**: 969–975.

27. C. L. Tien, "A hydrodynamic model for nucleate pool boiling," *Int. J. Heat Mass Transfer*, 1962; **5**: 540–553.

28. N. Zuber, "Nucleate boiling the region of isolated bubbles and the similarity with natural convection," *Int. J. Heat Mass Transfer*, 1963; **6**: 53–78.

29. H. K. Forster and R. Greif, "Heat transfer to a boiling liquid-mechanisms and correlations," *ASME J. Heat Transfer*, 1959; **81**: 45.

30. C. Y. Han and P. Griffith, "The mechanism of heat transfer in nucleate pool boiling," *Int. J. Heat Mass Transfer*, 1965; **8**: 887–914.

31. B. B. Mikic and W. M. Rohsenow, "A new correlation of pool boiling data including the effect of heating surface characteristics," *ASME J. Heat Transfer*, 1969; **91**: 245–250.

32. R. L. Judd and K. S. Hwang, "A comprehensive model for nucleate pool boiling heat transfer including microlayer evaporation," *ASME J. Heat Transfer*, 1976; **98**: 623–629.

33. R. J. Benjamin and A. R. Balakrishnan, "Nucleate pooling heat transfer of pure liquids at low to moderate heat fluxes," *Int. J. Heat Mass Transfer*, 1996; **39**: 2495–2504.

34. N. Zuber, "On the stability of boiling heat transfer," *ASME J. Heat Transfer*, 1958; **80**(2): 711–720.

35. Y. Haramura and Y. Katto, "New hydrodynamic model of critical heat flux applicable widely to both pool and forced convection boiling on submerged bodies in saturated liquids, *Int. J. Heat Mass Transfer*, 1983; **26**: 379–399.

36. S. G. Kandlikar, "A theoretical model to predict pool boiling CHF incorporating effects of contact angle and orientation," *ASME J. Heat Transfer*, 2001; **123**: 1071–1079.

37. K. Khanafer and K. Vafai, "A critical synthesis of thermophysical characteristics of nanofluids," *Int. J. Heat Mass Transfer*, 2011; **54**: 4410–4428.

38. G. Batchelor, "The effect of Brownian motion on the bulk stress in a suspension of spherical particles," *J. Fluid Mech.*, 1977; **83**: 97–117.

39. G. Son, V. K. Dhir, and N. Ramanujapu, "Dynamics and heat transfer associated with a single bubble during nucleate boiling on a horizontal surface," *ASME J. Heat Transfer*, 1999; **121**: 623–631.

40. J. L. Bi, X. P. Lin, and D. M. Christopher, "Effects of bubble coalescence dynamics on heat flux distributions under bubbles," *AICHE J.*, 2013; **59**: 1735–1745.

41. C. Gerardi, J. Buongiorno, L. Hu, and T. McKrell, "Study of bubble growth in water pool boiling through synchronized, infrared thermometry and high-speed video," *Int. J. Heat Mass Transfer*, 2010; **53**: 4185–4192.

42. S. Moghaddam and K. Kiger, "Physical mechanisms of heat transfer during single bubble nucleate boiling of FC-72 under saturation conditions-I," *Exp. Invest. Int. J. Heat Mass Transfer*, 2009; **52**: 1284–1294.

43. S. Moghaddam and K. Kiger, "Physical mechanisms of heat transfer during single bubble nucleate boiling of FC-72 under saturation

conditions-II," *Theor. Anal. Int. J. Heat Mass Transfer*, 2009; **52**: 1295–1303.

44. J. L. Bi, X. P. Lin, D. M. Christopher, and X. F. Li, "Analysis of coalescence phenomena on microheaters at two surface superheats," *Int. J. Heat Mass Transfer*, 2013; **67**: 798–809.

45. J. L. Bi, D. M. Christopher, X. P. Lin, and X. F. Li, "Effects of nucleation site arrangement and spacing on bubble coalescence characteristics," *Exp. Therm. Fluid Sci.*, 2014; **52**: 116–127.

46. H. K. Forster and N. Zuber, "Dynamic of vapor bubbles and boiling heat transfer," *AIChE J.*, 1955; **1**: 531–539.

47. T. Hibiki and M. Ishii, "Active nucleation site density in boiling systems," *Int. J. Heat Mass Transfer*, 2003; **46**: 2587–2601.

48. V. P. Carey, *Liquid Vapor Phase-transition Phenomena*, Hemisphere, New York, 1992., pp. 207.

49. I. C. Bang and S. H. Chang, "Boiling heat transfer performance and phenomena of Al_2O_3–water nano-fluids from a plain surface in a pool," *Int. J. Heat Mass Transfer*, 2005; **48**: 2407–2419.

50. H. D. Kim and M. H. Kim, "Effect of nanoparticle deposition on capillary wicking that influences the critical heat flux in nanofluids," *Appl. Phys. Lett.*, 2007; **91**: 014104.

Index